Measurement Challenges in Atmospheric Chemistry

ADVANCES IN CHEMISTRY SERIES **232**

Measurement Challenges in Atmospheric Chemistry

Leonard Newman, EDITOR
Brookhaven National Laboratory

Developed from a symposium sponsored
by the Division of Analytical Chemistry
of the American Chemical Society

American Chemical Society, Washington, DC 1993

Library of Congress Cataloging-in-Publication Data

Measurement challenges in atmospheric chemistry / Leonard
Newman, editor.

    p.   cm.—(Advances in chemistry series, ISSN 0065–2393; 232)
"Developed from a symposium sponsored by the Division of
Analytical Chemistry at the 199th National Meeting of the
American Chemical Society, Boston, Massachusetts, April 22–27,
1990."

    Includes bibliographical references and indexes.

    ISBN 0–8412–2470–6

    1. Atmospheric chemistry—Technique.   2. Aerosols—Mea-
surement.

    I. Newman, Leonard.   II. American Chemical Society. Division
of Analytical Chemistry.   III. American Chemical Society.
Meeting (199th: 1990: Boston, Mass.) IV. Series.

QC879.6.M4   1993
551.5′11—dc20   92–38528
                                     CIP

The paper used in this publication meets the minimum requirements of American National
Standard for Information Sciences—Permanence of Paper for Printed Library Materials, ANSI
Z39.48–1984. ∞

1993 Advisory Board

Advances in Chemistry Series

M. Joan Comstock, *Series Editor*

FOREWORD

The ADVANCES IN CHEMISTRY SERIES was founded in 1949 by the American Chemical Society as an outlet for symposia and collections of data in special areas of topical interest that could not be accommodated in the Society's journals. It provides a medium for symposia that would otherwise be fragmented because their papers would be distributed among several journals or not published at all.

Papers are reviewed critically according to ACS editorial standards and receive the careful attention and processing characteristic of ACS publications. Volumes in the ADVANCES IN CHEMISTRY SERIES maintain the integrity of the symposia on which they are based; however, verbatim reproductions of previously published papers are not accepted. Papers may include reports of research as well as reviews, because symposia may embrace both types of presentation.

ABOUT THE EDITOR

 LEONARD NEWMAN received his B.S. from the Polytechnic Institute of Brooklyn (now called Polytechnic University) and his Ph.D. from the Massachusetts Institute of Technology, both in chemistry. He is a senior scientist at Brookhaven National Laboratory and is head of the Environmental Chemistry Division. He has been at Brookhaven for 35 years, during which time he spent a year at the Royal Institute of Technology in Sweden as a visiting scientist. He is also an adjunct professor of environmental medicine at the New York University Medical Center and an adjunct professor in the School of Earth and Environmental Sciences at the Georgia Institute of Technology.

He has published more than 80 technical papers and given a comparable number of presentations at meetings and symposia. His early work was directed toward the development of spectrophotometric techniques for measuring complex ions in aqueous solution. He subsequently studied interactions and complex formation in organic media, with an emphasis on understanding the synergistic effects that influence the extraction of metals from aqueous media. In connection with the development of new nuclear fuel elements and reactor concepts, he studied interactions of metals with fused salts and developed techniques for measuring oxygen and carbon in liquid sodium.

In more recent years, he has been involved with studying atmospheric processes to develop an understanding of the transformations that lead to acid deposition and the formation of ozone. He has directed much attention toward the development of measurement techniques, especially those needed for obtaining information on sulfur and nitrogen chemistry. He has taken a special interest in measurement techniques used aboard aircraft—which have taken on a particular importance and have been used to study interactions in and about clouds—initially for understanding the formation of acid rain and now for addressing questions of the importance of clouds in global climate change.

He was one of the organizers of the symposium on gas–liquid chemistry of natural waters held at Brookhaven and the editor of the proceedings, which appeared in 1984. He has led a number of scientific groups to and has been a frequent lecturer in China. In 1989, he was one of the organizers of the International Conference on Global and Regional Environmental Atmospheric Chemistry held in Beijing, China, and he compiled and edited the proceedings.

CONTENTS

PREFACE

In THIS AGE OF ENVIRONMENTAL CONCERNS, it is important that we properly address the issues involved in the measurement of atmospheric constituents in order to come to realistic solutions expeditiously and economically. Nothing is more fundamental to this process than the ability to measure the substances of concern. Not only are these measurements necessary for documenting what is there, they are also important for describing and understanding the transport, transformation, and removal of substances from the atmosphere. We have become increasingly sophisticated in our approach to these questions, and, accordingly, we are continually facing the problem of not having the requisite tools to document and understand atmospheric processes. We are constantly in need of more sensitive, precise, and accurate methods of sampling and analyzing substances of interest. The purpose of this book is to educate the reader about the importance of the measurement of atmospheric constituents, give state-of-the-art descriptions of techniques, and point out specific problem areas in order to stimulate the community at large to address these problems.

This book is not intended to be an exhaustive treatise that discusses all substances of potential interest, but rather a discussion of selected substances that are of particular importance in describing the state and processes of the atmosphere and that have measurement problems that remain to be solved. The authors were asked to give an overview of the field, discuss the methods that are generally used, and, most importantly, discuss the measurement problems that still exist.

The first two chapters deal specifically with problems associated with sampling. Chapter 1 addresses sampling of labile, or reactive, substances. With these substances, the very act of sampling can change the chemical composition of the substance or of the materials with which they are in equilibrium; consequently, unique problems are posed for sampling and analysis. Gases must often be separated from particles during sampling, and this step requires a suitable matrix that can be used for subsequent chemical analysis. Consequently, Chapter 2 discusses diffusion-based collection procedures that are used for the separation process; diffusion coefficients of gases are greater than those of particles, so under laminar flow gases will diffuse to walls that are coated with reactive substances and will thereby be removed from the airstream.

Substances can be removed from the atmosphere by dry deposition to surfaces. A method for obtaining the parameters of dry deposition uses eddy correlation flux measurements that require chemical sensors with very fast responses. The problems of implementing this approach are discussed in Chapter 3. To understand large-scale transport and transformation properties, scientists frequently sample substances above ground level, where measurements are relatively uninfluenced by local sources. In Chapter 4, the problems of sampling from aircraft are discussed. In Chapter 5, measurements in the stratosphere, both by high-altitude aircraft and helium-filled balloons, are discussed.

The sampling and analysis of aerosols is a problem of long-standing interest. Our understanding of atmospheric aerosol chemical dynamics has been furthered by the development of specific instrumentation. The various capabilities and the problems associated with aerosol analysis are presented in Chapter 6. Aerosols persist in the atmosphere for relatively long periods, and a description of their sources, transport, and transformation has been furthered by the ability to analyze size-segregated fractions. The problems associated with sampling fine particles and with the analysis of the concomitant small masses are discussed in Chapter 7. Concerns about potential health effects and environmental damage originating from the acid content of aerosols have in large part prompted the development of methods for the measurement of the strong acid content of aerosols. The difficulties that still exist, especially in connection with sampling this reactive substance, are presented in Chapter 8.

The family of nitrogen compounds, by virtue of their number and complexity, has presented a series of challenges to the measurement community as the need to learn more about them has arisen because of their pivotal position in the chemistry of the atmosphere. A detailed description of the current ability to make measurements of known substances and the importance and possibility of the existence of as yet unmeasured species is presented in Chapter 9. In addition, the use of intercomparisons to verify and validate methods is discussed.

Hydrocarbons are a primary ingredient in the photochemical reactions that produce smog. Chapter 10 describes the methodologies used to determine the concentration of nonmethane hydrocarbon species in the atmosphere. Products of the oxidation of hydrocarbons are peroxy radicals. Both chemical and spectroscopic methods are used to measure them, and the principles and approaches of these methods are given in Chapter 11. Possibly the most important radical in the atmosphere for which a determination is required is the hydroxyl radical; the difficulty of measuring this radical has plagued the community for years. A review is given in Chapter 12, and the hurdles still to be surmounted are discussed in light of theoretical and experimental results.

A somewhat new and burgeoning field is the development of personal air monitors that permit the recording of an individual's direct and total exposure to pollutants. The associated technological issues and problems are described in Chapter 13. The criteria needed for the design of personal monitors are presented and should serve to guide the community in developing new methods for personal air monitoring.

Practicing atmospheric scientists should find the book useful as a resource, and scientists from other disciplines, such as analytical and physical chemistry, electrical and mechanical engineering, physics, and instrumentation, should find it a good introduction to the field.

Acknowledgments

M. Marsch and J. Williams were most important for the successful organization of the symposium and for expediting the collection of the chapters. My contribution was performed under the auspices of the U.S. Department of Energy under Contract DE–AC02–76CH00016, under the Office of Health and Environmental Research within the Office of Energy Research.

LEONARD NEWMAN
Brookhaven National Laboratory
Upton, NY 11973

October 20, 1992

Sampling of Selected Labile Atmospheric Pollutants

B. R. Appel

Air and Industrial Hygiene Laboratory, California Department of Health Services, 2151 Berkeley Way, Berkeley, CA 94704–9980

Nitric acid, particulate nitrate, and particulate organic carbon may be termed "labile" atmospheric pollutants; this name reflects the ease with which they undergo physical or chemical changes while an integrated sample is being collected. Manual sampling methods of varying accuracy are described for these species, together with sources of error. For nitric acid and particulate nitrate, the filter pack method, usually with Teflon and nylon filters in tandem, is the least accurate but simplest sampling procedure. More accurate techniques use diffusion denuders to separate gaseous HNO_3 from fine particulate nitrate. Most methods remain subject to interference from nitrous acid. Sampling techniques for particulate organic C remain relatively primitive. Tandem filter sampling permits correction for the error due to sorption of gaseous C on the filter medium. Denuder-based techniques, such as those described here for polyaromatic hydrocarbons, may hold the key to future development of improved samplers for particulate organic C as well.

T HE OBJECTIVE OF INTEGRATIVE ATMOSPHERIC SAMPLING is to measure accurately the concentrations of species of interest averaged over the sampling time. However, a substantial number of species are difficult to sample because of chemical or physical changes occurring during or after collection. Such pollutants are referred to as "labile". This chapter focuses on integrative sampling techniques for selected labile species, including particulate nitrate and its corresponding gas-phase species, nitric acid; particulate carbon; and particle-phase, polycyclic aromatic hydrocarbons. The terms "vapor-phase", "gaseous", and "gas-phase" are used interchangeably, as are the terms "particulate" and "particle-phase".

0065–2393/93/0232–0001$11.00/0

The need to determine accurately the phase-specific concentrations of these pollutants reflects several concerns: Compared to gaseous materials, particle-phase materials may penetrate more deeply into the human respiratory tract; particle-phase pollutants scatter light much more effectively than gaseous materials, and they thus have a greater contribution to visibility reduction; gaseous nitric acid has a much higher deposition velocity than particulate nitrates and can be a substantial contributor to the acidification of lakes, streams, forests, and vegetation.

In some instances specialized techniques were developed to determine concentrations of specific labile pollutants through the use of techniques that avoid (or nearly avoid) sample collection; these techniques are often referred to as "real-time" methods. These methods, as well as analytical methods to be applied to collected samples, are not discussed.

The literature cited is intended to illuminate the topics covered but in no sense represents an exhaustive review. Furthermore, numerous alternative sampling techniques are omitted in favor of more comprehensive treatments of what I judge to be the most generally useful strategies. The most important omission is probably impactor collection, a technique that may reduce sampling errors with some labile species (1).

It is useful to consider the general composition of atmospheric aerosols. Such composition varies markedly depending on sampler location (e.g., urban or rural), proximity of significant sources of aerosols and their gaseous precursors, and meteorology (e.g., wind speed, inversion height, and sunlight intensity). Anthropogenic particulate matter is concentrated in fine particles (i.e., less than 2.5 μm in diameter), whereas natural aerosols (e.g., wind-blown soil, sea salt, pollen, and spores) are concentrated in larger particles. Table I, taken from reference 2 and references cited therein, details average concentrations in the fine and coarse (2.5–15 μm) particle fractions obtained from a rural and an urban location. The materials that readily undergo phase and chemical changes are concentrated in the fine fraction. The major constituents shown for the fine fraction include carbonaceous materials, sulfate, ammonium, and nitrate.

Although Table I is generally self-explanatory, the carbonaceous material measurements require comment. Because of its chemical complexity, carbonaceous material is frequently characterized only on the basis of carbon measurements. These measurements attempt to divide the carbonaceous material into "organic C" and "elemental C". Carbon present in carbonate salts, frequently a minor contributor to the total particulate carbon, can be determined independently. Elemental carbon is among the most important pollutants in visibility reduction. Polycyclic aromatic hydrocarbons (PAHs) are relatively minor constituents of the particulate carbon but are of great interest in health effects studies. PAHs can also serve as model compounds in developing improved sampling techniques for semivolatile carbonaceous materials.

Table I. Average Aerosol Composition for Fine and Coarse Particles at a Rural, Forested Location (Great Smoky Mountains, Tennessee) and an Urban Location (Houston, Texas)

| Composition | Smoky Mountains | | Houston[a] | |
	Fine[b]	Coarse[b]	Fine	Coarse
Total mass	24,000 ± 3000	5600 ± 3000	42,500 ± 4250	27,200 ± 2700
SO_4^{2-}	12,000 ± 1300	NA[c]	16,700 ± 1380	1100 ± 200
NO_3^-	300 ± 300	NA	250 ± 260	1800 ± 260
NH_4^+	2280 ± 390	NA	4300 ± 390	<190
H^+	114	NA	67	<1
C (organic)	2220 ± 400	1200 ± 400	NA	NA
C (elemental)	1100 ± 800	<100	NA	NA
C (total)	3300 ± 600	1300 ± 600	7600 ± 500	3300 ± 500
Al	20 ± 18	195 ± 101	95 ± 60	1400 ± 420
Si	38 ± 10	580 ± 262	200 ± 60	3800 ± 1000
S	3744 ± 218	204 ± 187	NA	NA
Cl	<10	7 ± 4	19 ± 6	330 ± 21
K	40 ± 3	108 ± 30	120 ± 7	180 ± 21
Ca	16 ± 1	322 ± 73	150 ± 8	3100 ± 160
Ti	<6	18 ± 5	<8	48 ± 14
V	<4	<5	NA	NA
Mn	NA	NA	13 ± 2	23 ± 3
Fe	28 ± 2	118 ± 9	170 ± 9	730 ± 40
Ni	1 ± 0.5	1 ± 0.5	3 ± 1	5 ± 1
Cu	3 ± 0.7	<5	16 ± 2	14 ± 2
Zn	9 ± 1	<4	102 ± 6	68 ± 5
As	2.2 ± 1	<1	NA	NA
Se	1.4 ± 0.3	0.2 ± 0.2	NA	NA
Br	18 ± 1	5 ± 0.4	70 ± 4	39 ± 3
Pb	97 ± 5	14 ± 1	483 ± 23	127 ± 10

NOTE: All values are given in nanograms per cubic meter.
[a]Samples collected during the daytime.
[b]Fine and coarse particles were defined in this study as having aerodynamic diameters in the ranges 0–2.5 and 2.5–15 μm, respectively.
[c]Not analyzed.
SOURCE: Reproduced with permission from reference 2. Copyright 1986 Wiley.

Sampling errors associated with these species can result in measurements that are too high or too low and that lead to the frequently used terms "positive artifact" and "negative artifact". To ensure the accurate measurement of the species to be considered, researchers have focused on determining the sources and magnitude of such errors and on devising strategies that eliminate or minimize both positive and negative artifacts.

Sampling of Atmospheric Inorganic Nitrates

Inorganic nitrates in the atmosphere include the gaseous species HNO_3 and nitrate salts in suspended particulate matter. The sum of their concentrations is referred to as total inorganic nitrate (TIN). The composition of the particulate nitrate (PN) may be expressed as $M(NO_3)_x$, where M denotes NH_4 or a metal (especially an alkali metal or an alkaline earth element) and x denotes the valence state of M. Of these, the principal salts that have been identified are NH_4NO_3 and $NaNO_3$. The latter can be formed from the reaction of sea salt aerosol with HNO_3 (3, 4), a pathway consistent with the somewhat larger particle size associated with $NaNO_3$. Reaction of HNO_3 with suspended soil particles, a proposed pathway to coarse particulate nitrate, could involve other cations as well (5).

Most atmospheric studies do not determine the specific composition of the particulate inorganic materials being sampled, although composition has frequently been inferred from the balance between the principal anions and cations. The NO_3^- salts other than NH_4NO_3 are not significantly volatile, and their sampling and analysis should be less problematic although still subject to error. Specialized analytical techniques attempting to distinguish ammonium from other nitrate salts in atmospheric particulate samples have been devised (4) but are not reviewed here.

Ammonium nitrate, found primarily in the fine-particle fraction, results principally from the atmospheric reaction of HNO_3 with NH_3 according to the equilibrium

$$NH_4NO_3 \text{ (s)} \rightleftharpoons NH_3 \text{ (g)} + HNO_3 \text{ (g)} \qquad (1)$$

where $K = 9.5 \text{ ppb}^2$ at 20 °C. This value of K, which applies only below the deliquescence relative humidity (RH), increases sharply with increasing temperature. Above the deliquescence RH, K decreases with increasing RH (Figure 1) (6). Numerous studies, reviewed by Allen and Harrison (7), have compared observed atmospheric concentrations of HNO_3 and NH_3 with those expected from this equilibrium. Grosjean (8) assessed agreement between theory and observation both above and below the deliquescence RH, finding generally reasonable agreement for samples in Southern California, a location notable for elevated nitrate concentrations.

PN is often approximated by the NO_3^- retained on a filter, an impactor, or other nominal collector of particulate matter. However, inorganic nitrates are subject to particle–vapor transformations while present in the atmosphere as well as after collection on nonreactive sampling media. The following sections discuss the various pathways for such transformations, other sources of error, and alternative sampling strategies. Because the goal is to measure nitrates as they exist at the time of sampling, only transformations during and after the sampling process and other sampling artifacts are discussed. In the **Summary and Conclusions** section, the methods reviewed and their sources of error are presented in tabular form.

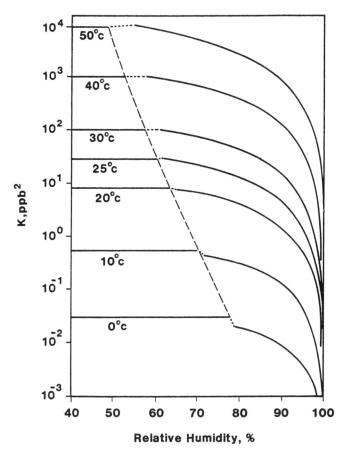

Figure 1. NH₄NO₃ dissociation constant dependence on temperature and relative humidity. (Reproduced with permission from reference 6. Copyright 1982.)

Pathways for Particle-to-Gaseous Nitrate Conversions.

It is generally assumed that if airborne NH_4NO_3, HNO_3, and NH_4 are present, they are in equilibrium immediately before collection. However, if, for example, the temperature increases before the end of the sampling period without partial pressures of HNO_3 and NH_3 being achieved at the sampler inlet at least equal to their new equilibrium values, dissociation and loss of NH_4NO_3 from the particulate sample should occur.

In addition to dissociation of NH_4NO_3, PN can be lost from the particle phase by reaction with both particulate and gaseous strong acids, yielding HNO_3. Such processes have been demonstrated in aqueous solutions and on inert filter surfaces. In aqueous systems including H^+, SO_4^{2-}, NH_4^+, and

NO_3^-, the partial pressure of HNO_3 depends on the nitrate-to-sulfate ratio (9). The same processes are expected to occur in acidic, suspended SO_4^{2-}- and NO_3^--containing particles before sampling. Laboratory and atmospheric sampling results with filter-collected samples support the reaction of nitrate salts with acidic particulate sulfates (reference 10 and references cited therein) and with gaseous HCl (10), liberating gaseous HNO_3.

Pathways for Real or Apparent Gaseous-to-Particle Nitrate Conversion. *Introduction.*

The retention of gaseous HNO_3 on many filter types is well recognized as a source of artifact PN (11, 12). Of those in common use, Teflon filters provide the least HNO_3 retention, less than 2% at 50% RH, and are the most widely used filter for PN sampling. With quartz-fiber filters, such retention increases with increasing filter alkalinity and RH (12, 13). Glass-fiber filters, which are relatively alkaline, can approximate efficient TIN samplers and retain both PN and HNO_3 with high efficiency under atmospheric conditions (14).

In addition to retention on sampling media, artifact PN can be formed by the retention of HNO_3 on previously collected particulate matter. Relatively few data are available that clearly document the magnitude of this problem, and unanswered questions remain.

Retention of Nitric Acid on Particulate Matter. Teflon filters preloaded with atmospheric particulate matter collected without particle-size segregation in Berkeley, California, showed greatly increased HNO_3 retention compared to clean Teflon filters (15). At 50–80% RH, retention of 300–500 $\mu g/m^3$ HNO_3 ranged from about 6 to 22%. Within this range, retention increased with increased particle loadings varying from 7 to 65 $\mu g/cm^2$.

The specific aerosol constituents responsible for HNO_3 retention were not established. The reaction of HNO_3 with Cl^- salts, yielding relatively nonvolatile nitrates, can be assumed to contribute; indeed, this reaction is the basis of a method for HNO_3 sampling (16). The reaction of HNO_3 with alkaline soil dust particles is also likely, as shown by a high correlation between nitrate and soil-related elements in coarse particles (5). Such reactions, including that with chloride salts, could occur both before and after particle collection, with only the latter causing a sampling error.

The mechanisms suggested for PN formation by reactions with aerosol constituents imply that the degree of HNO_3 retention on previously collected particulate matter would be greatly reduced if coarse particles were excluded from the sampler, and that the extent of HNO_3 retention with inert prefilters should generally decrease with decreased sampling volume.

Formation of Nitrate on Nylon Filters. *Introduction.*

Since first proposed by Spicer (17), nylon filters have played a prominent role in measuring atmospheric PN and HNO_3. Their prominence may be ascribed to a high efficiency for gaseous HNO_3 retention and relatively low retention of NO_2 and peroxyacetyl nitrate (PAN); Spicer demonstrated these features that were generally confirmed by others. However, as the state of the art improved, and with the finding of significant levels of HONO in the atmosphere, the selectivity of nylon filters has remained an area of active investigation. The change in the nylon polymer used to fabricate filters has also prompted some concern for the degree of generality possible with prior research results.

The nylon filter currently sold as 1-μm pore size Nylasorb (Gelman Sciences), probably the most widely used filter for HNO_3 sampling in the United States, was initially marketed by Ghia Corporation. Until 1985, these filters were fabricated from nylon 6, a polyamide formed from the homopolymerization of ϵ-caprolactam. More recently, Gelman, Sartorius, and other vendors have supplied filters fabricated from nylon 6,6, made by polymerization of adipic acid and hexamethylenediamine.

Potential interferents collected by nylon filters include organic nitrates (e.g., PAN), NO_2, and HONO. Each can yield NO_2^-, which can be oxidized to NO_3^- by O_3 or other oxidants. Although not related directly to nitrate sampling, SO_2 retention in varying degrees was also of concern.

Comparison of Nylon Filters from Different Suppliers. Nylasorb nylon 6 filters from Ghia Corporation exhibited an efficiency of 95 $\pm$ 11% for up to 3000 μg HNO_3 (sampling at 20 °C, 50% RH, and 20 L/min with 47-mm-diameter filters) (*14*). Comparisons of the measured atmospheric HNO_3 levels among different lots of Gelman Nylasorb and between Gelman Nylasorb and a nylon 6,6 filter from Sartorius Filters, Inc. showed no significant differences (*18*).

The retention of NO_2 on nylon 6 Nylasorb filters (Gelman Sciences), nylon 6 filters from Sartorius Filters, Inc., and nylon 6,6 filters from Schleicher & Schuell (S&S) was assessed; 0.5 ppm NO_2 was sampled at 75–80% RH and 21 °C for 6 h at 20 L/min with 47-mm filters. The results showed no more than 0.1% retention (*19*). Similarly, Perrino et al. observed with 47-mm Gelman Nylasorb filters (nylon type unstated) 0.4% retention, sampling 0.6 ppm NO_2 for 15 h at 1.5 L/min and 60–80% RH (*20*).

The degree of retention of SO_2 on these nylon filters was also assessed; 0.14 ppm SO_2 was sampled at 20 °C, 80% RH, and 20 L/min (total dosages for each filter, 2600 μg of SO_2) (*19*). At saturation, the artifact SO_4^{2-} values with the S&S filter, 56 $\pm$ 1 μg, was lower than that for the Sartorius and Gelman nylon 6 filters, which ranged from 65 to 70 μg.

Accordingly, the change in nylon polymer should not hamper atmospheric HNO_3 determinations. However, on the basis of the results with the acidic gas, SO_2, the capacity of nylon 6,6 filters for HNO_3 retention without breakthrough may be somewhat lower than that reported previously with nylon 6 filters.

Other Interferent Studies with Nylon Filters. Retention of HONO on Gelman Nylasorb (nylon 6) filters was tested; 0.3 to 1 $\mu g/m^3$ HONO was sampled in air (~50% RH) at 20 L/min with 47-mm filters for 16–23 h. At saturation, the filters contained 1.8 $\mu g/cm^2$ NO_2^-; this result indicates less than 50% efficiency for HONO retention (19). Sampling 440 $\mu g/m^3$ HONO at 20 L/min with 47-mm Gelman Nylasorb filters (nylon type unstated), Sickles and Hodson reported about 25% retention efficiency, with a similar saturation value, and noted that such retention was reversible. Substantial oxidation of HONO to NO_3^- on the filter surface by O_3 was demonstrated in laboratory and atmospheric trials (21). Perrino et al. (20) observed with 47-mm Gelman Nylasorb (nylon type unstated) filters an efficiency for HONO that increased from about 25 to about 90% as the flow rate decreased from 12 to 2 L/min, with saturation at ~50 $\mu g/$filter. Nitrite to nitrate conversion on the nylon filter surface increased from 13% at 45 ppb O_3 to 93% at 200 ppb O_3.

Because atmospheric HONO exhibits nighttime and early morning maxima (22), in contrast to the midday maxima for HNO_3 and O_3, little midday interference from HONO would be expected in HNO_3 measured with short-term samplers. However, with long-term (e.g., 24-h) sampling, substantial interference from atmospheric HONO might be observed with techniques involving nylon filters, depending on details of the sampling strategy.

PAN is a potential interferent with daytime concentration maxima that can coincide with the HNO_3 maxima. However, early work by Joseph and Spicer (23) and later work by Fahey et al. (24) indicated that PAN is not a significant interferent in HNO_3 measurements relying on nylon-filter collection. The change in nylon-filter composition introduces a small degree of uncertainty to this conclusion.

Talbot et al. reported that O_3 can react with unwashed nylon filters to produce increased extractable NO_3^- (25). The artifact HNO_3 was equivalent to about 0.1 ppb (0.25 $\mu g/m^3$) under the conditions used; this result suggests that the artifact would only be significant at background locations.

Nitrite oxidation to NO_3^- following removal from the sampler can probably be minimized by freezing nylon (and other) filter samples immediately after collection and analyzing the aqueous filter extracts as soon as possible.

Common Pathways for Loss of Nitric Acid (Without Artifact Particulate Nitrate Formation). *Introduction.* Loss of HNO_3 can result from sorption on the surfaces of sampler inlets and inlet lines; this situation is common to all collection procedures for atmospheric HNO_3. Such inlet losses are rarely determined, so it is difficult to assess the accuracy of atmospheric HNO_3 measurements or to pinpoint the source of bias between methods in intermethod comparisons (e.g., *see* reference 26). In addition to sampler-associated losses, NO_3^- can be lost from samples during storage.

HNO₃ Losses on Tubing and Inlet Surfaces. Several laboratory studies have examined the degree of loss of HNO_3 at low concentrations in air when it passes through tubing of varying composition. All types of tubing showed decreasing loss of HNO_3 as the surfaces became conditioned. These studies concluded that Teflon and Pyrex glass tubing provide the least HNO_3 retention; Teflon is slightly better than glass (27, 28). A comparison of tubing fabricated from three types of Teflon, polytetrafluoroethylene (PTFE), fluorinated ethylene–propylene (FEP), and perfluoroalkoxy (PFA), was made by sampling successive <1-μg doses of HNO_3 in air. The results showed PTFE tubing to produce severe losses of HNO_3. Loss with FEP tubing varied from zero to moderate, depending on the batch tested. A single batch of PFA tubing exhibited no measurable loss (19). The results suggest that differing porosities account for the results with the three types of Teflon; FEP and PFA products are prepared by extrusion molding of the molten polymers, whereas PTFE products are prepared by compression and sintering of a powdered polymer of relatively high melting point, a process that results in a substantial, and easily varied, void content in the finished product.

A laboratory evaluation (29) of losses of HNO_3 within seven different cyclones or impactors used as inlets for HNO_3 measurements suggested that large differences should be observed between atmospheric samplers using these devices; initial transmission efficiencies ranged from 28 to more than 100%. Loss of HNO_3 within the devices varied inversely with residence time as well as with the total dose sampled, and loss increased with RH. The extensive loss of HNO_3 observed with PTFE tubing was not reflected in the results with a cyclone fabricated from this polymer; this result probably reflects the much lower surface-to-volume ratio with the cyclone and, perhaps, differing Teflon porosity. A comparison of atmospheric HNO_3 results between samplers equipped with these same inlets showed much better agreement than that predicted by the laboratory findings (29). The data suggested that atmospheric sampling passivated the inlet walls for HNO_3 sampling much more quickly than sampling HNO_3 in clean, humidified air. Further work is needed to resolve the discrepancy between laboratory and atmospheric results.

HNO₃ Losses During Sample Storage. The loss of up to 90% of the NO_3^- from atmospheric particulate-matter-loaded glass and quartz-fiber filters has been documented during 2- to 18-month storage at room temperature (30, 31). The storage container used probably influenced strongly the degree of loss; in trials with atmospheric samples collected with Teflon filters stored in plastic petri dishes (which had relatively tightly fitting lids), no loss of nitrate was detectable in 2 months of storage at room temperature (32). Loss of nitrate may be partially offset by bacterial oxidation of NH_4^+ to NO_3^-, a process that is evident in aqueous extracts of atmospheric particulate samples. Low-temperature storage immediately following sampling is the prudent strategy for all nitrate samples.

Filter Pack Method for Atmospheric Particulate Nitrate and Nitric Acid and Associated Errors. *Introduction.* The sampling method chosen to determine atmospheric PN and HNO_3 is usually the simplest technique affording an acceptable degree of accuracy for a given application. The most widely used technique for atmospheric PN and HNO_3 uses two filters in tandem (a "filter pack"); the first retains PN, while the second traps HNO_3. The well-recognized problem of artifact PN on most filter types has led to the use of Teflon prefilters as the preferred medium; this use can nearly eliminate HNO_3 retention on the filter itself. However, the apparent PN and HNO_3 measured with such a filter pack still reflects the net result of opposing sources of error. Factors minimizing one error source may enhance others. Therefore, concentrations of the nitrate species measured in this way are approximations. Even the observation of relatively good agreement between such results and those for reference procedures may reflect approximate cancellation of errors under the conditions encountered.

Methodology. Teflon filters exhibit the lowest HNO_3 retention of those in common use (*12*). Fluoropore (Millipore Corporation) and Zefluor PTFE filters (Gelman Sciences) of 0.5- to 2-μm pore size, are frequently used as prefilters in the filter pack method (FPM). With 47-mm-diameter filters, flow rates with the FPM from 10 to 20 L/min are typical. The 2-μm pore size Zefluor filter exhibited 97% efficiency for 0.035-μm particles at a face velocity of 33 cm/s, with a pressure drop of 1.0 cm Hg (1.3 kPa) (*33*).

The after-filters in common use with the FPM include nylon (*17*) and NaCl-impregnated cellulose (e.g., Whatman 41) filters (*13, 16*). The latter, impregnated by the user, have very high efficiency and capacity (*15, 16*), low cost, and relatively low air-flow resistance. However, the high chloride ion concentration in aqueous filter extracts has hampered NO_3^- determination by ion chromatography for some analysts. Nylon filters (e.g., 1-μm pore size Nylasorb from Gelman Sciences) are relatively expensive and can exhibit significant and varying NO_3^- blanks together with varying air-flow resistance. Filters impregnated with strongly alkaline tetra-*n*-butylammonium hydroxide have also been used as after-filters (*34*). Retention of NO_3^- of less than 5% was demonstrated. However, the probable partial retention of PAN and efficient retention of HONO make this and other strong-base-impregnated after-filters less desirable options for atmospheric sampling of HNO_3 under most circumstances.

Loss of HNO_3 to Filter Holder Surfaces. The walls of filter holders as well as the prefilter support screens are likely sites for HNO_3 retention. Polycarbonate filter holders (Nuclepore Corporation) are frequently used because these can be easily assembled into two or more holders in tandem. Significant loss of HNO_3 on these filter holders has not been reported but would be expected when HNO_3 is sampled at very low concentrations (unless the holders are conditioned by prior sampling). Using an all-Teflon filter holder to minimize losses, Goldan et al. (28) observed significant losses of HNO_3 by sampling with a nylon filter preceded by a Teflon prefilter. These workers sampled dosages from 2 to 10 μg of HNO_3 to simulate sampling in a clean air environment, and they observed losses up to 40% in dry air and 55% in moist air. The lost HNO_3 was not recoverable from the prefilter.

Discussion of Atmospheric Results.   Depending on the aerosol composition and concentration, the concentration of strongly acidic species in the particle and gas phases, and the meteorological conditions, the potential concentration of artifact HNO_3 formed by conversion of the deposited PN may be relatively small or very large. In Southern California, the concentrations of particle-phase strong acids are usually relatively low, and HNO_3 and NH_3 levels can be relatively high with large diurnal temperature changes. Particulate nitrate levels during summertime periods can exceed 30 μg/m³ and can be similar in concentration to HNO_3. Consequently, a large potential error, as a percent of TIN, PN, or HNO_3 (or in μg/m³), exists with the FPM because of NH_4NO_3 dissociation following collection. Recoveries of PN from 2- to 8-h samples collected with Teflon prefilters can be as low as 10% when sampling includes periods at or above 30 °C. At nighttime, however, under conditions of lower ambient temperatures and high RH, little or no net loss of PN is usually found, a finding consistent with the decrease in dissociation constant for NH_4NO_3 under these conditions (14). In a major intermethod comparison performed in Southern California, FPM HNO_3 results by all participants averaged 20% higher than those by a spectroscopic technique (26).

At the opposite extreme, for atmospheric sampling in the northeastern United States, where the acidity of atmospheric particulate matter from acidic sulfates can be relatively high (35), or at rural or background sites with very low TIN, a large proportion of the TIN can be present as HNO_3 when sampled. For example, at a rural site in southwestern Ontario, Canada, nearly 70% of the TIN appeared to be HNO_3 (36). In these cases, the maximum potential artifact HNO_3, as a percent of TIN or of HNO_3, is reduced. The Canadian FPM HNO_3 results did not exceed those with the reference procedure, a tunable diode laser system; indeed, the filter results were lower. Similarly, at background sites, with measured TIN levels less than a few hundred parts per trillion, the potential artifact HNO_3 concentration has been argued to be reduced for the same reason. At these background sites, other sources of error (e.g., contamination of blanks and samples) were judged to be more significant than loss of NO_3^- from the prefilter (28, 37) in both PN and HNO_3 measurement.

If the prefilter contains relatively little PN, then retention of HNO_3 on the particulate matter of the prefilter might dominate over loss of PN. This situation is most likely to occur at locations providing a large percentage of the TIN as HNO_3, accompanied by lower ambient temperatures and higher RH.

Alternative Integrative Sampling Strategies. *The Denuder Difference Method. General Description.* First proposed by Shaw et al. (38), the denuder difference method (DDM) provides a strategy that can measure HNO_3 and fine-particle nitrate (FPN) free of a number of potential errors to which the FPM is subject. The DDM samples with two filter packs in parallel, with each using a cyclone or impactor to exclude particles larger than 2 or 3 μm in aerodynamic diameter. Between the inlet and filter pack of one sampler is a diffusion denuder, designed to remove HNO_3 with high efficiency. Each filter pack consists of a Teflon and a nylon filter as described in the preceding section. The nylon filter in the sampler with the denuder serves to trap (as HNO_3) any FPN that volatilizes from the Teflon filter. Thus, one sampler collects atmospheric HNO_3 plus FPN, and the second collects FPN only. The difference in the total of the NO_3^- recovered from the Teflon plus nylon filters of each sampler, together with the measured air volume, provides a measure of the HNO_3 concentration. Alternatively, if particulate SO_4^{2-} concentrations are not required, the Teflon filters can be omitted and the nylon filters used as efficient collectors of both FPN and HNO_3. This strategy provides improved precision, decreases filter and analytical costs, and has been recommended elsewhere (39).

The restriction of samplers to fine particles and gases is necessitated by the frequent presence of coarse PN. Such particles may be collected by impaction on the walls of the denuder, thereby decreasing the recovered PN and increasing the calculated HNO_3. An addition filter sampler, operated without restriction to fine particles, is needed to collect TIN and thereby to obtain coarse nitrate (CPN):

$$CPN = TIN - (HNO_3 + FPN) \tag{2}$$

Design Parameters. In a diffusion denuder an airstream, in laminar flow, contacts a surface (e.g., the inner wall of a tube) coated with a sorptive or reactive trapping medium for the gaseous pollutant of interest. Fine particles, which have diffusion coefficients more than 4 orders of magnitude smaller than those for gases, penetrate the denuder with high efficiency. Forrest et al. (40) reported losses of 0.2–2.2% for particles between 0.3 and 0.6 μm and about 4–5% for 1- to 2-μm particles.

The theoretical efficiency of a diffusion denuder for a gaseous pollutant can be calculated for various geometries with the assumption of a high probability for reaction and complete retention of the pollutant upon contact with the coated surface. The Gormley–Kennedy equation (*41*) provides for a denuder constructed from cylindrical tubes:

$$C/C_0 = 0.82 \exp(-15\Delta) + 0.098 \exp(-89\Delta)$$
$$+ 0.033 \exp(-228\Delta) \tag{3}$$

where C is the average concentration leaving the tube, C_0 is the average concentration entering the tube, $\Delta = DL\pi/4Q$, D is the diffusion coefficient (cm^2/s), L is the aggregate total length of the coated section in each tube (cm), Q is the flow rate (cm^3/s), and $(1 - C/C_0)$ is the fractional denuder efficiency. This expression is valid if the flow rate is sufficiently low to ensure laminar flow (i.e., the Reynolds number is less than 2000). The Reynolds number is determined from the equation

$$N_{Re} = dV \rho/\eta \tag{4}$$

where N_{Re} is the Reynolds number, d is the internal diameter of each tube (cm), V is the velocity of gas in the tube (cm/s), ρ is the gas density (g/cm^3), and η is the gas viscosity ($g/cm \cdot s$).

Diffusion coefficient values for HNO_3 ranging from 0.12 to 0.15 cm^2/s have been reported. Trapping media for HNO_3 include MgO (*38*), Na_2CO_3 (*42*), Na_2CO_3–glycerol (*43*), NaCl (*44, 39*), NaF (*45*), and anodized aluminum (*46*).

A typical design uses multiple tubes operating in parallel. For example, 11 tubes, each coated for 30 cm with MgO, provide a calculated efficiency of 85.7% at a total flow rate of 20 L/min (with $D = 0.15$ cm^2/s). A laboratory determination at RH = 30–80% and T = 21–32 °C yielded an efficiency of 87.8 ± 5.1% ($n = 5$), a value consistent with the calculated result (*47*). Experimental results can be corrected for denuder penetration. Denuders for atmospheric nitrate use are normally designed for ≥95% efficiency.

To develop laminar flow prior to HNO_3 collection, the initial section of each tube remains uncoated. The uncoated length required can be calculated (reference *48*; *see* also Chapter 2). For example, a denuder with 24 6-mm-i.d. tubes operating at a total flow of 20 L/min requires a 6-cm uncoated section in each tube.

Accuracy and Sources of Error.   A comparison of DDM HNO_3 results by separate laboratories using the Teflon–nylon and "nylon filter only" strategies showed agreement, on average, within 15% (*49*). Further validation of this alternative would be required when sampling with increased concentrations of particle-phase acids. Additionally, nylon filters (1-μm pore size Nylasorb) exhibit a greater tendency to plug up at high particle loadings than 2-μm pore size Zefluor filters (J. Horrocks, private communication).

The DDM method has been extensively compared to other techniques, including spectroscopic methods (e.g., reference 26). For example, Figure 2 compares 4- to 6-h DDM atmospheric results (without Teflon filters) to those obtained with a tunable diode laser (TDL) technique; values are about 30% higher with the DDM (49). The TDL method uses a PFA Teflon inlet line and an in-line PTFE Teflon filter to introduce a particle-free air sample into an optical cell at subambient pressure. The TDL HNO_3 results averaged about 16% less than those by Fourier transform infrared spectroscopy, a long-path in situ method considered to be the reference procedure (26). Therefore, the above DDM results were inferred to be too high by, on average, 14%.

The potential sources of error and imprecision in the DDM include the following:

1. Retention, on nylon filters as well as on the denuder, of atmospheric species other than HNO_3 that form NO_3^-.

2. Volatilization of NH_4NO_3 during passage through the denuder, which would cause low FPN values and correspondingly high calculated HNO_3 concentrations.

3. Poor precision at low concentration because the DDM is a difference method. The limit of detection is, therefore, relatively high. For example, with Nylasorb nylon filters exhib-

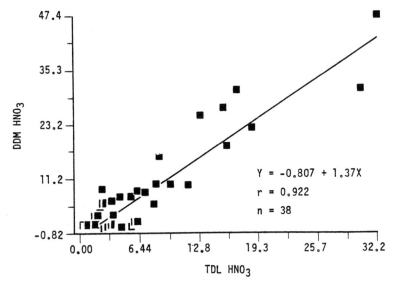

Figure 2. TDL HNO_3 results plotted against DDM HNO_3 results ($\mu g/m^3$). (Reproduced with permission from reference 49. Copyright 1988.)

iting a NO_3^- field blank of 3.0 ± 0.3 μg per 47-mm filter, the limit of detection by the DDM for 5-h samples at 20 L/min was 0.2 μg/m^3, based on twice the standard deviation of the blanks (49).

Nitrate formation from retention of HONO on nylon filters was discussed in the **Formation of Nitrate on Nylon Filters** section. Efficient removal of HONO has been demonstrated on carbonate-coated denuders (42), so potential interference with all DDM methods using strongly alkaline denuders is likely (but not on denuders coated with NaCl or NaF, as discussed in following sections). Positive interference in HNO_3 measurements due to HONO is likely with the DDM (as well as with the FPM) during nighttime and early morning periods; the magnitude varies with details of sample handling as well as ambient HONO levels.

To examine volatilization of NH_4NO_3 while within a denuder, several laboratories evaluated the extent of such loss by using laboratory-generated aerosols. None found evidence supporting significant losses under normal denuder conditions. For example, Larson and Taylor (50) followed the rate of evaporation of 0.4-μm aqueous droplets of NH_4NO_3. They concluded that for a denuder residence time of 1.0 s the error due to evaporation and loss of the nitrate salt averaged $1.8 \pm 1.5\%$. Typical denuders for HNO_3 use residence times of 0.2–0.3 s, decreasing the anticipated error. Similarly, Forrest et al. (40) found insignificant loss of smaller than 3-μm-diameter NH_4NO_3 particles by such evaporation. Pratsinis et al. (51) developed a model for a dissociating aerosol and concluded that for NH_4NO_3 particles no larger than 0.1 μm in diameter evaporation during a 0.2-s residence time *was* significant and that such evaporation increased markedly with temperature. However, size distribution measurements for NH_4NO_3, including a recent report (52) using techniques that appeared to minimize volatilization following collection, support generally larger particle sizes for this aerosol. Thus, on the basis of the Pratsinis model as well as experimental results, volatilization of NH_4NO_3 within a denuder should not be a significant factor.

Annular Denuder Method (ADM).   General Description.   By replacing open-bore tubes with a 1- to 2-mm annular space between concentric cylinders, Possanzini et al. (53) developed a denuder that was about a factor of 30 more efficient per unit length compared to an open-bore denuder tube. Typically, annular denuder tubes are used with two or three sections in tandem (Figure 3) and operate at 10–20 L/min. As with the DDM, it is important to remove coarse PN to minimize error in nitrate measurements because of its possible impaction on the annulus walls or elsewhere within the denuder. The system shown in Figure 3 uses a Teflon-coated glass cyclone and glass denuders available from University Research Glassware (Carrboro, North Carolina).

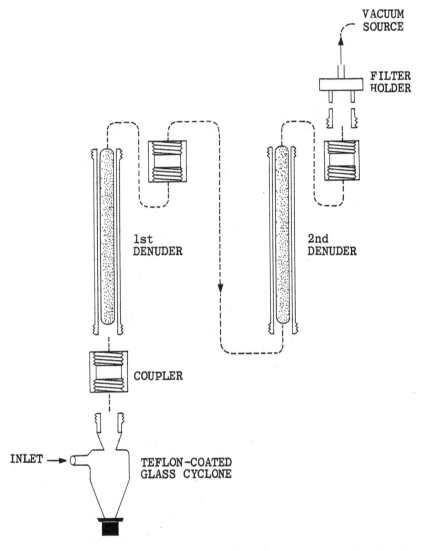

Figure 3. Schematic diagram of an annular denuder system. (Reproduced with permission from reference 22. Copyright 1990.)

In contrast to the DDM, in the ADM HNO_3 (together with other acidic gas) concentrations are obtained directly by analyzing aqueous extracts of the denuder coating. In principle, therefore, greater precision at low concentration should be possible for atmospheric HNO_3 with the ADM compared to the DDM methods. However, the method is labor-intensive, requiring extraction and recoating of the denuders at the sampling site. The greater potential of the method for operator error (e.g., sample loss) compared to the DDM may diminish the observed precision and accuracy. Furthermore, the commercially available glass denuders are both expensive and fragile. A Teflon–nylon filter pack or a nylon filter alone downstream of the denuder provides a measure of FPN analogous to the denuded sample in the DDM.

Design Parameters. With the assumption that the walls of the annulus are coated with a perfect collector for HNO_3, the efficiency of an annular denuder can be calculated (53):

$$c/c_0 \approx 0.819 \exp\left(-22.53\frac{\pi DL(d_1 + d_2)}{4F(d_2 - d_1)}\right) \tag{5}$$

where c is the average gas concentration leaving the denuder, c_0 is the gas concentration entering the denuder, D is the diffusion coefficient of the pollutant, L is the length of the coated annulus, F is the flow rate, and d_1 and d_2 are the internal and external diameter of the annulus, respectively.

Accuracy, Limit of Detection, and Sources of Error with CO_3^{2-}–Glycerol Coated Denuders.

1. **Retention Efficiency of Acidic Gases.** Acidic gases (e.g., HNO_3, HONO, HCl, or SO_2) are efficiently trapped by a denuder coating prepared by using 1% Na_2CO_3–1% glycerol in a 1:1 (v/v) mixture of methanol–water (19, 43, 54). For example, when sampling 10 to 20 $\mu g/m^3$ HNO_3 in clean air at 17 L/min, 50% RH, and 20 °C and at 80% RH and 13 °C, the collection efficiency of a single, 21-cm, CO_3^{2-}–glycerol coated annular denuder was 99% or greater (19).

2. **Interferents.** The principal sources of error in HNO_3 determination with the ADM and CO_3^{2-}–glycerol coated denuders, include (1) greater than 98% retention of HONO on the first denuder (55) followed by partial conversion of the resulting NO_2^- to NO_3^- during sampling and following extraction, and (2) retention, with low efficiency, of other NO_2^-- and NO_3^--yielding species, including NO_2 and PAN.

   Relatively little retention of NO_2 on CO_3^{2-}–glycerol coated denuders has been observed. For example, Appel et al. observed 0.2% at 80% RH at room temperature (22). However, greater retention is possible, because NO_2 can be converted to HONO by reaction with H_2O on inlet surfaces, and HONO is retained with high efficiency by this coating. Because atmospheric NO_2 levels can be much higher than HNO_3 concentrations, such retention cannot be ignored. The inlet sec-

tion of commercial annular denuders is Teflon-coated for about
2.5 cm ahead of the annulus, but Jenkins et al. (56) noted that
PTFE-coated glass was indistinguishable from glass in pro-
moting HONO formation from NO_2.

PAN can be hydrolyzed quantitatively to NO_2^- in alkaline
solution. Thus, at least partial retention of PAN on alkali-
coated denuders is likely. Ferm and Sjodin (42) reported up
to 10% retention of PAN on an open-tube denuder coated with
Na_2CO_3. Facile conversion of NO_2^- to NO_3^- on carbonate-
impregnated filters by oxidation with O_3 has been reported
(21), so similar conversion on alkali-coated denuders is likely.
The use of glycerol in the alkaline denuder coating has been
shown to decrease substantially the extent of nitrite oxidation
(55). Slow oxidation of NO_2^- to NO_3^- in extracts of the denuder
coating is also probable, a factor that can probably be minimized by low-temperature storage and by minimizing delays
before analysis.

The use of tandem denuders, each with the same coating,
permits at least partial correction for these sources of error in
HNO_3 measurement. Species retained with low efficiency on
carbonate coatings (e.g., NO_2 and PAN) are collected in about
equal amounts by successive denuders. If it is assumed that
oxidation during and subsequent to sampling occurs equally
on successive denuders as well as in their corresponding extracts, the NO_3^- from the second denuder extract can be subtracted from that in the front denuder extract to obtain a
measure of atmospheric HNO_3. However, this technique does
not correct for the efficient retention (and subsequent oxidation) of atmospheric or artifact HONO.

The measurement of FPN from the NO_3^- recovered from
a backup Teflon–nylon or nylon filter with the ADM is subject
to some of the same potential sources of error as with the
DDM. No loss of FPN passing through the annular denuder
is expected on the basis of particle-loss measurements (55).
Although HONO is efficiently removed ahead of the filter or
filters, nearly all of the PAN would reach the nylon filter.
Retention of PAN on nylon filters has been reported to be
insignificant. Thus, relatively little error in FPN is expected
with the ADM and CO_3^{2-}–glycerol coated denuders.

3. **Uncertainty in Correcting Atmospheric Data.** The level of
uncertainty with this method can be illustrated with results
from atmospheric trials (57). After 4- to 6-h samples were
collected in Southern California with CO_3^{2-}–glycerol coated
annular denuders during a late fall period, the NO_3^- recovered

from the rear denuder averaged 14% of that on the front denuder. If the nitrate on the rear denuder was due to incomplete retention of HNO_3, in spite of the laboratory results showing greater than 99% removal efficiency by the front denuder (19), atmospheric HNO_3 concentrations should have been corrected by +14%. However, we judged the partial retention of nitrite-yielding species, followed by oxidation, to be the most probable source of the observed NO_3^- on the rear denuder. Thus, subtraction of this NO_3^- from that on the front denuder is the preferred strategy to calculate atmospheric HNO_3 concentrations.

The simultaneous collection of HONO and HNO_3 during nighttime and early morning periods on the front denuder, followed by partial oxidation of the resulting NO_2^- to NO_3^-, remains the most significant source of error with the ADM and CO_3^{2-}–glycerol denuder coatings. In Southern California, 4- to 6-h average HONO concentrations up to ~15 ppb (28 $\mu g/m^3$ as NO_2^-) were recently observed (22). Much lower HONO concentration maxima are the norm at other locations.

The efficiency of HNO_3 collection with a single annular denuder should be confirmed by each user with a known source of HNO_3, regardless of denuder coating used. On the basis of laboratory studies, Ferm and Sjodin (42) reported a technique that can be used to correct simultaneously for penetration of HNO_3 through the first of tandem denuders as well as for artifact NO_3^- due to atmospheric species retained with low efficiency.

An Improved ADM Strategy for HNO₃. An alternative strategy that is reported to eliminate interference from HONO uses an NaCl-coated denuder. Removal of HNO_3 with 97% efficiency on an initially NaCl-coated denuder at 15 L/min was reported, without collection of HONO, and with less than 1.5% retention of NO_2 (39, 44). An NaF coating (45) would also provide efficient HNO_3 collection while minimizing retention of interferents and eliminating interference with NO_3^- determination by ion chromatography due to high chloride concentrations. However, NaF is highly toxic, a disadvantage for field operations. Variability in its performance was also observed (44). Nevertheless, this approach has been recently used by U.S. Environmental Protection Agency (EPA) personnel in atmospheric sampling (58). The NaCl- (or NaF-) coated denuder can be followed by a second denuder with the same coating to confirm that the combined effects of penetration and artifact NO_3^- formation are negligible. Whether any NO_3^- observed on the second denuder should be added or subtracted from that on the first denuder remains uncertain. Alternatively, the front denuder can be followed by one CO_3^{2-}–glycerol coated annular denuder to remove HONO ahead of a Teflon–nylon (or nylon alone) filter pack to determine FPN free of error from partial HONO retention on nylon filters. If simultaneous measurement of HONO and other acidic gases is desired, then two CO_3^{2-}–glycerol coated denuders to permit correction for partial retention of PAN and NO_2 can be operated downstream of an NaX-coated denuder (X denotes Cl or F).

In addition to greatly increasing the sampling and analytical requirements, if other acidic gases are to be measured the NaCl-coated denuder may show decreased efficiency at low RH. Forrest et al. (13) observed a sharp decrease in HNO_3 collection efficiency with NaCl-coated filters at 25% RH. Similar studies with denuders have not been reported, however. A comparison of atmospheric HNO_3 results with the ADM using NaX-coated denuders with those by reference procedures has not been reported, so the accuracy of the method remains to be confirmed.

Summary and Conclusions. The methods discussed and the potential errors associated with specific sampler components are summarized in Table II. Measurement of the labile atmospheric pollutants HNO_3 and particulate NO_3^- is subject to a multiplicity of errors with all methods that rely on the collection of a sample. The method selected will frequently be that providing an acceptable degree of uncertainty and may be influenced by the need to determine other species simultaneously. In general, the filter pack method is the least accurate but simplest procedure. A comparatively new technique, collection of HNO_3 with an NaCl- or NaF-coated annular denuder followed by HONO removal with a CO_3^{2-}–glycerol coated denuder and fine particulate NO_3^- collection with a backup nylon filter, appears to address many of the systematic sources of error and may prove to be the integrative sampling method that provides the greatest accuracy. However, its validation is incomplete. Regardless of the inherent accuracy of the method chosen, all inlet and sampler surfaces ahead of collection media should be treated as potential sources of HNO_3 loss until proven otherwise.

Because of the labor-intensive nature of annular denuder sampling, alternative techniques deserve further scrutiny. The DDM, relying on Teflon–nylon filter packs (or nylon filters alone) and an anodized aluminum annular denuder ahead of one of a pair of modified dichotomous samplers (46), merits evaluation for the influence of potential interferents (e.g., HONO). The DDM with NaCl-coated denuders would probably be nearly free of interference from HONO, but this conclusion requires evaluation. Because coarse particulate NO_3^- must be excluded to achieve improved accuracy for HNO_3 and fine particulate NO_3^-, supplementary samplers are needed to obtain improved estimates of total as well as coarse particulate NO_3^-.

Sampling of Particulate Organic Compounds and Organic Carbon

Introduction. *Sampling Problems.* The vapor pressures of organic compounds in the atmosphere range from more than 1 atm (101 kPa) (i.e., completely gaseous under ambient conditions) to nearly zero. Semivolatile materials, those with vapor pressure (v.p.) in the range 10^{-4} to 10^{-8} torr (10^{-2} to 10^{-6} Pa), can have significant concentrations in both the particle and vapor phases. Although it is relatively easy to determine the total concentrations (i.e., particle plus vapor phase) for such semivolatile materials, the accurate measurement of the average concentrations existing in one phase at the moment of sampling is difficult. The determination of particulate organic compounds (as total C or as individual semivolatile species) may be subject to positive errors from sorption of species that are in vapor phase at the sampler inlet. Sorption may occur on a sampling medium, usually a filter, as well as on previously collected particulate matter. Conversely, volatilization may occur following collection, resulting in a negative artifact.

The composition of the carbonaceous species retained by sorption on sampling media has rarely been evaluated. Fung (59) reported that the carbonaceous materials retained by sorption on quartz-fiber filters were relatively polar organic compounds, including phthalate esters and nitrogen-containing heterocyclic compounds. Additional studies of this type would be very helpful.

Sorption on previously collected particulate matter as well as on filter media may be inferred from a comparison of sampler results. Filter samples collected for 14 h were compared to seven successive 2-h samples for recovered carbonaceous materials taken in parallel atmospheric sampling in Southern California (60). The collected organic materials were characterized by their solubility in solvents of varying polarity. The mean ratio of calculated-to-observed 14-h average concentrations was 0.7 ($n = 12$) for nonpolar organic compounds and 1.7 for polar organic compounds. Such results are consistent with sorption of nonpolar organic compounds on previously collected particles (more significant in the 14-h samples) and sorption of polar organic compounds on blank Gelman AE glass-fiber filters (which would give an upward bias to the 14-h results calculated from 2-h samples). The latter results are consistent with the more recent work of Fung. However, the nature and significance of sorption on previously collected particles requires additional, and more direct, experimental results.

Experiments attempting to measure the extent of loss of organic C from the particle phase following collection have also been performed. The simplest experiment consists of passing carbon-free air through a filter loaded with freshly collected atmospheric particulate matter and determining the decrease in organic C. Such experiments (61, 62) have failed to show sig-

Table II. Summary of Manual Techniques and Sources of Error in Sampling Particulate Nitrate and Nitric Acid

Method	Sampler Element	Purpose(s)	Possible Sources of Error
Filter pack	Cyclone or impactor	Remove coarse particles able to react with HNO_3	Partial retention of HNO_3, especially at high RH and longer residence times
	Tandem filter holder	Support tandem filters without leakage	Slight retention of HNO_3 even with Teflon holders; significant at very low concentrations
	PTFE prefilter	Collect FPN and other particles and transmit HNO_3 to collection filter	Partial retention of HNO_3 on particles (enhanced if coarse particles are collected); volatilization or reaction of FPN yielding HNO_3
	Quartz prefilter	Same as PTFE prefilter	Partial retention of HNO_3 on filter and particles (enhanced if coarse particles are collected); volatilization or reaction of FPN yielding HNO_3
	Nylon after-filter	Collect HNO_3	Collection of volatilized PN; retention of HONO; retention of unidentified species yielding NO_3^-
	NaCl-impregnated cellulose after-filter	Collect HNO_3	Collection of volatilized FPN; possibly low HNO_3 efficiency at very low RH

Method			Comment
Denuder difference	Cyclone or impactor	Prevent loss of CPN to the denuder	Same as with filter pack method
	Denuder (all coatings)	Remove HNO_3 and transmit FPN to after-filter(s)	Incomplete removal of HNO_3
	Denuders with highly alkaline coatings (e.g., CO_3^{2-} or MgO)	Same for all coatings	Incomplete removal of HNO_3; efficient retention of HONO
	Anodized Al denuder	Same for all coatings	Incomplete removal of HNO_3 (degree of interferent retention unknown)
	Teflon prefilter plus nylon after-filter following denuder	Collect FPN (Teflon plus nylon) and other aerosol constituents (Teflon)	None identified (assuming HONO removed by denuder)
	Holder, Teflon prefilter plus nylon after-filter without denuder	Collect HNO_3 + FPN (Teflon + nylon) and other aerosol constituents (Teflon)	Slight loss of HNO_3 to holder; retention of HONO and unidentified species yielding NO_3^-
Annular denuder	Cyclone or impactor	Same as with denuder difference method	Same as with filter pack method
	Denuder with CO_3^{2-} or glycerol–CO_3^{2-} coating	Remove HNO_3 and transmit FPN to after-filters	Incomplete retention of HNO_3; retention of HONO and unidentified NO_3^--yielding species
	Denuder with NaCl or NaF coating	Same for all coatings	Incomplete removal of HNO_3 (degree of interferent retention appears to be small)

nificant loss of organic C and are in marked contrast to findings with specific organic compounds [e.g., PAH compounds (63)].

The problems associated with sampling particulate organic C in the atmosphere crudely parallel those with sampling nitrates. However, the very large number of carbonaceous species, the range of vapor pressures involved, and the ubiquitous nature of carbon severely complicate development of improved sampling techniques based on total (or organic) carbon measurements.

Theory of Vapor–Particle Partitioning. A theoretical treatment of the volatilization of semivolatile particles following collection has dealt with a pure, single-component aerosol in equilibrium with its vapor at the sampler inlet (1). Pure compounds assumed to be retained by adsorption on the surface of less-volatile particulate matter (e.g., PAHs) have been treated more empirically by Yamasaki et al. (64) and Bidleman et al. (65) and reviewed by Pankow (66). In both cases, the driving force for volatilization is a decrease in vapor concentrations below the equilibrium value for each species. This decrease can be caused by a combination of (1) atmospheric dilution or diurnal concentration change, (2) the pressure drop across the aerosol deposit on the sampling medium, (3) the pressure drop across a sampling medium that allows particles below the surface (e.g., a "depth" collector such as a quartz- or glass-fiber filter), and (4) the pressure drop caused by sampler inlets and inlet tubing, impactor jets, and so forth. Zhang and McMurry (1) concluded that for single-component aerosols, losses from impactors should be less than those from filter samplers.

Qualitative evidence of the possible influence of pressure drop can be seen by comparing recoveries of carbonaceous material, as carbon, while varying filter sampling flow rates. For example, samplers operating at a face velocity (i.e., flow rate per unit area of the macroscopic filter surface) of 11 cm/s yielded an average of 30% greater total carbon, in $\mu g/m^3$, compared to a standard high-volume sampler operated at 50 cm/s (67).

Efforts to overcome the problems of sampling semivolatile carbonaceous materials have dealt with particulate organic C (62, 67, 68) and, more successfully, with the measurement of individual organic compounds (e.g., reference 63). The partitioning of PAHs and pesticides between particle and vapor states has served as the framework for developing and testing theories and sampling techniques. A discussion of these theories and techniques follows. However, the use of PAH compounds as models for semivolatile carbonaceous materials may be inappropriate for those that can approach single-component aerosols.

With PAH as well as other organic compounds, oxidation reactions, including reactions with O_3 and NO_2, may be significant in altering recoveries (e.g., reference 69) of specific compounds. Chemical reactivity of carbonaceous species is not, however, considered in this chapter, in part because, on a total C basis, such reactions would have reduced significance.

Table III lists the empirical formulas, molecular weights, and, where available, boiling points and vapor pressures for PAH compounds. This chapter discusses, in addition to particulate C, both theory and application in sampling such compounds as a guide in developing improved particulate C samplers.

Table III. Physical Properties of Selected PAH Compounds

Compound	Empirical Formula	Molecular Weight	Boiling Point (° C)[a]	Vapor Pressure (torr) for Crystalline Solid at 20 °C	
				Bidleman et al.[b]	Pupp et al.[c]
Phenanthrene	$C_{14}H_{10}$	178.2	338	6.1×10^{-4}	
Anthracene	$C_{14}H_{10}$	178.2	340	4.1×10^{-4}	
Fluoranthene	$C_{16}H_{10}$	202.3	383	4.2×10^{-5}	
Pyrene	$C_{16}H_{10}$	202.3	393	7.3×10^{-5}	
Benz[a]anthracene	$C_{18}H_{12}$	228.3	435[d]		
Cyclopenta[c,d]pyrene	$C_{18}H_{10}$	226.3			
Perylene	$C_{20}H_{12}$	252.3			
Benzo[b]fluoranthene	$C_{20}H_{12}$	252.3	481		
Benzo[j]fluoranthene	$C_{20}H_{12}$	252.3	~480		
Benzo[k]fluoranthene	$C_{20}H_{12}$	252.3	481	3.7×10^{-8}	1.1×10^{-7}
Benzo[a]pyrene	$C_{20}H_{12}$	252.3	496	9.2×10^{-8}	7.1×10^{-7}
Benzo[e]pyrene	$C_{20}H_{12}$	252.3	493	9.6×10^{-8}	7.1×10^{-7}
Benzo[g,h,i]perylene	$C_{22}H_{12}$	276.3		1.1×10^{-8}	
Dibenz[a,h]anthracene	$C_{22}H_{14}$	278.4			
Coronene	$C_{24}H_{12}$	300.4	525	1.7×10^{-10}	

[a]Reference 79. Boiling points are at 760 torr (1 torr = 133 Pa).
[b]Reference 65.
[c]Calculated from the experimental data of Pupp et al. (86).
[d]Sublimes.

Theory of Vapor–Particle Partitioning of PAH Compounds in the Atmosphere.

Yamasaki et al. (64) treated semivolatile PAHs as inherently vapor-phase materials, existing in the particle phase only because of their tendency to sorb on nonvolatile materials. Under this assumption, the proportion in each phase should vary with the available surface area for sorption and with the ambient temperature. They assumed that sorption followed a Langmuir adsorption isotherm, requiring that there be only a low fractional coverage of the particulate matter with semivolatile material. In this case, competition for sorption sites can be ignored. At equilibrium, the rate of sorption equals the rate of evaporation. They derived this relationship:

$$\log ([\text{PAH}_{vap}][\text{TSP}]/[\text{PAH}_{partic}]) = -A/T + B \qquad (6)$$

where $[\text{PAH}_{vap}]$ is the concentration of the PAH in vapor phase, $[\text{TSP}]$ is the concentration of total suspended particulate matter in a sample, $[\text{PAH}_{partic}]$ is the concentration of PAH sorbed on the particulate matter, A and B are constants, and T is temperature (in kelvins).

Testing of this relationship required the ability to measure the PAH concentrations in the vapor and particle phases. Yamasaki et al. used glass-fiber (GF) filter-collected PAH as an approximation of PAH_{partic}. Polyurethane foam (PUF) downstream of the filter in a high-volume sampler was used to trap, and, thereby, operationally define the atmospheric concentrations of vapor-phase PAH. Figure 4 shows typical semilog plots for $[PAH_{vap}][TSP]/[PAH_{partic}]$ against $1/T$. The mean temperature for 24-h sampling periods was used for these plots; the variability of these temperatures was expected to increase the scatter in the resulting linear regressions. Nevertheless, regression fits to equation 6 gave $r^2 \geq 0.8$ for most PAH compounds (r, linear correlation coefficient). The slopes of these lines correspond to the constant A. Thus, this simple approach seemed consistent with atmospheric behavior for a series of PAH compounds.

This approach was developed further by Bidleman and co-workers (65, 70–72) and included the use of filter-collected particles and PUF traps to operationally define the required parameters. For example, the predicted influence of temperature on the vapor–particle partitioning of fluoranthene

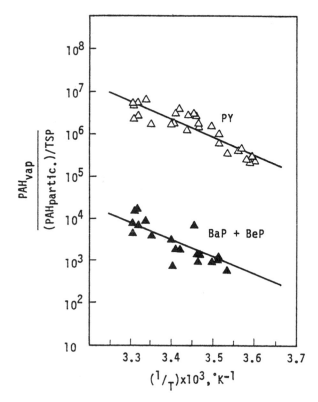

Figure 4. Plot of $PAH_{vap}/(PAH_{partic}/TSP)$ against $1/T$. (Reproduced from reference 64. Copyright 1964 American Chemical Society.)

(FL) may be calculated from the regression parameters obtained by Keller and Bidleman (71). At 0 °C and an assumed 60 μg/m^3 [TSP], 47% of FL should be in the particle state, compared to only 2.2% at 25 °C.

Sorption Media for Collecting Vapor-Phase PAH Compounds.

Analogous to sampling TIN with the filter pack method, sampling with a quartz- or glass-fiber filter followed by a sorbent can provide a relatively accurate measure of the total (i.e., vapor plus particle phase) concentration of specific PAH compounds. Such methodologies were reviewed by Lee et al. (73). Selection of a sorption medium for collection of vapor-phase PAHs is influenced by air-flow resistance, collection efficiency, breakthrough volume, the effort required to purify the medium for use, and the efficiency with which the PAH can be recovered from the sorbent.

The sorbents most frequently used for collection of vapor-phase PAH compounds include Tenax GC (74–76), PUF (70, 77), and XAD–2 (77, 78). Comparisons of retention efficiencies and breakthrough volumes indicate that Tenax GC and XAD–2 are more efficient than PUF for more volatile PAH compounds (e.g., anthracene). Lee et al. (73) noted XAD–2 resin to be preferable to Tenax GC because of its higher capacity, especially for lower boiling compounds.

Gas/Particle Ratios for Atmospheric PAH Compounds Estimated from Filter and Sorbent Sampling.

Yamasaki et al. (79) sampled ambient air in Osaka, Japan, using a 20 × 25 cm GF filter followed by two cylinders of PUF; face velocity through the filter was 33 cm/s. For 28 24-h periods, on average, 92% of the recovered benzo[a]pyrene (BaP) and 100% of the benzo[ghi]perylene (BghiP) was on the GF filter. The four-ring PAHs, including benzo[a]anthracene (BaA), chrysene, and triphenylene, averaged 62% recovery from the filter, the balance being recovered from the PUF.

A test of the recovery of 25 μg of PAHs ranging from phenanthrene (PH) to BghiP was performed by passing clean air through the spiked filter followed by two PUF disks for 24 h (the temperature was unstated). Except for PH, PAHs volatilized from the GF filter were 90–95% trapped and recovered from the PUF plugs. The losses of BghiP and BaP from the filter were 0 and 3%, respectively (79).

Galasyn et al. (80) made a similar evaluation with Pallflex quartz filters backed by PUF disks sampling in New Hampshire. For winter samples collected with an average temperature of 9 °C, the percentages found on PUF for BaA and BaP were 10 and 0%, respectively. During summer, these percentages were about 75 and 0%, respectively.

Improved Sampling Techniques for Atmospheric PAHs.   *Filter Versus Impactor Collection.*   The relative degree of volatilization of PAH compounds from filters during sampling can be inferred by comparison of

results in which filter and impactor samplers operated in parallel. Katz and Chan (*81*) compared the total PAHs collected with an Andersen cascade impactor operated at 0.57 m^3/min to that retained with the high-volume filter method using a glass-fiber filter operating at 1.5 m^3/min (face velocity 64 cm/s). In general, markedly higher PAH concentrations were recovered from the impactor. Although this result is seemingly consistent with the theory and conclusions of Zhang and McMurry (*1*) for single-component aerosols formed by condensation, it is unclear if this model should be relevant to PAH compounds.

A Multiple Prefilter Approach.   Van Vaeck et al. (*75*) used what they termed an "integrated gas phase–aerosol sampling system" to distinguish the material volatilized after collection from the initially gas-phase PAHs. The sampler consisted of a filtration system followed by a Tenax cartridge to collect vapor-phase PAHs; the entire flow passed through this cartridge at 18.3 L/min. The sample volume was kept at least an order of magnitude below the breakthrough volume for the most-volatile PAH considered, anthracene. Accordingly, quantitative retention was expected on the sorbent for all vapor-phase PAH compounds of interest. The unique element of their design was the filtration system, a standard 20 × 24 cm glass-fiber filter in a holder consisting of two stainless steel plates with ten circular apertures and Viton O-ring seals; this design effectively created 10 filter samplers. Using solenoid valves and a timer, they programmed the unit so that a fresh filter disk was introduced 10 times per 24-h day into the sampling train ahead of the Tenax cartridge, which was changed once a day. The total air volume to which any filter-collected aerosol was exposed was reduced by a factor of 100–300 in comparison to a second filter sampler run for 24-h periods. As a consequence, the Tenax cartridge behind the short-term filters was thought to be relatively free of organic compounds volatilized from the prefilter. Accordingly, the PAHs on this Tenax cartridge were used to measure those in the gas phase when sampled, PAH_g. The concentration of PAH recovered from the filter plus sorbent from the long-term sampler was used as a measure of the total gas plus particle-phase concentrations, PAH_t. Particle-phase concentrations for specific PAHs, PAH_p, were obtained by difference: $PAH_p = PAH_t - PAH_g$.

This approach remains subject to criticism, however, if sorption of initially gas-phase PAHs onto glass-fiber filters or on previously collected particulate matter is significant. The very strategy that reduces positive errors in measuring initially gas-phase PAH concentrations might cause significantly enhanced negative errors in such measurements because a fresh prefilter surface must be saturated with vapor-phase PAHs every 2.4 h. The data of Van Vaeck et al. do not permit assessment of the possible error from this source. However, an earlier study (*67*) showed less than 1% retention of vapor-phase PH on a glass-fiber filter. The error due to sorption onto the prefilter of vapor-phase PAHs and other nonpolar organic compounds similar in v.p. to PH is, therefore, likely to be small. The error due to sorption on previously collected particulate matter remains to be assessed but would be minimized by the frequent filter changes.

For samples collected at various times of the year, results with the relatively volatile PAHs, PH through pyrene (PY), were highly variable, showing as little as ~20% retention on the prefilters. However, BaP and other five-ring PAH compounds were found only on the filter.

Diffusion Denuders for Specific Organic Compounds. Diffusion denuder techniques have been applied successfully to sampling individual organic materials, including pesticides (82), nicotine (83, 84), and PAH compounds (63). The methodology and results relating to PAH compounds are detailed here.

Diffusion coefficients have been measured for comparatively few materials, but they can be calculated (67). Values for vapor-phase naphthalene and anthracene of 0.058 and 0.048 cm^2/s at 20 °C, respectively, are nearly a factor of 3 lower than that for HNO_3 and other inorganic gaseous pollutants. Thus, their removal efficiency is lower per unit length of denuder.

Figure 5 shows the sampler components used by Coutant et al. (63) for sampling semivolatile PAH compounds; these components include a denuder, a filter, and a cylinder of PUF. A second unit, lacking the denuder, sampled total concentrations of the PAH compounds. The denuder consisted of a parallel bank of seven stainless steel tubes, each 61 cm long with i.d. of 1.5 cm. The inner surface of each tube was coated with high-vacuum silicone grease (Dow Corning) before each sampling trial. Denuder efficiency was tested with naphthalene vapor; sampling was done with a single tube at 2 L/min. The observed efficiency, 89%, compares to 92% predicted by the Gormley–Kennedy equation (equation 3). Capacity for the single tube was "several tens of micrograms" without significant decline in efficiency.

A prefired Pallflex 2500 QAST quartz filter (104-mm diameter) downstream of the denuder retained particle-phase PAHs together with other particulate matter at a face velocity of about 3 cm/s. The PUF retained vapor-phase PAH compounds that penetrated the denuder, together with those volatilized from the filter following collection in the particle state. The presence of a denuder is expected to enhance the extent of volatilization of PAHs from the filter relative to the unit without a denuder. The PAH concentrations volatilized during sampling from the unit without a denuder were referred to as the normal artifact, A_n. The corresponding concentrations from the unit with the denuder included $A_n + A_e$, where A_e indicates "excess artifact" caused by the presence of the denuder. The possibility that particle-phase PAH compounds could volatilize and be retained while within the denuder was not addressed.

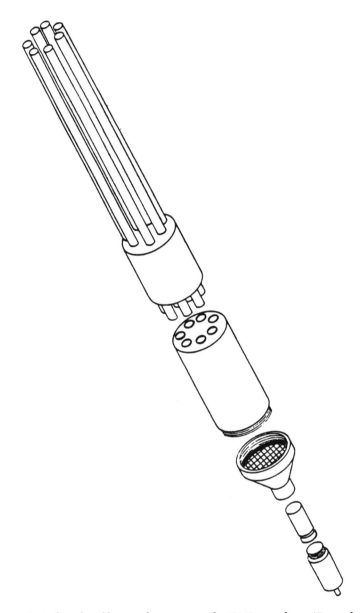

Figure 5. A denuder–filter–sorbent system for PAH sampling. (Reproduced with permission from reference 63. Copyright 1988.)

The apparent particle-phase and vapor-phase PAH concentrations and the sampling artifacts were determined from the system of simultaneous equations:

$$F_{nd} = P - A_n \tag{7}$$

$$F_d = P - A_n - A_e \tag{8}$$

$$PUF_{nd} = V + A_n \tag{9}$$

$$PUF_d = 0.11\,V + A_n + A_e \tag{10}$$

where F_{nd} and F_d are PAH levels recovered from filters without and with the denuder, respectively; P is the true particle-phase PAH concentration; V is the true vapor-phase PAH concentration; A_n is the normal volatilization artifact; A_e is the excess volatilization artifact; PUF_{nd} and PUF_d are PAH levels recovered from PUF without and with the denuder, respectively; and $0.11V$ is a correction for the 11% penetration of vapor-phase PAH through the denuder, with subsequent collection only on PUF assumed. (The degree of penetration for naphthalene was assumed relevant to higher molecular weight PAH as well.) This approach is analogous to that for HNO_3 and fine-particle NO_3^- by the DDM after correction for significant penetration through the denuder. (The values of the normal and excess artifact cancel out when P and V are calculated.)

Table IV summarizes results for eleven PAH compounds, including median values and ranges for the percent of the total amount of each compound found in the vapor phase at the moment of sampling. Also given are corresponding values for the "artifact" (i.e., A_n) for each compound. The three-

Table IV. Summary of PAH Vapor-Phase and Artifact Levels

Chemical	Ranges		Median	
	Vapor	*Artifact*	*Vapor*	*Artifact*
Phenanthrene	25–87	13–68	39	60
Anthracene	5–71	14–92	26	71
Fluoranthene	27–64	7–62	43	47
Pyrene	5–80	16–83	43	47
Benz[a]anthracene	19–67	8–45	26	13
Chrysene	5–65	17–50	28	38
Benzo[e]pyrene	NA	NA	NA	NA
Benzo[a]pyrene	NA	NA	NA	NA
Indeno[1,2,3-c,d]pyrene	NA	NA	NA	NA
Benzo[g,h,i]perylene	NA	NA	NA	NA
Coronene	NA	NA	NA	NA

NOTE: Ranges and median values are expressed as percentages of the total amounts of each compound found. NA denotes not applicable: no evidence of vapor or artifact was found.
SOURCE: Reproduced with permission from reference 63. Copyright 1988.

and four-ring PAHs showed substantial vapor-phase components as well as volatilization of the initially particle-phase fraction during sampling. The five-ring and higher PAHs were recovered exclusively from the particle phase. Thus, for example, the median percentage of PY in the vapor phase was 43%, indicating that 57% was in the particle state when sampled. The 47% of the total PY shown as the median artifact indicates that 82% of the initially particulate PY was volatilized after passage through the denuder (i.e., from the filter) during sampling. These results suggest that the use of filter–sorbent collection to estimate atmospheric particle-to-vapor ratios may cause substantial errors for more volatile PAHs as well as other materials.

The implications of such denuder results for the approach of Yamasaki (64) remain to be addressed. However, the findings of Coutant et al. (63) cannot be accepted uncritically. Their conclusions rest on the assumption that volatilization of particle-phase PAHs within the denuder is insignificant. Additional control studies on such methodology are essential.

The variability observed in the percentages shown in Table IV for the three- and four-ring PAHs was thought to reflect the variability in both the temperature and the timing of these variations with respect to the sampling cycle. Lesser volatilization was found in winter samples. Coutant et al. recommended that 24-h filter sampling for PAH compounds should be started and finished during early morning hours so that only samples collected during the first 8 to 10 h would be subjected to rising temperatures within the sampler, conditions favoring enhanced volatilization of previously collected PAHs.

Improved Samplers for Particulate Organic Carbon. *Tandem Filter Techniques To Correct for a Positive Sampling Artifact.* As detailed in the preceding sections, data indicate, although not unequivocally, that the determination of specific particle-phase organic materials, namely, PAH compounds, can be subject to large negative sampling artifacts. However, there is no similar indication that positive sampling errors are encountered with such materials. The situation is somewhat reversed for the sampling of atmospheric, particle-phase carbonaceous materials determined as carbon. Volatilization during sampling must be inferred from such indications as the influence of face velocity, an interpretation that has been challenged recently (vide infra). By contrast, the sampling medium typically used, quartz-fiber filters, can easily be shown to retain vapor-phase carbon, with atmospheric CO_2 being only a minor contributor. In a typical experiment (62), atmospheric sampling is done with two or more quartz filters sampling in tandem, and the amounts of carbon recovered from backup filters are compared to the total carbon (particulate plus sorbed) on the front filter. The concentration of sorbed carbon on the backup filter, expressed per volume of air sampled or relative to the front filter carbon, increases with decreasing air volume; this observation is consistent with sorption of carbonaceous materials on a limited number of sites on the filter. For air volumes ranging from 2.4 to 7.2 m^3, the ratios ranged from 0.51 to 0.13.

In experiments in which ambient air was sampled by three quartz filters in tandem, Cadle et al. (*61*) proposed to measure the sorption efficiency, *E*, of a clean filter from the equation

$$E = 1 - A_3/A_2 \qquad (11)$$

where A_2 and A_3 are the amounts of adsorbed organic C on the second and third filters, respectively. Using 47-mm Gelman Microquartz filters for 24-h samples at 20 L/min (face velocity 23 cm/s), they found, on average, *E* = 34% (four trials). In similar experiments (*62*) with 47-mm Pallflex 2500 QAO quartz-fiber filters (24 trials and sample volumes ranging from 2.4 to 60 m^3 at 20 L/min and a face velocity of 26 cm/s), the mean ratio A_3/A_2 was 0.88 ± 0.17. The variability in results prevented a useful estimation of average absorption efficiency with equation 11.

The current state of the art in filter sampling particulate C is to use a dual-filter strategy to provide an approximate correction for sorption of vapor-phase organics on the particle collection filter medium. The carbon determined on a backup filter is assumed to be equal to that retained by sorption on the front filter composed of the same material. Two techniques are in use. The simplest involves two quartz-fiber filters sampling in tandem and subtracting the carbon recovered from the backup quartz filter from the organic C determined on the front quartz filter (*Q* – *Q* strategy). This approach is supported by the near equivalence of the carbon on the second and third quartz filters (*62*). However, this approach cannot correct for any hypothetical vapor-phase, organic C sorbed efficiently by the front quartz filter. Additional support for this approach is discussed in the **Diffusion Denuder Techniques** section.

In the alternative method (*85*), a second sampler, consisting of a Teflon particle filter followed by a quartz filter, is operated in parallel with a quartz-filter sampler. The carbon recovered from the quartz backup filter of this second sampler is subtracted from the organic C determination of the first sampler (*Q* – *TQ* strategy). The rationale for this alternative approach is to minimize the extent of sorption of vapor-phase carbon on the first (particle-collection) filter, which, otherwise, might significantly diminish the amount retained on the backup quartz filter. The need for an additional sampler and very accurate air volume measurements is a disadvantage.

Comparisons of the two strategies permit differing conclusions. One comparison (*62*) found the *Q* – *Q* strategy to be more accurate at sample volumes less than 10 m^3. However, the experimental technique prevented an unequivocal conclusion. The elimination of the face velocity dependence of particulate organic C concentrations was used by McDow and Huntzicker (*85*) as an indication of the accurate measurement of this material by filter sampling. They compared corrections for vapor adsorption on quartz-fiber filters by using the $Q - TQ$ and $Q - Q$ strategies. The organic C on both front and backup quartz filters exhibited decreasing concentrations with increased face velocity. Rather than arguing for an increase in volatilization of particulate organic C from the front filter, they argue that this decrease reflects diminished absorption of vapor-phase C with increased face velocity. Their findings showed the $Q - TQ$ correction strategy to eliminate 90% of the face velocity dependence of the front quartz-filter organic C compared to 45% with the $Q - Q$ strategy. This comparison was performed at constant sample volume; face velocities were varied by altering the exposed filter areas. Additional comparisons of this type should be performed, but with varying sample volumes.

The dual-filter technique for particulate C would not correct for sorption onto previously collected particulate matter. To the degree that this sorption is significant, dual-filter techniques would probably provide decreasing accuracy with increased particle loading.

Diffusion Denuder Techniques. In principle, a diffusion denuder–filter–sorbent system analogous to that described for PAH compounds can be designed to measure particulate organic C as well as the significance of positive and negative sampling errors. The denuder must be able to retain, with very high efficiency, the carbon-containing materials that are, otherwise, retained by sorption on the filter or on the particulate matter thereon. The carbon recovered from the sorbent following the filter in such a system would provide an upper limit to that lost from the filter during sampling. Evaluation of potential denuders for this application provided efficiencies up to ~60% (62) and, in work by Fitz, 78% (68).

The experimental system used by Fitz is shown in Figure 6. Dual, 47-mm filters sampled ambient air in parallel at 15 L/min from a common cyclone inlet with one of the samplers preceded by a denuder composed of quartz-filter strips. The denuder, which provided an average residence time of ~2 s, was designed for greater than 99% removal efficiency of a material with a diffusion coefficient at least that of anthracene if 100% retention of species contacting the filter strips is assumed. The filter material, Pallflex QAST quartz fiber, was identical for the denuder and dual filters. The experimental results for 18 12-h sampling periods are shown in Table V. The organic C (OC) on the front filter following the denuder was consistently lower by about 12% (~2 µg/m³). The OC on the back filter following the denuder averaged 22% of that on the undenuded side; this value suggests an average denuder efficiency of 78%. Elemental C (EC) values were about equal, a situation consistent with the absence of significant loss of fine particles passing through the denuder.

Fitz concluded that the OC on the back filter is primarily due to irreversible adsorption of gas-phase organics rather than to volatilization of par-

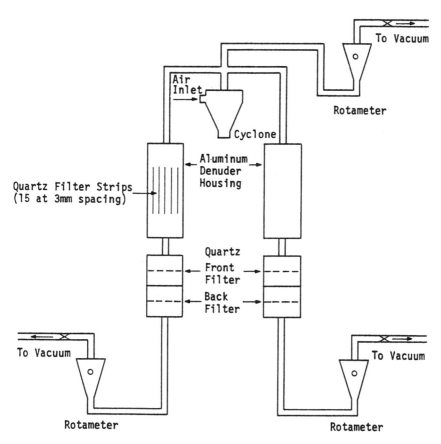

Figure 6. A denuder–dual filter system for investigating the organic positive artifact on quartz-fiber filters. (Reproduced with permission from reference 68. Copyright 1990.)

ticulate C from the front filter and re-adsorption. Accordingly, the front filter OC should be corrected by subtracting the backfilter OC. Corrected organic particulate C values are nearly identical for the denuded and undenuded side: 14.0 and 13.8 $\mu g/m^3$, respectively.

Fitz's data provide additional support for the $Q - Q$ correction strategy. If a portion of the vapor-phase C were efficiently sorbed on quartz filters, such materials should be efficiently removed by the denuder. As a consequence, the OC on the front filter following the denuder would be lower than that on the undenuded side. Accordingly, the $Q - Q$ strategy should undercorrect for sorption on the undenuded side. The near equivalence of the corrected results from the two sides argues that strongly sorbed OC is not significant.

This experiment does not address possible sorption of OC on the particulate matter, volatilization within the denuder, or the degree of volatili-

Table V. Comparison of Organic and Elemental Carbon Analyses With and Without the Quartz-Fiber Filter Denuder ($n = 18$)

| Sample | No Denuder | | Denuder | | Difference in Means (No Denuder − Denuder) ($\mu g/m^3$) | Standard Error of Means ($\mu g/m^3$) |
	Mean ($\mu g/m^3$)	Standard Deviation ($\mu g/m^3$)	Mean ($\mu g/m^3$)	Standard Deviation ($\mu g/m^3$)		
OC front	16.57	4.33	14.56	3.41	2.01	0.26
EC front	1.53	0.52	1.51	0.61	0.02	0.05
OC back	2.73	0.78	0.60	0.29	2.13	0.15
EC back	0.01	0.05	0.01	0.04	0.00	0.01

SOURCE: Adapted from reference 68.

zation following filter collection. The addition of a relatively efficient inorganic sorbent [e.g., a fluidized bed of activated Al_2O_3 (62)] downstream of the dual filters shown in Figure 6 might prove useful in the assessment of the negative artifact and the attempt to more closely approach the elusive goal of a true particulate organic C sampler.

Conclusions. Compared to nitrate sampling, the sampling of particulate organic C remains relatively primitive, with room for much creative research effort. Although there is general agreement about the significance of a positive artifact due to sorption of vapor phase C on filter media, the existence of a negative artifact due to volatilization of particulate C remains unclear. Dual-filter techniques permit correction of the error due to sorption on filter media. Denuder-based sampling techniques may hold the key to improved particulate C samplers and provide an estimate of the negative artifact. The quartz-filter denuder developed by Fitz can serve as the starting point for developing such improved samplers.

Glossary

ADM	annular denuder method
BaA	benzo[*a*]anthracene
BaP	benzo[*a*]pyrene
BeP	benzo[*e*]pyrene
BghiP	benzo[*ghi*]perylene
CPN	coarse particulate nitrate
DDM	denuder difference method
EC	elemental carbon
FEP	fluorinated ethylenepropylene (Teflon)
FL	fluoranthene
FPM	filter pack method
FPN	fine particulate nitrate
GF	glass fiber
OC	organic carbon
PAH	polyaromatic hydrocarbon
PER	perylene
PFA	perfluoroalkoxy (Teflon)
PH	phenanthrene
PN	particulate nitrate
PTFE	polytetrafluoroethylene (Teflon)
PUF	polyurethane foam
PY	pyrene
RH	relative humidity
S&S	Schleicher and Schuell
TDL	tunable diode laser

TIN   total inorganic nitrate
v.p.   vapor pressure

References

1. Zhang, X. Q.; McMurry, P. H. *Atmos. Environ.* **1987**, *21*, 1779–1789.
2. Finlayson-Pitts, B. J.; Pitts, J. N., Jr. *Atmospheric Chemistry: Fundamentals and Experimental Techniques*; John Wiley: New York, 1986; p 787.
3. Robbins, R. C.; Cadle, R. D.; Eckhardt, D. L. *J. Met.* **1959**, *16*, 53–56.
4. Yoshizumi, K.; Hoshi, A. *Environ. Sci. Technol.* **1985**, *19*, 258–261.
5. Wolff, G. T. *Atmos. Environ.* **1984**, *18*, 977–981.
6. Stelson, A. W.; Seinfeld, J. H. *Atmos. Environ.* **1982**, *16*, 983–992.
7. Allen, A. G.; Harrison, R. M. *Atmos. Environ.* **1989**, *23*, 1591–1599.
8. Grosjean, D. *Sci. Total Environ.* **1982**, *25*, 263–275.
9. Tang, I. N. *Atmos. Environ.* **1980**, *14*, 819–828.
10. Appel, B. R.; Tokiwa, Y. *Atmos. Environ.* **1981**, *15*, 1087–1089.
11. Spicer, C. W.; Schumacher, P. M. *Atmos. Environ.* **1977**, *11*, 873–876.
12. Appel, B. R.; Tokiwa, Y.; Haik, M.; Kothny, E. L. *Atmos. Environ.* **1984**, *18*, 409–416.
13. Forrest, J.; Tanner, R.; Spandau, D.; D'Ottavio, T.; Newman, L. *Atmos. Environ.* **1980**, *14*, 137–144.
14. Appel, B. R.; Tokiwa, Y.; Haik, M. *Atmos. Environ.* **1981**, *15*, 283–289.
15. Appel, B. R.; Wall, S. M.; Tokiwa, Y.; Haik, M. *Atmos. Environ.* **1980**, *14*, 549–554.
16. Okita, T.; Morimoto, S.; Izawa, S.; Konno, W. *Atmos. Environ.* **1979**, *10*, 1085–1089.
17. Spicer, C. W. *Atmos. Environ.* **1977**, *11*, 1089–1095.
18. Appel, B. R.; Tokiwa, Y.; Kothny, E. L.; Wu, R.; Povard, V. Final Report to the California Air Resources Board, Contract No. 4–147–32, 1986.
19. Appel, B. R.; Tokiwa, Y.; Povard, V.; Kothny, E. L. Final Report to the California Air Resources Board, Contract No. A4–074–32, 1987.
20. Perrino, C.; De Santis, F.; Febo, A. *Atmos. Environ.* **1988**, *22*, 1925–1930.
21. Sickles, J., II; Hodson, L. *Atmos. Environ.* **1989**, *23*, 2321–2324.
22. Appel, B. R.; Winer, A. M.; Tokiwa, Y.; Biermann, H. W. *Atmos. Environ.* **1990**, *24A*, 611–616.
23. Joseph, D. W.; Spicer, C. W. *Anal. Chem.* **1978**, *50*, 1400–1403.
24. Fahey, D. W.; Hubler, G.; Parrish, D.; Williams, E.; Norton, R.; Ridley, B.; Singh, H.; Liu, S.; Fehsenfeld, F. *J. Geophys. Res.* **1986**, *91*, 9781–9783.
25. Talbot, R. W.; Vijgen, A. S.; Harriss, R. C. *J. Geophys. Res.* **1990**, *95D*, 7553–7561.
26. Hering, S. V. *Atmos. Environ.* **1988**, *22*, 1519–1539.
27. Bowermaster, J.; Shaw, R. W., Jr. *J. Air Pollut. Control Assoc.* **1981**, *31*, 787.
28. Goldan, P. D.; Kuster, W. C.; Albritton, D. L.; Fehsenfeld, F. C.; Connell, P. S.; Norton, R. B.; Huebert, B. J. *Atmos. Environ.* **1983**, *17*, 1355–1364.
29. Appel, B. R.; Povard, V.; Kothny, E. L. *Atmos. Environ.* **1988**, *22*, 2535–2540.
30. Smith, J. P.; Grosjean, D.; Pitts, J. N., Jr. *J. Air Pollut. Control Assoc.* **1978**, *28*, 930–933.
31. Dunwoody, C. L. *J. Air Pollut. Control Assoc.* **1986**, *36*, 817–818.
32. Appel, B. R.; Hoffer, E. M.; Tokiwa, Y.; Haik, M.; Wesolowski, J. J. Final Report for Year One, EPA Grant No. R806734–01–0, 1980.
33. Liu, B. Y. H.; Pui, D. Y. H.; Rubow, K. L. In *Aerosols in the Mining and Industrial Work Environment*; Marple, V. A.; Liu, B. Y. H., Eds.; Ann Arbor Science: Ann Arbor, MI, 1983; Vol. III, Chapter 70, pp 989–1038.

34. Huebert, B. J.; Lazrus, A. L. *Geophys. Res. Lett.* **1978**, *5*, 577–580.
35. Tanner, R. L.; Cederwall, R.; Garber, R.; Leahy, D.; Marlow, W.; Meyers, R.; Phillips, M; Newman, L. *Atmos. Environ.* **1977**, *11*, 955–966.
36. Anlauf, K.; Fellini, P.; Wiebe, H.; Schiff, H.; Mackay, G.; Braman, R.; Gilbert, R. *Atmos. Environ.* **1985**, *19*, 325–333.
37. Stevens, R. K.; Shaw, R. W., Jr.; Appel, B. R.; Forrest, J. Discussion of reference 28 and response. *Atmos. Environ.* **1983**, *17*, 2561.
38. Shaw, R. W., Jr.; Stevens, R. K.; Bowermaster, J.; Tesch, J.; Tew, E. *Atmos. Environ.* **1982**, *16*, 845–853.
39. Allegrini, I.; De Santis, F. *CRC Crit. Rev. Anal. Chem.* **1989**, *21*, 237–253.
40. Forrest, J.; Spandau, D.; Tanner, R.; Newman, L. *Atmos. Environ.* **1982**, *16*, 1473–1485.
41. Gormley, P. G.; Kennedy, M. *Proc. R. Irish Acad., Sect. A* **1949**, *52*, 163–169.
42. Ferm, M.; Sjodin, A. *Atmos. Environ.* **1985**, *19*, 979–983 and references cited therein.
43. Vossler, T.; Stevens, R. K.; Paur, R.; Baumgardner, R.; Bell, J. *Atmos. Environ.* **1988**, *22*, 1729–1736.
44. Febo, A.; De Santis, F.; Perrino, C. *Proceedings of the Fourth European Symposium on Physico-Chemical Behavior of Atmospheric Pollutants*; Angeletti, G.; Restelli, G., Eds.; Reidel: Dordrecht, 1986; p 121.
45. Slanina, J.; Lamoen-Doornenbal, L. V.; Lingerak, W. A.; Meilot, W.; Klockow, D.; Niessner, K. *Int. J. Environ. Anal. Chem.* **1981**, *9*, 589.
46. John, W.; Wall, S.; Ondo, J. *Atmos. Environ.* **1988**, *22*, 1627–1635.
47. Appel, B. R.; Tokiwa, Y.; Hoffer, E. M.; Kothny, E. L.; Haik, M.; Wesolowski, J. J. Final Report to the California Air Resources Board, Contract A8–111–31, 1980.
48. Ferm, M. *Atmos. Environ.* **1979**, *13*, 1385–1393.
49. Appel, B. R.; Tokiwa, Y.; Kothny, E. L.; Wu, R.; Povard, V. *Atmos. Environ.* **1988**, *22*, 1565–1573.
50. Larson, T. V.; Taylor, G. S. *Atmos. Environ.* **1983**, *17*, 2489–2495.
51. Pratsinis, S.; Xu, M.; Biswas, P.; Willeke, K. *J. Aerosol Sci.* **1989**, *20*, 1597–1600.
52. Wall, S.; John, W.; Ondo, J. *Atmos. Environ.* **1988**, *22*, 1649–1656.
53. Possanzini, M.; Febo, A.; Liberti, A. *Atmos. Environ.* **1983**, *17*, 2607–2610.
54. Brauer, M.; Koutrakis, P.; Wolfson, J.; Spengler, J. *Atmos. Environ.* **1989**, *23*, 1981–1986.
55. Allegrini, I.; De Santis, F.; Di Palo, V.; Febo, A.; Perrino, C.; Possanzini, M.; Liberti, A. *Sci. Total Environ.* **1987**, *67*, 1–16.
56. Jenkins, M. E.; Cox, R.; Williams, D. *Atmos. Environ.* **1988**, *22*, 487–498.
57. Appel, B. R.; Tokiwa, Y.; Povard, V. Final Report to the California Air Resources Board, Contract A732–098, 1988.
58. Stevens, R. K.; King, F.; Bell, J.; Whitfield, J. Presented at the 81st Annual Meeting of the Air Pollution Control Association, Dallas, TX, 1988; paper 88–57.3.
59. Fung, K. "Artifacts in the Sampling of Ambient Organic Aerosols". Presented at the 1988 EPA/APCA International Symposium: Measurement of Toxic and Related Air Pollutants, Research Triangle Park, NC, 1988.
60. Appel, B. R.; Hoffer, E. M.; Kothny, E. L.; Wall, S. M.; Haik, M. *Environ. Sci. Technol.* **1979**, *13*, 98–104.
61. Cadle, S.; Groblicki, P.; Mulawa, P. *Atmos. Environ.* **1983**, *17*, 593–600.
62. Appel, B. R.; Cheng, W.; Salaymeh, F. *Atmos. Environ.* **1989**, *23*, 2167–2175.
63. Coutant, R. W.; Brown, L.; Chuang, J. C.; Riggin, R. M.; Lewis, R. G. *Atmos. Environ.* **1988**, *22*, 403–409.
64. Yamasaki, H.; Kuwata, K.; Miyamoto, H. *Environ. Sci. Technol.* **1982**, *16*, 189–194.

65. Bidleman, T. F.; Billings, W. N.; Foreman, W. T. *Environ. Sci. Technol.* **1986**, *20*, 1038–1043.
66. Pankow, J. F. *Atmos. Environ.* **1987**, *21*, 2275–2283.
67. Appel, B. R.; Tokiwa, Y.; Kothny, E. L. *Atmos. Environ.* **1983**, *17*, 1787–1796.
68. Fitz, D. *Aerosol Sci. Technol.* **1990**, *12*, 142–148.
69. Pitts, J. N., Jr. *Science (Washington, D.C.)* **1980**, *210*, 1347–1349.
70. You, F.; Bidleman, T. F. *Environ. Sci. Technol.* **1984**, *18*, 330–333.
71. Keller, C. D.; Bidleman, T. F. *Atmos. Environ.* **1984**, *18*, 837–845.
72. Foreman, W. T.; Bidleman, T. F. *Environ. Sci. Technol.* **1987**, *21*, 869–875.
73. Lee, M. L.; Novotny, M. V.; Bartle, K. D. *Analytical Chemistry of Polycyclic Aromatic Compounds*; Academic Press: New York, 1981; Chapter 4.
74. Cautreels, W.; Van Cauwenberghe, K. *Atmos. Environ.* **1978**, *12*, 1133–1141.
75. Van Vaeck, L.; Van Cauwenberghe, K.; Janssens, J. *Atmos. Environ.* **1984**, *18*, 417–430.
76. Jones, P. W.; Giammar, G. D.; Strup, P. E.; Stanford, T. B. *Environ. Sci. Technol.* **1976**, *10*, 806.
77. Chuang, J. C.; Hannan, S. W.; Wilson, N. K. *Environ. Sci. Technol.* **1987**, *21*, 718–804.
78. Otson, R.; Leach, J. M.; Chaung, L. T. K. *Anal. Chem.* **1987**, *59*, 1701–1705.
79. Yamasaki, H.; Kuwata, K.; Miyamoto, H. *Bunseki Kagaku* **1978**, *27*(6), 317–321, Japanese with English abstract.
80. Galasyn, J. F.; Hornig, J. F.; Soderberg, R. H. *J. Air Pollut. Control Assoc.* **1984**, *34*, 57–59.
81. Katz, M.; Chan, C. *Environ. Sci. Technol.* **1980**, *14*, 838–843.
82. Johnson, N. D. "Evaluation of a Diffusion Denuder Based Gas/Particle Sampler for Chlorinated Organic Compounds," Presented at the 1986 EPA/APCA Symposium on Measurement of Toxic and Related Air Pollutants, Research Triangle Park, NC, 1986.
83. Koutrakis, P.; Fasano, A.; Slater, J.; Spengler, J.; McCarthy, J.; Leaderer, B. *Atmos. Environ.* **1989**, *23*, 2767–2773.
84. Eatough, D.; Benner, C.; Bayona, J.; Richards, G.; Lamb, J.; Lee, M.; Lewis, E.; Hansen, L. *Environ. Sci. Technol.* **1989**, *23*, 679–687.
85. McDow, S. R.; Huntzicker, J. J. *Atmos. Environ.* **1990**, *24A*, 2563–2571.
86. Pupp, C.; Lao, R.; Murray, J.; Pottie, R. *Atmos. Environ.* **1974**, *8*, 915–925.

RECEIVED for review March 20, 1991. ACCEPTED revised manuscript June 24, 1992.

Automated Measurement of Atmospheric Trace Gases

Diffusion-Based Collection and Analysis

Purnendu K. Dasgupta

Department of Chemistry and Biochemistry, Texas Tech University, Lubbock, TX 79409

The theoretical and practical aspects of automated diffusion-based collection and analysis systems are described. The design of thermodenuders, wet denuders, and diffusion scrubbers is discussed. Theoretical considerations include the estimation of collection efficiency, inlet length necessary for laminar flow development, and particle transmission through the system as well as the choice of a given design for the intended application. Practical collection and analysis systems are described on the basis of literature accounts of thermally cycled denuders, ion-exchange and porous-membrane-based diffusion scrubbers, and wet denuders, both of simple tubular and annular geometry.

T HE DETERMINATION OF ATMOSPHERIC TRACE GASES has represented an area of active endeavor since the onset of humanity's interest in the chemistry of the atmosphere. Present-day practice ranges from sampling the air with a liquid absorber contained in a bubbler to making direct spectroscopic measurements with path lengths of several kilometers, with obvious attendant differences in cost, sensitivity, and reliability. Generally applicable direct spectroscopic techniques in present use relying on absorptiometry include tunable diode laser spectroscopy (*1*), differential optical absorption spectroscopy (*2*), and Fourier transform infrared spectroscopy (*3*). In general, these methods are relatively free from interferences and thus represent the techniques of choice as reference procedures. However, their widespread

0065–2393/93/0232–0041$13.50/0
© 1993 American Chemical Society

application is deterred by their bulk and cost. Occasionally, compact absorptiometric instruments of adequate sensitivity are possible for a specific gas, as is the case for ozone (4). Other examples of dedicated affordable instruments for specific gases typically involve luminometry, for example, pulsed fluorometry for SO_2 (5) and chemiluminometry for the determination of NO upon reaction with O_3 (6) or for the determination of O_3 upon reaction with C_2H_4 (7). In general, however, if the gases of interest can be reproducibly collected in a liquid (typically aqueous) absorber, a great variety of analytical alternatives (e.g., colorimetric–fluorometric–electrochemical detection following wet chemical manipulations or ion–liquid chromatography) become applicable. In most cases, such approaches allow good specificity and respectable limits of detection (LODs) at a modest cost.

Continuous Gas–Liquid Contactors

Collection of an atmospheric trace gas for subsequent wet analysis has been classically accomplished by using a bubbler or an impinger. Usually, the liquid volume is not the limiting factor in the analytical procedure; if the gas collection efficiency is not seriously sacrificed at the same sampling rate, the ability to utilize a smaller liquid volume translates into a greater analyte concentration and thus a lower attainable LOD. Several alternatives are superior to bubblers and impingers in this respect. Moreover, these alternatives permit continuous liquid and gas flow, allowing continuous analysis of the liquid stream. The earliest example of such a device is a multiturn glass coil in which the sampled air causes the simultaneously pumped liquid absorber to form a film on the interior walls of the coil. A gas–liquid separator follows, and typically the isolated liquid stream is fed to an air-segmented continuous-flow analyzer. Fully automated ambient air analyzers based on this principle were reported more than two decades ago (8), and more modern adaptations have been reported, most notably for the continuous measurement of H_2O_2 and HCHO (9, 10).

A continuous gas scrubber can be designed around the Venturi principle as well (11). Nebulizing the absorber liquid by using the sample air as propellant is among the most efficient and ingenious strategies utilized to design continuous gas–liquid contactors. The nebulized liquid is then collected by providing either an impaction site (12) or a hydrophobic membrane (13). All of these gas–liquid contactors involve a flow regimen that is far from laminar. Consequently, a significant (and variable, depending on the design of the device) fraction of the simultaneously sampled atmospheric aerosol is scrubbed as well, and it finds its way in the liquid effluent. If the analyte measured is not significantly present in the aerosol and does not interact with any of the aerosol constituents, no particular problems are posed. If it does either, however, a different strategy is needed because removal of the aerosol from the sample gas by prefiltration is generally unacceptable because of filter-induced artifacts.

Diffusion Denuders

Indeed, it is often impossible to distinguish between an analyte originating from the aerosol and the same analyte originating from the gas phase after incorporation into a liquid absorber: the inability to distinguish between particulate NH_4^+ and NH_3 (g) or between particulate NO_3^- and HNO_3 (g), once collected into an aqueous solution, can be cited as examples. In such cases, diffusion-based collection has proven to be the only approach for reliably measuring a gas in the presence of an aerosol.

Because the diffusion coefficient of a gas molecule is typically 4 orders of magnitude larger than the smallest atmospheric aerosol of significance (in terms of mass contribution), it is possible to effect essentially quantitative removal of analyte gas molecules by diffusion in an appropriately designed system. The simplest form of a diffusion-based gas–aerosol discriminator of this type is a tube with its interior walls coated with some substance that serves as an efficient sink for the gas molecules. Under laminar flow conditions and with a vertical deployment of the tube to avoid gravitational settling of the aerosol, the aerosol transmission efficiency can be nearly quantitative (*14, 15*). These devices were suggested originally by Townsend (*16*), and the initial interest was to remove certain gases to better study the associated particles. The removal of gases by means of diffusion thus eventually led to the term "diffusion denuder". In 1949, Gormley and Kennedy (*17*) provided a mathematical treatment concerning the diffusion from a stream flowing through a cylindrical tube. Several recomputations of the original Gormley–Kennedy solution have been made, and the expression of Bowen et al. (*18*) is regarded the most accurate (*19*). The following expression (equation 1) agrees with the results of Bowen et al. up to four significant figures.

$$1 - f = 0.81905\ e^{-3.6568\mu} + 0.09753\ e^{-22.305\mu}$$
$$+\ 0.0325\ e^{-56.961\mu} + 0.01544\ e^{-107.62\mu} \tag{1}$$

where f is the fraction collected by the denuder and μ is a dimensionless quantity given by

$$\mu = \pi DL/Q \tag{2}$$

where D is the diffusion coefficient of the gas, L is the length of the tube, and Q is the volumetric flow rate. For most denuder systems designed to collect a gas, f is high, and only the first term on the right side of equation

1 needs to be used. These principles are regarded as sufficiently sound to permit reliable measurements of diffusion coefficients of gases (20, 21).

The first reported application of a denuder device with subsequent chemical analysis involved the speciation of gaseous and particulate fluoride. This study used a denuder system containing three concentric tubes (22). Few details were given as to why such a design was chosen; denuders of annular geometry were not reinvestigated until the 1980s. Using the simple single-tube geometry, Crider et al. (23) reported the first continuous gas–aerosol discriminating analytical instrument. Since this time, reported designs have included multiple tubes operated in parallel (24), two concentric tubes (with the inner wall of the outer tube and the outer wall of the inner tube being suitably coated to serve as the sink surface) held together by one or two tripoint welds with the air being sampled through the annular space (25), and even a set of 12 concentric tubes operated in the multiple annular geometry (26). Stevens et al. (27) similarly described a concentric denuder with three glass annuli, protected on the exterior by a stainless steel sheath so as to render the assembly less fragile. Koutrakis et al. (28) described the design of a compact serial acid- and alkali-coated denuder, complete with an inlet impactor, for the measurement of acidic aerosols and gases. Another reported diffusion-based sampler utilizes flow in the laminar–turbulent transition region (29, 30). A decisive judgment on the advantages and disadvantages of the transition-flow reactor, including its general applicability to a variety of analytes, awaits further studies.

Three reviews describing applications of diffusion denuders have been published. The doctoral dissertation of Ferm (31) reflects considerable experience with single-tube denuders for the measurement of a variety of species. The review by Ali et al. (32) is extensive; it provides an excellent historical and theoretical background and summarizes the literature based on the type of analyte gas determined. The focus of the most recent review, by Cheng (19), is diffusion batteries used for size discrimination of aerosols as well as diffusion denuders. Various physical designs are discussed in some detail in that review.

Diffusion denuders have become a relatively common tool for the present-day atmospheric analytical chemist. In a comparison of methods for nitrogen species analysis conducted in urban Los Angeles in 1985, some 8 out of 18 instruments deployed to measure HNO_3 used a diffusion denuder (33). This study also indicated the uncertainties associated with prefilters and filter-based measurements. Despite the increasing popularity of diffusion denuders, the use of a typical diffusion denuder is very labor-intensive. A typical application involves washing and coating the active surfaces of a denuder tube, field sampling, returning to the laboratory, washing and removing the coating under noncontaminating conditions, analyzing the wash solution for the collected analyte, and then beginning the cycle anew. Efforts have been and continue to be made to fully automate diffusion-based collection and analysis, at least for specific analytes. These devices are discussed in the remainder of this chapter; the cited reviews give the general application of diffusion denuders.

Operating and Design Considerations

Some considerations common to all diffusion-based collection–analysis systems are discussed in the following sections.

Laminar Flow Development. In a typical application, it is not generally possible to sample isokinetically. Equations governing the collection efficiency (e.g., equation 1) apply only under laminar flow conditions. It is considered necessary therefore to leave a length of the tubing surface at the entrance deliberately uncoated to permit laminar flow to fully develop before actual collection occurs. For a simple tube, the minimum inlet length, L_i, necessary to fully develop laminar flow (within about 98%) is given by (34):

$$L_i = 0.05 \, dN_{Re} \tag{3}$$

where d is the diameter of the tube and N_{Re}, the Reynolds number, can be expressed as

$$N_{Re} = 4Q\rho/(\pi d\eta) \tag{4}$$

where Q is the volumetric flow rate and ρ and η are the density and the viscosity of the sample gas, respectively. Equations 2 and 4 can be combined to

$$L_i = 0.2 \, Q\rho/(\pi\eta) \tag{5}$$

This equation shows that at a given location (ρ and η constant), L_i is solely dependent on the sampling rate Q. For dry air at 20 °C and 1 atm and with a density of 1.2 g/L and a viscosity of 1.8×10^{-4} P, equation 5 can be rewritten

$$L_i = 6.2 \, Q \tag{6}$$

where L_i is expressed in centimeters and Q in liters per minute. Inlet length considerations for an annular denuder are not fundamentally different. For most annular geometries in present use, the annular gap is small relative to the radius of curvature; this situation permits the parallel-plate approximation. For a set of parallel plates separated by the distance x, Schlichting

(35) stated that the inlet length necessary for laminar flow development can be approximated by

$$L_i = 0.04 \, xN_{Re} \tag{7}$$

For an annular denuder,

$$x = (d_o + d_i)/2 \tag{8}$$

where d_o and d_i are the inner diameter of the outer tube and the outer diameter of the inner tube (rod), respectively, and

$$N_{Re} = 2Q\rho/[\pi\eta(d_o + d_i)] \tag{9}$$

Combining equations 7 through 9 gives

$$L_i = 0.04 \, Q\rho(d_o - d_i)/[\pi\eta(d_o + d_i)] \tag{10}$$

When the inlet length is expressed in terms of number of "gap widths", the difference between the flow in a tube and the flow in an annulus of narrow gap differs only by 25% [(0.05 − 0.04)/0.05]. This situation is an indication that the growth of the laminar boundary layers from the wall to the center of the channel is similar in both cases. Because duct friction coefficients, a measure of momentum transfer, do not vary by more than a factor of 2 for ducts of regular cross sections when expressed in terms of hydraulic diameters, the use of the inlet length for tubes or parallel plates can be expected to be a reasonable approximation for the inlet lengths of other cross sections under laminar flow conditions. In the annular denuder, the dimensionless inlet length for laminar flow development, L', can be expressed as

$$L' = 2L_i/(d_o - d_i) \tag{11}$$

For a given flow rate, L' is inversely related to the gap width or to the diameter (d_o or d_i) at constant gap width (36).

Because it may not be possible to provide an inlet length sufficient for full development of laminar flow in some applications, the consequences of not meeting this requirement may be worthy of examination. Although the lack of a fully developed laminar flow profile may make it impossible to compute diffusion coefficients from the overall fraction of the gas that penetrates the denuder, most practitioners are solely interested in obtaining a good collection efficiency for the analyte gas. Over the inlet region, analyte gas collection efficiency is expected to increase slightly relative to predictions based on diffusive mass transfer under laminar flow conditions. According to Mercer (37) and Friedlander (38), the length required for parabolic flow to develop is short compared to the length required for the particle concentration boundary layer to develop. As such, the entrance length should have little influence on the extent of particle deposition on the denuder. However, the existence of some uncoated inlet length is important, because otherwise particles deposited in the entrance region will become part of the sample analyzed. On the other hand, leaving long, uncoated lengths of a glass tube, from which denuders are commonly constructed, is hardly a desirable state of affairs. Such a length can easily acquire a static charge (particularly at low relative humidities) and can contribute to significant loss of particles (39); the loss, predictably, is even higher when the surface is poly(tetrafluoroethylene) (PTFE), known to be particularly susceptible to acquiring an electrostatic charge (40). On the basis of these conflicting considerations, the prudent course of action appears to be to require several centimeters of uncoated entrance length, although this length may not be sufficient to fully develop laminar flow. However, this discussion may be irrelevant to issues of leaving any uncoated lengths at all in denuders that are manually washed (with wash liquid being subsequently analyzed). In such cases, typically both the uncoated and coated portion of the denuder is washed (41).

Collection Efficiency. *Single-Tube Denuders.* For the appropriate design of a diffusion-based collection device for an intended application, the ability to estimate the collection efficiency a priori is of considerable help. Although the theoretical soundness of the Gormley–Kennedy equation (equation 1) is not questioned, it is based on the assumption that the uptake probability of the analyte gas at the wall is unity; that is, the wall is truly a "perfect sink", and every collision results in uptake. This assumption is unrealistic. In recent years, this issue has been reexamined. McMurry and Stolzenburg (42) showed for a liquid-coated denuder how the uptake probability (discussed by the authors in terms of the "mass accommodation coefficient") can be evaluated from collection efficiency measurements. Murphy and Fahey (43) utilized the mathematical solution originally developed for hemodialyzers by Cooney et al. (44); this treatment assumes a constant uptake probability that may be less than unity. To use the Murphy–Fahey approach, however, this probability must be precisely known.

When the denuder active surface is an inert porous membrane such that the analyte molecule must diffuse across the pores to be trapped by an absorber liquid, only a fraction of the membrane surface is porous, and the pores may also be tortuous. Consequently, collision at the membrane surface is not synonymous with uptake. Corsi et al. (45) developed a numerical solution for the collection efficiency observed for such a membrane-based diffusion denuder, hereinafter referred to as a diffusion scrubber (DS). Both groups of researchers dealing with the issue of less than unity uptake probability reached the conclusion that this value must be very much less than unity before a significant departure from Gormley–Kennedy behavior becomes experimentally discernible. The collection efficiency determined by Dasgupta et al. (46) for a hydrophobic porous PTFE membrane tube (2-μm pores, 50% surface porosity, 0.4-mm-thick wall) collecting SO_2 (dilute H_2O_2 flowing as scrubber liquid outside the tube) indeed agrees within 1% with values computed directly from the Gormley–Kennedy equation at sampling rates lower than 2 L/min (89 to 100% collection efficiency). At higher sampling rates, however, experimental collection efficiencies were lower than Gormley–Kennedy values. Overall, for most practical single-tube denuders, whether membrane-based or not, the Gormley–Kennedy equation provides a reliable starting estimate for device design. A gradual decrease of collection efficiency during sampling due to surface saturation is possible with typical sorbent-coated denuders (47); with automated diffusion denuders–scrubbers this decrease would not appear to be an important issue inasmuch as the collecting medium is either continuously or cyclically renewed.

Annular Denuders.   Computing the collection efficiency of annular denuders—which are becoming increasingly popular because of their ability to maintain near-quantitative collection efficiencies at high sampling rates within a compact design—is less straightforward. Possanzini et al. (25) introduced the annular geometry to the present-day practice of diffusion-based sampling and suggested an empirical equation to calculate the collection efficiency f:

$$1 - f = Ae^{-\alpha\Delta} \tag{12}$$

where A and α are empirical constants and Δ is given by

$$\Delta = \mu(d_o + d_i)/(d_o - d_i) \tag{13}$$

Although the precise values of A and α were said to be dependent on the exact physical design of the device, under conditions where the radius of curvature is large compared to the annular gap ($d_o \gg d_o - d_i$), A and α were regarded to be constants. For SO_2 as the test gas in tetrachloromercurate(II)-coated glass annular denuders ($Q = 1$ to 40 L/min, $\Delta = 0.064$ to 0.272, $f = 0.81$ to 1.00), they reported $A = 0.82$ and $\alpha = 5.63$. In a later study dealing with the collection of NO_2 (g) with KI-impregnated annular denuders coated with polyethylene glycol, the same value for A but a different value for α was reported. The much lower value for α, 1.12, was attributed to poor sink efficiency (48). Ali et al. (32) have criticized the drawbacks of such empirical approaches. On the basis of the original work of Gormley (49) on diffusion between parallel plates, they suggested that the parallel-plate approximation of an annular denuder will lead to a form of equation 12 where $A = 0.91$ and $\alpha = 3.77$. The differences in the actual collection efficiencies resulting from the different values of A and α can be substantial unless the denuder is operating at nearly quantitative collection efficiency.

More recently, Winiwarter (50) has built on the work of Stephan (51) by dealing with heat transfer in annular tubes and applying a Fickian diffusion model to compute collection efficiencies in an annular denuder. The resulting differential equations were solved by the Runge–Kutta method. Collection efficiencies thus computed for HNO_3, HCOOH, and CH_3COOH agreed well with the experimental data obtained by Rosenberg et al. (52). Winiwarter provided tabulated data and graphs to enable computation of collection efficiencies for specific annular denuder dimensions and operating conditions. Coutant et al. (26) also developed a numerical method; they used the parallel-plate approximation. Their model takes into account the nonunity uptake probability, which can be input as a separate parameter. The software (written in C programming language for IBM and compatible PCs) is available from the authors and permits calculation of collection efficiencies for multiple concentric tubes as well.

None of these approaches are applicable to denuders of the annular geometry where only one of the surfaces in contact with the sampled gas is effective in uptake. Membrane-based DS devices of the reverse geometry, introduced originally by Tanner et al. (53), utilize a membrane tube concentrically suspended within an outer jacket tube. The gas is sampled through the annular space, and the analyte gas is captured by collection through the pores in the membrane. With this design, only the inner surface of the annulus is effective as a sink. Solutions to the parallel problem for heat transfer are available. Lundberg et al. (54) presented an extensive set of numerical solutions for cases where either the outer or the inner surface of the annulus is the sink. These results can be directly used, with the limitation that the treatment assumes the probability of uptake to be unity. The fraction collected, f (referred to as the parameter $\theta_{mi}^{(3)}$ by Lundberg et al.), is calculated as a function of the dimensionless axial position $\overline{X}$, where $\overline{X}$ in this application would be given by

$$\overline{X} = \mu \frac{d_o + d_i}{4(d_o - d_i)} \qquad (14)$$

Figure 1 graphically depicts the numerical data relevant to our application listed by Lundberg et al. Different sets of curves of f vs. $\overline{X}$ are provided for individual values of d_i/d_o. Discrete data were provided in the numerical tables of the original work; to produce the continuous traces in Figure 1, a cubic spline fitting was used.

The experimental collection efficiencies for H_2O_2 (estimated Graham's law diffusion coefficient is 0.18 cm²/s) for an aqueous scrubber Nafion (perfluorinated ionomer) membrane DS ($L = 40$ cm, $d_o = 0.5$ cm, and $d_i = 0.06$ cm) have been experimentally determined by two independent ap-

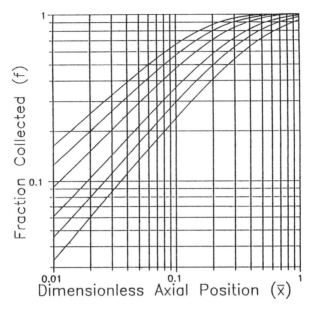

Figure 1. Collection efficiency (f) for a denuder of annular geometry in which only the inner surface of the annulus is a sink as a function of axial position $\overline{X}$ (see equation 14). The curves are based on numerical data in reference 54. From top to bottom, the traces correspond to d_i/d_o = 1.0, 0.5, 0.25, 0.1, 0.05, and 0.02.

proaches (40). Here the sink surface is expected to be efficient in uptake, and f has been observed to be 0.944 ± 0.017, 0.838 ± 0.013, 0.732 ± 0.006, and 0.638 ± 0.002 at $Q = 0.5$, 1.0, 1.5, and 2.0 L/min, respectively. The corresponding values for $\overline{X}$ are 0.86, 0.43, 0.29, and 0.22, respectively. The d_i/d_o value for this DS is 0.12; interpolating between the curves for $d_i/d_o = 0.1$ and 0.25 in Figure 1 reveals that the theoretical predictions are in excellent agreement with the experimental data.

With porous membrane DS devices of this geometry and for thin membranes with low-tortuosity pores (i.e., where the diffusion distance within the pores is very small compared to the radial diffusion distance in the DS), good predictions for the collection efficiencies can be obtained if the nominal $\overline{X}$ and d_i/d_o values are both multiplied by the fraction of the surface that is porous. For example, with a diffusion scrubber based on such a membrane tube ($L = 40$ cm, $d_o = 0.5$ cm, $d_i = 0.045$ cm, fractional porosity 0.4), the corrected $\overline{X}$ values for H_2O_2 as sample gas are 0.32, 0.16, 0.11, and 0.08, respectively for $Q = 0.5$, 1.0, 1.5, and 2.0 L/min, and the corrected d_i/d_o value is 0.036. The collection efficiencies predicted from Figure 1 (interpolating between d_i/d_o values of 0.02 and 0.05) are in good agreement with the experimental results (*40*): 0.661 ± 0.017, 0.371 ± 0.013, 0.310 ± 0.001, and 0.234 ± 0.024 at the respective flow rates.

The efficient sink criteria cannot be overemphasized, however. When the experimental data for formaldehyde for either of the two different membrane scrubbers are considered, the theoretical predictions significantly exceed the experimental collection efficiencies. The uptake of formaldehyde at an aqueous interface is controlled by its rate of hydration to methylene glycol, a process that is acid- or base-catalyzed. The collection efficiency significantly increases in going from pure water to 0.1 M H_2SO_4 as a scrubber liquid (*55*), but the uptake probability still remains a controlling factor in determining the collection efficiency. Obviously, in such cases theoretical predictions merely establish an upper limit.

Particle Transmission.   Coarse particles ($>\sim 2.5$ μm) are present in all atmospheric samples. If a diffusion-based collection system is not equipped with an impactor or cyclone at the front end to remove the coarse particles, some coarse particle deposition will occur in the system. This is especially true for annular denuders (in which the inner member must have some physical means of attachment to the outer member and such structural supports are obligatorily present in the flow path) and DS devices with T-type air inlets. Larger particles are often derived from crustal sources and display significant acid-neutralizing capacity. If total acidity is to be determined in the atmospheric sample, significant errors can be introduced unless such particles are first removed. PTFE-coated cyclones as an integral part of the sample inlet appear to be the best choice for removing the coarse particles (*27, 56*). Relative to a more polar surface like glass, the adsorption of sticky gases like HNO_3 is minimized on a PTFE or PTFE-coated surface. Although such a surface is more prone to promote electrostatically induced fine-particulate deposition (vide infra), the residence time within the cyclone is typically below 1 s, and significant losses of fine particles have not been observed (*27, 56*). In this connection, maintenance requirements of any instrument with a coarse-particle removal system at the inlet need to be pointed out. Although the rest of the system may well be configured to be completely automated, without periodic cleaning of the cyclone the resulting data may be subject to significant error. Because of diurnal temperature variations, errors may accrue, for example, from cyclic deposition and evaporation of NH_4NO_3 as well as from loss of acid gases due to gas–particle interactions with deposited coarse particles.

For fine particles, despite the fact that the major rationale behind diffusive sampling of a gas is to achieve discrimination from the concurrently present atmospheric aerosol, relatively little attention has been paid to actually characterizing the particle transmission through these systems. A summary of existing data has been presented (*40*). The only thorough characterization of particle losses that has appeared in the literature concerns the Harvard–Environmental Protection Agency annular denuder system (39, 57). The system is equipped with an integral inlet cyclone or impactor. The first study (57) suggested that total particle losses are $< \sim 3\%$ for the particle size 1.50 to 2.77 μm. The more recent and more extensive study (39) dealt with particles over a much greater size range and with different degrees of charge: neutral, with Boltzmann charge, or with a single charge. Further, the glass denuder surfaces were either uncoated or coated with NaCl or citric acid. The loss increased in going from the neutral to the charged aerosol and was somewhat higher for an uncoated denuder surface compared with the coated surface, presumably because of the greater tendency of the un-coated surface to acquire a static charge. Typical losses in a single-stage uncoated denuder in the 0.10- to 0.86-μm size range ranged from 0.9 to 8% for an aerosol with Boltzmann charge distribution; an additional 1.8 to 5.3% was lost on essential system components. Although the overall extent of the loss is still reasonably small, it may not be insignificant if the intent is to measure the aerosol composition.

The effect of electrostatic charging cannot be overemphasized—dramatic increases in particle deposition occurred when the denuder was dried with a vigorous flow of compressed air. With PTFE tubes, much more susceptible to acquiring a static charge, as much as 3 to 15% of the aerosol, 0.1 to 1.09 μm in diameter and with a Boltzmann charge distribution, can be lost under dry conditions merely by passage through a 70-cm-long straight PTFE tube (40). Clearly, long inlet lines must be avoided to measure particle compo-sition along with gases. Unfortunately, tubing materials that are considered inert are also more susceptible to acquiring a static charge. The choice of the sample conduit material is also dictated by the adsorption characteristics it displays toward the analyte gases of interest. For a variety of polar gaseous analytes, for example, HNO_3, HCl, and H_2O_2, glass inlet lines are unac-ceptable. Even when the conduit is electrically conductive, for example, stainless steel, the extent of aerosol deposition is acutely dependent on the charge on the aerosol (58); little control can be exercised over this situation.

In any fully automated instrument operating unattended, it is also highly desirable that the instrument periodically zeros and calibrates itself. Thus, aside from the sample air, zero and calibrant gases must be accessible to the collection system inlet by some automated valving arrangement. Typi-cally, PTFE or perfluoroalkoxy (PFA) solenoid valves are used to switch sample streams. However, virtually all available commercial valves have right-angled passageways inside the valve, and deposition of ambient par-ticulate matter occurs at the bend. Over a period of time, this buildup compromises sample integrity. Some valves also become sufficiently warm upon energization to cause significant loss of labile analyte gases like H_2O_2 (55). A three-position electropneumatic slider valve with no sharp angular bends (59) minimizes these problems. An altogether different and possibly superior approach is to provide a T-connection at the very beginning of the inlet line. Zero or calibration gases are introduced through this T-arm as desired by appropriate valves connected to these sources. The zero or calibration gas flow is made greater than the sample flow. Thus, during zero or calibration periods, no ambient air is drawn into the system; rather, excess zero or calibration gas is vented through the inlet.

Annular Geometry: Preferred in All Applications?   The annular geometry is rapidly replacing the single-tube geometry in most applications. Clearly, the annular design is capable of collecting a greater analyte mass per unit time. For automated collection–analysis systems that operate on a cyclic sorption–desorption protocol (e.g., thermodenuders and vide infra), the collected analyte is generally desorbed as a plug within a fixed time window. Greater collected mass therefore translates to better LODs or temporal resolution or both. On the other hand, for a wet denuder in which a flowing scrubber liquid continuously wets the active surfaces and the effluent is then taken for analysis, the advantages of an annular geometry, in view of its more complex construction, are less clear-cut unless the scheme involves concentration of the denuder liquid effluent prior to analysis. This situation arises because most continuous flow-through analytical detectors are concentration-sensitive rather than mass-sensitive. The ability of the annular geometry to collect a greater analyte mass is associated with a larger active surface area. With the reasonable assumption that the minimum necessary scrubber liquid flow rate to effectively present a continuously renewed collection surface is directly proportional to the active surface area, most of the advantage of the annular geometry disappears. For the example of the isoefficient denuder trio cited by Possanzini et al. (25), the collection efficiency of a single-tube denuder—$L = 50$ cm, $d = 0.3$ cm at $Q = 1.7$ L/ min—is identical to that of two annular denuders—$L = 10$ cm, $d_i = 1.0$ cm, $d_o = 1.3$ cm at $Q = 3.8$ L/min and $L = 20$ cm, $d_i = 3.0$ cm, and $d_o = 3.3$ at $Q = 20$ L/min. The differences in the analyte mass collected per unit time is reflected in the ratio of the flow rates, 1:2.23:11.8. The corresponding ratio of the active surface area, $\pi(d_o + d_i)L$, is 1:1.52:8.4. The collected mass per unit surface area ratio, the superiority factor, is then obtained by dividing the first set of numbers by the second and is 1:1.45:1.20. This degree of improvement is hardly worth the added complexity of the annular design. Higher sampling rates are sometimes considered desirable to minimize inlet losses. However, a high flow denuder is not essential for this purpose. An inlet manifold can be set up with a high flow rate and small residence time, and the denuder, with its own aspiration source, is then connected with a very short conduit to this manifold.

These considerations pertain largely to automated systems. In systems utilizing manual collection and analysis, the annular geometry does have advantages, as evident from its current widespread use.

Automated Collection–Analysis Systems

The existing systems can be broadly classified into two groups: systems that deal with a gaseous analyte, and systems that deal with the analyte in the liquid phase. To date, the first group is composed solely of thermodenuders, devices that rely on thermally cycled sorption–desorption steps. The second group includes wet diffusion denuder–scrubbers and devices that are cyclically sorbent-coated and washed.

Thermodenuders

Sampling ambient air through sorbents like diphenylphenylene oxide (Tenax) and thermally desorbing the adsorbed compounds for gas chromatographic analysis are among the most established and useful practices for the determination of organic compounds in ambient air. Thermodenuders represent the equivalent approach with diffusion-based sampling.

Braman et al. (47) reported the first such device for the measurement of atmospheric NH_3 and HNO_3, and application to ambient air analysis was reported by McClenny et al. (60). Tungsten(VI) oxide (WO_3, often referred to as tungstic acid) is coated on the inside walls of porous glass (Vycor) or quartz tubes by electrically heating a concentrically suspended tungsten wire. The polymeric blue W(IV) oxide initially formed is converted to the preferred yellow W(VI) form by heating the tube to 500 °C while passing oxygen through it. In the automated instrument (60), the sampler inlet allowed, by appropriately connected solenoid valves, the choice of sample air, calibrant, and a purge gas (20% O_2 in He, used during the desorption step). The sample air is drawn through a WO_3-coated denuder tube containing an integral heating coil and then through a catalyst tube containing a gold catalyst, also provided with a heating coil. A typical operational cycle involves (1) 10–50 min of sampling of ambient air ($Q = 1$ L/min); (2) integration of the baseline signal of a chemiluminescence-based NO_x monitor for 10 min (this is considered blank, and step 1 continues during this time; (3) ceasing sampling, switching to purge gas, and connecting the converter tube exit to the NO_x monitor by appropriate valving, controlled electrical heating of both the denuder and the converter, and recording of the resultant signal from the NO_x monitor (10 min); and (4) cooling (10 min) before the cycle is begun anew. During step 1, sampled NH_3 and HNO_3 are taken up by the denuder. When it is heated in step 3, HNO_3 first comes off as NO_2, and at a higher temperature NH_3 comes off without decomposition. In the presence of the heated Au catalyst, both are converted to NO and are thus measured by the NO_x monitor. The individual signals from HNO_3 and NH_3 are temporally separated; although they are not baseline-resolved, the separation is considered adequate. The LOD for NH_3 or HNO_3 was 70 parts per trillion by volume (pptrv) for a 20-min sample. Collection of particulate ammonium and nitrate with a packed column of WO_3-coated sand and analysis by similar thermal desorption techniques are also mentioned in these papers (*47, 60*), although few details are given.

Although the tungstic oxide thermodenuder in principle is an attractive means of measuring two analyte gases of considerable interest, field studies have indicated that unknown interferences can seriously compromise the reliability of the data generated for either NH_3 or HNO_3. In intercomparison studies, the data generated by this technique have been in error by as much as a factor of 2 to 6. [*See* Fox et al. (*61*), Roberts et al. (*62*), and Appel et al. (*63*)]. In a more detailed subsequent study, Roberts et al. (*64*) made a number of important observations regarding this technique as applied to the measurement of background levels of NH_3. They found that neither the blue W(IV) oxide nor the yellow W(VI) oxide is best suited for the purpose; rather, the blue-green oxide, presumably of an intermediate oxidation state, is the most desirable form. Alkylamines like ethylamine were also found to be taken up by such a denuder. Although such amines are desorbed at a temperature significantly lower than NH_3 under idealized test conditions, significant amounts of the amines are converted to NH_3 on the active surface of the denuder at relative humidity levels greater than 25%. Consequently, it is doubtful that such compounds can actually be differentiated from NH_3 under actual measurement conditions. Most importantly, continued reliable performance of the denuder is very susceptible to overheating; the recovery of NH_3 decreases and the peak becomes ill-defined for an overheated denuder. With frequent calibration to ensure accuracy, the authors measured NH_3 levels in isolated regions in the Colorado mountains and coastal California to be 200 ± 80 and 360 ± 170 pptrv, respectively; these are among the lowest NH_3 concentrations reported for NH_3 in continental air. Despite the apparent applicability, Roberts et al. did not endorse the technique. Because desorption of NH_3 occurs at temperatures near those that alter the surface, they concluded that the performance is susceptible to slow evolutionary and occasionally catastrophic failure and that other alternatives should be sought. Much more favorable results were later reported with a molybdenum oxide denuder of annular geometry (*65*); this denuder is described in a later section.

Interestingly, much of the early interest in developing automated thermodenuders centered on the determination of aerosol-phase analytes rather than gases. Lindqvist (*66*) described a system for the determination of aerosol H_2SO_4 in which the analytical and regeneration step of the denuder was fully automated but the transfer of the denuder tube between the collection system and the analyzer was manual. As discussed subsequently, this step is not difficult to automate. The strategy behind the collection of aerosols by a denuder typically involves the removal of all gases that may pose an interference by an appropriate predenuder. The aerosol is then thermally converted to a gas (either a phase transition or decomposition into gaseous products), and this gas is taken up by a denuder that is thermally desorbed during the analysis step. In Lindqvist's instrument, two serial glass (Pyrex) predenuders (one coated with oxalic acid to remove NH_3, and the other coated with alumina impregnated with $AgNO_3$, H_3BO_3, $NaHCO_3$, and tartaric acid) designed to remove SO_2, H_2S, CH_3SH, CH_3SCH_3, and CH_3SSCH_3 are used. The principal denuder is a 73.3- × 0.45-cm quartz tube, first solution-coated with an acetone solution of $Mn(NO_3)_2$ and $PdCl_2$ and later heated to 800 °C in sequential H_2, air, and He streams to obtain a coating of MnO_2–PdO. During collection, this tube is maintained at 138 °C to volatilize the aerosol H_2SO_4, which is collected as sulfate. During analysis, the denuder is supplied with a small flow of helium and heated to 800 °C, whereupon the sulfate decomposes to SO_2. The exit gas from the denuder is merged with a small flow of H_2 that then flows through a quartz tube heated to 1000 °C. The resulting H_2S is collected on a silver wool trap. When this step is complete, the silver wool is flash-heated to 600 °C, and the liberated H_2S is fed to a gas chromatograph equipped with a photoionization detector. The entire analytical cycle requires 12 min. Samples of 60 min provide an LOD of 60 ng/m^3 (15 pptrv gas-phase equivalent) if immediately analyzed. Blanks increase upon storage, and the attainable LOD increases by an order of magnitude upon 24 h of storage; this observation clearly underscores the advantages of a fully automated system with no elapsed time between collection and analysis. Under ambient conditions, the relative standard deviation (rsd) of the measurement procedure is reported to be 17%.

Lindqvist subsequently described a similar system for the determination of HNO_3 (67). HNO_3 was collected on an $Al_2(SO_4)_3$-coated quartz denuder. The thermally desorbed NO_x was determined by gas chromatography–photoionization detection. Subsequently Tanner et al. (68) simplified and fully automated the overall system configuration. A 51- × 0.4-cm quartz denuder tube was solution-coated with 20% w/v $Al_2(SO_4)_3$ and used at a very low Q (0.1 L/min). The desorption step involved heating to 500 °C for 1 min; the liberated NO_x was determined by a chemiluminescence monitor. Although laboratory results were attractive, field intercomparisons with a number of other methods indicated low and variable results; the reasons for this discrepancy could not be identified with certainty.

The Netherlands Energy Research Foundation (ECN) has made many significant contributions to the progress of atmospheric analytical chemistry, particularly in the area of automated diffusion denuder-based analytical systems. As early as 1981, the ECN group, in collaboration with Klockow and Niessner, who originally developed the thermoanalytical strategy for the measurement of strong acids, described a serial seven-section denuder system to perform measurements of HNO_3, NH_3, H_2SO_4, ammonium nitrate, and ammonium sulfate (69). The first section was coated with NaF and was reported to retain HNO_3 (g), the second section was coated with H_3PO_4 and retained NH_3 (g), the third section was coated with NaOH to retain SO_2 (a potential interferent), the fourth section was coated with NaF and heated to 130 °C to convert H_2SO_4 into the vapor phase and retain it as well as to dissociate NH_4NO_3 and to retain the resulting HNO_3, the fifth section was coated with H_3PO_4 to retain NH_3 evolved from the fourth section, the sixth section was coated with NaF and heated to 230 °C to dissociate ammonium sulfates and retain the resulting H_2SO_4, and the seventh section contained an H_3PO_4-coated tube to collect the NH_3 evolved in the sixth section. The system was not automated; it would be indeed difficult to automate such an involved scheme.

Slanina et al. (70) subsequently described a fully automated computer-controlled thermodenuder system for the measurement of sulfuric acid and ammonium sulfate aerosol. As in Lindqvist's work, sulfur gases are removed by two serial glass predenuders, one coated with K_2CO_3 and the other with activated carbon. The main denuders are 50- × 0.6-cm quartz tubes solution-coated initially with $Cu(NO_3)_2$, dried and then fired at 900 °C in a nitrogen stream to obtain a mixed Cu–CuO coating. Two sequential Cu–CuO denuder tubes are used: The first is maintained at 120 °C and the second at 240 °C for the respective collection of aerosol sulfuric acid and ammonium sulfates. During analysis, the tubes are heated, one at a time, to 800 °C, and the liberated SO_2 is measured by a flame photometric sulfur analyzer. The tubes are then regenerated while still hot by injecting a small volume of clean air, and the cycle is started anew. The instrument is equipped with a second pair of Cu–CuO denuder tubes that are used for sampling while the first pair goes through the analytical cycle. Provisions for periodic automated calibration with SO_2 as a calibration standard are also provided. The system provides an LOD of 20 ng/m^3 for a 1-h sample and 100 ng/m^3 for a 5-min sample for either H_2SO_4 or $(NH_4)_2SO_4$.

Although this thermodenuder collection–analysis system represents an interesting and ingenious advance in the art of atmospheric analysis, it did not meaningfully solve the intended measurement problem. The premise of the thermal differentiation step is that H_2SO_4 and $(NH_4)_2SO_4$ aerosols exist as an external mixture. In light of what is presently known about the heterogeneous oxidation of SO_2, this premise is incorrect. Internal mixtures, that is, aerosols containing H_2SO_4 in various stages of neutralization with NH_3, cannot be thermally differentiated to hypothetical constituent components in the described manner. On the basis of subsequent findings, it is doubtful that $(NH_4)_2SO_4$ and NH_4HSO_4, even when discretely present in the same sample, can be thermally differentiated. Further, the thermal decomposition of more volatile ammonium salts (e.g., NH_4NO_3) can produce negative interferences in H_2SO_4 measurements with thermal denuders (41). To be fair, science advances on faltering steps; I have been just as guilty of making the identical erroneous assumption (71, 72).

In any case, the development of the automated H_2SO_4–$(NH_4)_2SO_4$ determination system was clearly an important step toward the subsequently reported thermodenuder system for the measurement of SO_2 (73). In this instrument, a quartz predenuder, solution-coated with $NaHSO_4$–Ag_2SO_4, removes interfering sulfur gases, notably H_2S. This removal is followed by a Cu–CuO-coated denuder as described in a preceding section. During the analysis step, the sulfate resulting from the collected SO_2 is thermally decomposed to SO_2 and measured with a pulsed fluorescence instrument. Under field conditions, this setup permitted an LOD of 40 pptrv of SO_2 for a 30-min sample (Q = 0.5 L/min). An LOD almost an order of magnitude better was attainable with the flame photometric detector in the laboratory. The reproducibility of the technique was measured with parallel triplicate denuders and ranged from 2 to 5% rsd.

Continuing applications of thermodenuders typically utilize the annular geometry. Keuken et al. (74) described a vanadium pentoxide coated quartz annular denuder system. The denuder itself and the overall setup of the system are shown in Figures 2 and 3. The denuder is coated by an aqueous slurry of V_2O_5 and dried at 70 °C. The denuder (d_o = 2.8 cm, d_i = 2.5 cm, and L = 25 cm) exhibited essentially quantitative efficiency for collecting NH_3 at Q = 10 L/min. Following the collection period (10–20 min), a motorized screw drive moves a preheated oven (700 °C) around the denuder (Figure 3) and stops when the denuder is half covered (position 1). After 5 min, the oven is moved again, this time to cover the entrance half of the denuder as well (position 2). During sampling, ammonia is essentially removed completely by the first half of the denuder. Unknown nitrogen-containing gaseous interferences are, on the other hand, removed inefficiently and are sorbed approximately to the same extent on the first and the second half of the denuder. When the denuder is heated, both NH_3 and any other nitrogen-containing compounds are converted to NO_x and measured in turn by a chemiluminescence-type NO_x monitor. Thus, the peak obtained in position 1 corresponds to the interferences only, and the peak obtained in position 2 corresponds to both NH_3 and the interferences. The NH_3 signal is determined by difference. For a 10-min sample, the LOD is ~150 pptrv with a typical rsd of 5% under ambient measurement conditions. In parallel measurements with a H_3PO_4-coated single-tube denuder, a wet annular denuder (vide infra) and a tunable diode laser spectrometer, very good agreement was reported.

Earlier, a less conventional annular geometry was reported by Langford et al. (65) for the determination of NH_3. Following their previous work with tungstic oxide coated denuder tubes (62, 64), they sought a more rugged and reproducible means of fabricating metal oxide coated denuders. These goals were met in a geometry in which a W or Mo rod is concentrically placed within a quartz tube (d_i = 0.32 cm, d_o = 0.40 cm, and active L = 30.5 cm); this geometry is analogous to that of the DS described by Tanner

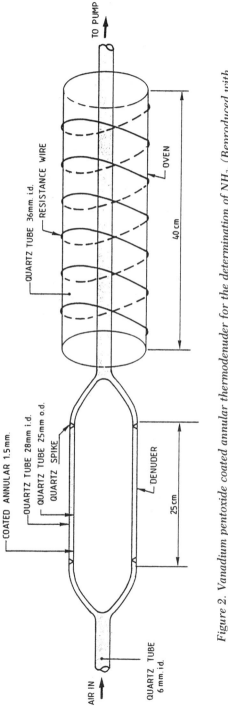

Figure 2. Vanadium pentoxide coated annular thermodenuder for the determination of NH₃. (Reproduced with permission from reference 74. Copyright 1989 Pergamon Press.)

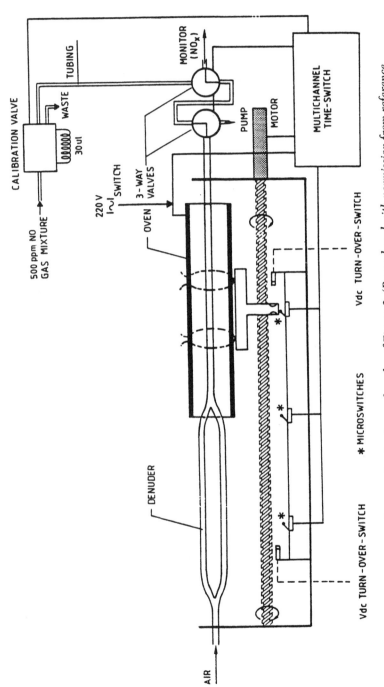

Figure 3. Instrument schematic for the the thermodenuder of Figure 2. (Reproduced with permission from reference 74. Copyright 1989 Pergamon Press.)

et al. (53) and reduces the overall thermal mass of the system. Heating under a flow of O_2 at 300 to 400 °C produces Mo(IV) oxide within 15 min, and the desired Mo(VI) oxide is produced upon further heating to 450 °C. (The W rod produces similar coatings of the corresponding (IV) and (VI) oxides but at higher temperatures.) After 1 h, the flow is reversed and the assembly is heated for another hour to produce a more even coating. The entire procedure requires 3 h and produces an optimum oxide coating 0.25 mm thick. Sampling is typically conducted for 30 min at $Q = 0.9$ L/min.

Detailed studies of the desorption behavior of NH_3 from all four oxide surfaces [W(IV), W(VI), Mo(IV), and Mo(VI)] were reported (65). The WO_2, MoO_2, and WO_3 surfaces all desorb NH_3 as NH_3 at respective temperatures of ~280, 300, and 350 °C and thus require the usual heated Au converter for the conversion of NH_3 to NO prior to measurement by a chemiluminescence-type NO_x monitor. In contrast, the MoO_3 surface shows two distinct desorption steps at 350 and 380 °C; the first corresponds to the desorption of NH_3 as NH_3, and the second peak has been shown to be NO, the denuder surface itself acting as the necessary oxidative converter. At the low NH_3 levels of interest, the second peak amounts to ~30% of the total NH_3 sorbed (the figure is essentially constant for a given denuder, but individual denuders require separate calibrations), and this situation allows sufficient sensitivity for the system to be operated without an additional oxidative converter. The MoO_3 surface also takes up NO_2 and HNO_3, which are both desorbed as NO at 200 and 200 to 250 °C, respectively. The feasibility of thermal separation of sorbed NO_2 and HNO_3 was not investigated in this work; rather, using a program that rapidly heats the denuder to 250 °C and then gradually raises the temperature to 400 °C, a complete separation of the NO_2–HNO_3 and the NH_3 peaks (the rejection ratio of the NH_3 peak from NO_2 is greater than 5000:1) is obtained. A flow of N_2 is used during the desorption and the cooling steps.

Like the WO_3 denuder, the MoO_3 denuder is unable to distinguish between NH_3 and amines; however, the predominance of the former in most measurement situations may make this limitation of little consequence. The LOD is estimated to be 10–20 pptrv of NH_3; during extensive field studies, concentrations as low as 50 pptrv were measured and constitute the lowest reported gaseous NH_3 concentrations thus far. At low levels virtually every surface shows some uptake of NH_3; to avoid this situation, the entire train is maintained at 50 °C (including the denuder, during the collection step, to prevent adsorption on the quartz walls). This practice does cause some concern about interference from the potential evaporative dissociation from aerosol NH_4NO_3; however, side-by-side studies with citric acid coated simple tubular denuders indicate that the evaporation kinetics is slow, and the observed extent of artifactual NH_3 is likely less than 3%. Overall, the MoO_3 thermodenuder represents an elegant combination of good design and clever use of chemistry, and the studies clearly reflect the attention to detail bestowed by its inventors.

Klockow et al. (75) described two separate automated thermodenuder systems for the simultaneous determination of HNO_3 and NH_4NO_3. One problem in the determination of HNO_3 is that most denuder coatings tried for this purpose also take up some other nitrogen-containing compounds, presumably organic in nature. Although such compounds are not collected with high efficiency, the NO_x evolution from these species during the thermal desorption step cannot be temporally separated from that of HNO_3 by any reasonable programmed temperature steps, and the resulting interference can be significant. The first system devised attempted to solve this problem by using a train of three serial annular quartz denuders similar to that described earlier (74). All three denuders are solution-coated with $MgSO_4$. The serial denuder train incorporates valves at the connecting point between each denuder such that during the desorption step each denuder can be separately heated and its effluent can be analyzed by a chemiluminescence-type NO_x monitor. During the collection step ($Q = 5$ L/min), the first two denuders, operating at room temperature, separately collect HNO_3 plus unknown interference and the unknown interference only. The third denuder is operated at 150 °C to volatilize NH_4NO_3 and collect the resulting HNO_3. During the desorption step, each tube is heated in turn to 700 °C, and the evolved NO_x is detected. The difference between the signals from tubes 1 and 2 is taken to be HNO_3, and the signal from tube 3 is taken to be NH_4NO_3. The system is provided with provisions for automated detector calibration with cylinder NO. A second train of three sequential denuders is provided to alternate the collection–analysis cycles between the two trains and thus improve temporal resolution. With a 30-min sample, LODs of 40 pptrv and 100 ng/m^3 were reported for HNO_3 and NH_4NO_3, respectively.

The second system described in this paper (75) utilizes single-tube denuders operating at $Q = 0.66$ L/min. The first and third denuders are made of glassy carbon (50 × 0.6 cm) and are coated with a methanolic suspension of $BaSO_4$ and a water–methanol solution of NaF, respectively. The middle denuder is a glass tube (50 × 1.5 cm) containing an inserted rolled sheet of activated-carbon-impregnated filter paper. According to Klockow et al., the $BaSO_4$-coated glassy carbon denuder does not collect the unknown organic nitrogen-containing interference(s) to a significant extent and collects HNO_3 selectively. The nitrogen-containing interferents are removed by the activated carbon denuder, and the NH_4NO_3 is then volatilized and collected in the NaF-coated denuder maintained at 140 °C. During the analysis step denuders 1 and 3 are sequentially heated to 700 °C, the liberated NO_x is measured, and the denuders respectively yield measures of HNO_3 and NH_4NO_3. For 24-h samples, respective LODs of 20 pptrv of HNO_3 and 60 ng/m^3 of NH_4NO_3 were reported.

The general advantages of thermodenuder systems include facile pre-concentration, automation, and the ease with which they can be coupled to a variety of detectors designed to handle gas-phase samples. The fact that all flowing fluids are in the gas phase simplifies overall system design. On the negative side, thermodenuders frequently have high analytical blanks, they consume a large amount of electrical power, and substantial cool-down times are required for a new sampling cycle to begin unless parallel sampling trains are used instead.

Wet Effluent Denuders–Scrubbers

Wet effluent diffusion-based collection devices can be subdivided into three groups—conventional coated denuders, in which the coating–washing–analysis steps are fully automated; membrane-based diffusion scrubbers; and wet diffusion denuders. There is only one report of an example of the first type. Bos (76) describes a system in which a single-tube glass denuder is coated in situ with a methanolic solution of citric acid, the denuder is dried with clean air, ambient air is sampled for a preset period of time, the denuder surface is washed down with water, and the washings are sent (for the determination of the collected NH_3) to an air-segmented flow analysis system relying on the indophenol blue chemistry, and then the cycle is begun anew. Such an obviously complex arrangement cannot be replicated from the very limited description provided in the brief paper by Bos (76). Should it be possible to operate such instruments reliably over long periods of time, they clearly may have many applications. No further reports on this technique have appeared in the decade since Bos published his work, and the method is not in present use.

Diffusion Scrubbers

Diffusion scrubbers are membrane-based denuders in which the sample air flows on one side of a membrane and a suitable scrubber liquid flows on the other side. The analyte gases of interest are collected in the scrubber liquid, and the effluent is subjected to analysis. The simplest geometry is that of a conventional single-tube denuder. Air is sampled through a tubular membrane while the scrubber liquid is pumped in a countercurrent fashion through an external jacket tube surrounding the membrane tube.

Ion-Exchange Membrane DS Devices. The first DS reported (77) was based on a perfluorosulfonate cation-exchange membrane tube (Nafion, wet internal diameter 700 μm, wall thickness 75 μm, 30 cm long) contained in a glass jacket. Dilute H_2SO_4 was pumped through the jacket, and the DS was used for sampling NH_3. Ion-exchange membranes are hydrophilic, and during the sampling the tube is moist on the interior wall as well. Ammonia is captured at the membrane surface as NH_4^+, which migrates across the membrane to the scrubber liquid. Nafion is a membrane of considerable structural strength and inertness (aside from its ion-exchange properties). Small neutral polar molecules, for example, H_2O_2 and HCHO, and small cations show good transport properties across Nafion. It should be possible to sample low-molecular-weight amines, for example, CH_3NH_2, but as the cations become larger, more hydrophobic, or both, their diffusion coefficients in the membrane decrease markedly. Acid gases, which lead to characteristic anions, cannot generally be sampled with Nafion because the Donnan barrier inhibits the transport of anions across the cation-exchange membrane.

However, acid gases constitute a singularly important class of analytes to the atmospheric chemist. It is logical to attempt their collection with a DS based on an anion-exchange membrane. The major problem is the lack of the availability of suitable anion-exchange membrane tubes. Phillips and Dasgupta (78) studied a DS based on a PTFE membrane tube into which vinylbenzyl chloride is radiation-grafted and then quaternized (79). With a DS built from a 25-cm-long (3000 μm i.d., 100-μm wall) tube, the collection of HNO_3 was studied with on-line UV detection. The D of HNO_3 determined from the dependence of f on Q was in good agreement with other D_{HNO_3} values in the literature. The scrubber liquid was a solution of K_2SO_4 and sulfamic acid; the latter was incorporated to minimize the interference from NO_2 and HONO in the determination of HNO_3. (A DS based on an ion-exchange membrane cannot be used with pure water as the scrubber liquid if the collected analyte species is an ion that is bound on the ion-exchange sites. The scrubber liquid must contain ionic species of suitable displacing power in sufficient concentration to displace the collected analyte ion from the ion-exchange site.) Although this work proved in principle the applicability of an anion-exchange membrane based DS, it did not represent a practical means of measuring atmospheric HNO_3; the direct UV detection method provided neither the necessary sensitivity nor selectivity. Other problems may occur in dealing with the simultaneous collection and analysis of a variety of acid gases with an anion-exchange membrane based DS. Anion-exchange sites tend to catalyze the oxidation of sulfite to sulfate and nitrite to nitrate because of the greater affinity of the ion exchanger for the more oxidized form. The first oxidation does not present a major problem; collected SO_2 is often deliberately oxidized to sulfate before determination. However, the second oxidation does complicate the differentiation of HONO from HNO_3.

Ion-exchange membranes are hydrophilic. They present an active surface that has the same uptake probability for the analyte gas as may be obtained with a wet denuder with the same aqueous scrubber liquid. Thus, with scrubber liquids that are efficient sinks, these denuders typically exhibit Gormley–Kennedy behavior in terms of collection efficiency. The transport of the analyte from one side of the membrane to the other takes place in the condensed phase. For typical analyte species, the relaxation time, given by t^2/D_m, t being the membrane thickness and D_m being the analyte diffusion coefficient in the membrane matrix, is of the order of 0.75 to 3 min for 50- to 100-μm-thick membranes when the analyte species is not retained in the membrane by electrostatic forces. This condition may be the determining factor in controlling the response times of continuous analyzers based on such membrane-based collectors. Nonaqueous or mixed aqueous scrubber liquids can also be used with an ion-exchange membrane, should their use be deemed beneficial for the collection of certain analytes. Transport across an ion-exchange membrane tube does not rely on passage through pores; the membranes are therefore not easily "fouled". However, in designing a simple DS based on tubular ion-exchange membranes, it is difficult to achieve the concurrent goals of having a membrane tube large enough in diameter to sample air under conditions of low N_{Re}, thin enough to have rapid transport of analytes across it, and yet structurally strong enough to resist deformation as the scrubber liquid is pumped in an annular space necessarily small to minimize the integration volume. Other available hydrophilic membrane tubes, for example, cellulosic tubes used for dialysis, are likely to have the same limitations. Nevertheless, recent work (*40, 80*) shows that successful designs can indeed be achieved with ion-exchange membranes, and response times may actually be better for uncharged analytes or analytes that are charged similarly to the membrane matrix rather than for analytes that are retained on the membrane by ion exchange.

Porous Membrane DS Devices.   The applicability of a simple tubular DS based on a porous hydrophobic PTFE membrane tube was demonstrated for the collection of SO_2 (dilute H_2O_2 was used as the scrubber liquid, and conductometric detection was used) (*46*). The parameters of available tubular membranes that are important in determining the overall behavior of such a device include the following: First, the fractional surface porosity, which is typically between 0.4 and 0.7 and represents the probability of an analyte gas molecule entering a pore in the event of a collision with the wall. Second, wall thickness, which is typically between 25 and 1000 μm and determines, together with the pore tortuosity (a measure of how convoluted the path is from one side of the membrane to the other), the overall diffusion distance from one side of the wall to the other. If uptake probability at the air–liquid interface in the pore is not the controlling factor, then items 1 and 2 together determine the collection efficiency. The transport of the analyte gas molecule takes place within the pores, in the gas phase. This process is far faster than the situation with a hydrophilic membrane; the relaxation time is well below 100 ms, and the overall response time may in fact be determined by liquid-phase diffusion in the boundary layer within the lumen of the membrane tube, by liquid-phase dispersion within the analytical system, or by both. Consequently, the response times of porous membrane DS based analyzers are unlikely to be limited by the membrane response time.

The third important membrane parameter is the pore size; it is typically 0.02 to 5 μm. Together with the contact angle of the scrubber liquid on the membrane surface (which is a combined measure of the hydrophobicity of the membrane and the surface tension of the liquid), this parameter determines the pressure at which the scrubber liquid will undesirably seep through the membrane. Liquids with low surface tensions (e.g., methanol) cannot be used with porous membranes—they readily leak through the membrane. With water, the pressure necessary to force it through a 0.4-mm-thick PTFE membrane with 2-μm pores is ~10 psi; the necessary pressure is more than an order of magnitude greater for a much thinner, similar membrane with 0.02-μm pores. In general, membranes with >2-μm pore size leak much too easily and are unusable for DS applications. Although the first porous membrane scrubber was demonstrated with PTFE membranes with 2-μm pores (46), the pressure tolerance toward liquid leakage for such a membrane continually decreases with use, presumably because of soiling that compromises the hydrophobicity of the membrane surface. Consequently, the membrane eventually leaks. Although this problem has been solved independently by two research groups as described in the following section, such solutions were dictated by tubular membranes then available—a series of inert porous polypropylene membrane tubes with many attractive features (0.2-μm pores, surface porosity 0.7, diameters 0.2–6 mm) have since become available (Accurel, Enka AG, Wuppertal, Germany) and may permit the fabrication of attractive porous membrane DS devices of the simple tubular geometry.

A few other characteristics of a porous membrane DS should be pointed out here. The collection efficiency is dependent on the pores being open and available. Deposition of particulate matter on the membrane, small as it may be, can reduce the membrane collection efficiency during continued use. For essentially the same reason, solutions containing significant amounts of dissolved solids cannot be used as scrubber liquids because of the evaporative deposition of the solid in the pores. On the other hand, should analyte-bearing particles deposit on the membrane (e.g., if an $(NH_4)_2SO_4$ particle is deposited during the determination of NH_3), the membrane behaves as a filter, and there is little probability that the contents of the particle will be incorporated into the scrubber liquid (although, admittedly, it will interact with the incoming sample gas). The fate of the particle, if deposited on a wetted membrane surface, would obviously be quite different.

Annular Geometry DS Devices.   The leakage problems associated with a 2-μm-pore PTFE membrane tube were solved by Tanner et al. (53) by adopting the annular geometry. Air flows through the annular space of a jacket tube while liquid flows through the concentrically placed membrane tube. The membrane tube is filled with a solid PTFE filament to reduce the interior holdup volume. The design is shown in Figure 4. This configuration results in very little liquid pressure on the membrane and minimizes the probabilities of leaks. Tanner et al. (53) used the porous membrane DS to collect H_2O_2 after prereacting the sample air with NO to remove O_3 (and thereby eliminate O_3-induced artifact H_2O_2 formation). Field and laboratory results were comparable with parallel impinger collection. Manual collection of the scrubber effluent and off-line analysis was used in this work; nevertheless, Tanner et al. clearly recognized that the DS-based approach can provide a clear advantage only if the scrubber effluent is coupled to a continuous analysis system.

Work in Tanner et al.'s laboratory with the 2-μm-pore PTFE membranes indicated, on the other hand, long-term reproducibility problems aside from issues of leakage. They believed that minor changes in pore structure may be occurring in this extremely flexible and pliable membrane when kept under tension; these changes may be reflected in changes in collection efficiency. In our laboratory, thin-walled membranes of minimum tortuosity (straight pores) of very small pore size were therefore sought to provide good and reproducible collection efficiency and immunity from leaks. The only membrane that met these criteria (Celgard X-20, porous polypropylene, surface porosity 0.4, pore size 0.02 μm, wall thickness 25 μm) was available in a maximum internal diameter of 400 μm. The amount of air that can be drawn through such a membrane is obviously limited, and thus the annular geometry, similar to that described by Tanner et al. (53), was also adopted (55).

A glass jacket tube (to maintain a linear configuration) is lined with a 5-mm i.d. PTFE tube and terminates in polypropylene T-joints at either end. A 40-cm length of a 400-μm i.d. microporous Celgard tube, filled with a 300-μm i.d. nylon monofilament fishing line and provided with PTFE connecting tubes at the termini is suspended concentrically inside the jacket tube by appropriate seals at the T-fittings. Air flows in and out through the perpendicular arms of the T-joints, and the scrubber liquid is pumped through the membrane countercurrent to the airflow. To minimize particle deposition and permit laminar flow development, the membrane begins ~7 cm from the T-port. The results of some 10 person-years of effort to develop automated instruments programmed with periodic zero and calibrate functions based on such a DS and intended for the measurement of HCHO, H_2O_2, and SO_2 have been described (55) and are summarized briefly here.

Each instrument is dedicated to a specific analyte and shares the following common features. The sample inlet allows microprocessor-programmed selection of sample, zero, and calibrant gas. Rather than being programmed for continuous aspiration of the sampled air, the instrument is programmed to alternate between sample and zero with periodic (every 4

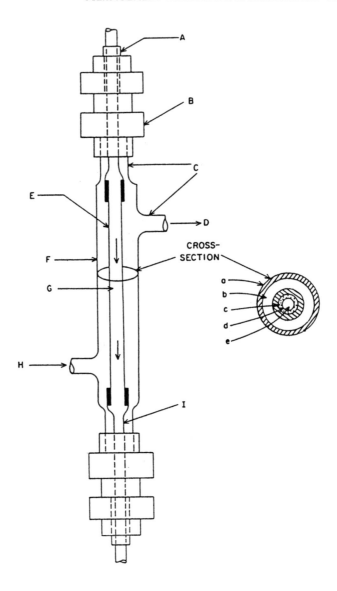

Figure 4. Diffusion scrubber device for H₂O₂ removal from air, consisting of the following: A, Teflon tubing adaptor; B, Teflon Swagelok union; C, 6-mm o.d. glass tube; D, air outlet; E, porous Teflon tube (2 mm i.d. × 30 cm, 70% porosity); F, 8-mm i.d. glass tube; G, scrubber solution with concentric Teflon displacement rod; H, air inlet; and I, Teflon tube connection to porous Teflon tube. Cross section inset: a, 8-mm o.d. (6-mm i.d.) glass tube; b, air stream; c, porous Teflon tube; d, H₂O scrubber solution; and e, Teflon displacement rod. (Reproduced from reference 53. Copyright 1986 American Chemical Society.)

h or longer) calibrant–zero cycles. The sample periods are not quite long enough to reach plateau response; the signal for each analytical cycle is therefore represented by a peak followed by a return to the baseline, which establishes the liquid-phase zero. The scrubber liquids used are 0.1 M H_2SO_4, water, and 5 μM HCHO for the HCHO, H_2O_2, and SO_2 instruments, respectively, and they are pumped through the DS at flow rates of 40–85 μL/min. Total evaporative loss of the scrubber liquid ranges up to a maximum of 6 μL/min for completely dry air ($Q = 2$ L/min). Appropriate reagents are added via T-joints at microflow rates to the scrubber effluent. In each case a fluorescent product is formed and monitored with a filter fluorometer. For formaldehyde, the reaction with acetylacetone and ammonium acetate to produce diacetyldihydrolutidine in an in-line heated reactor is used (81). For H_2O_2, the peroxidase-mediated oxidation of 4-hydroxyphenylacetate followed by the introduction of ammonia through a permeative membrane reactor is used (82). For S(IV), the reaction with alkaline 9-N-acridinylmaleimide in a water–N,N-dimethylformamide medium to produce the corresponding fluorescent sulfonate is used (83). The operational and performance characteristics are shown in Table I; typical calibration runs and field data are shown for the H_2O_2 instrument in Figures 5 and 6. Various calibration sources have been developed for providing calibrant gases to the DS instruments (84–87). No meaningful interferences for any of the techniques were found except for that from O_3 for H_2O_2—this interference does not occur in the laboratory with clean calibrant gases and a clean DS, but it occurs with ambient air and clearly increases with the degree of soiling of the entire inlet system. Although none of the instruments underwent a field intercomparison study after their development was complete, the HCHO and H_2O_2 instruments have been used in various inter-

Table I. Typical Operating Parameters and Performance Specifications

Analyte	Sample (Calibrate)/ Zero Periods (min)	Lag Time[a] (min)	Rise Time[b] (min)	Detection Limit[c] (pptrv)	Linearity Limit (ppbv)
HCHO	3/7	6.3	1.7	100	800[d]
H_2O_2	1/4	1	0.8	30	30[e]
SO_2	3/6	1.6	1.9	175	300[d]

NOTE: All values were obtained with 40-cm-long scrubbers (jacket tube i.d. 5 mm) at sampling rates of 2.0 L/min.
[a]Time between the valve switching from zero to calibrate gas and the onset of change at instrument output.
[b]Time required for signal output to change from 10 to 90% of the plateau value.
[c]Based on three times the noise level when zero air is being sampled. Experiments were conducted in all cases at concentrations less than or equal to three times the quoted detection limit.
[d]Highest concentration tested; linearity may extend further.
[e]Response was less than linear at higher concentrations.

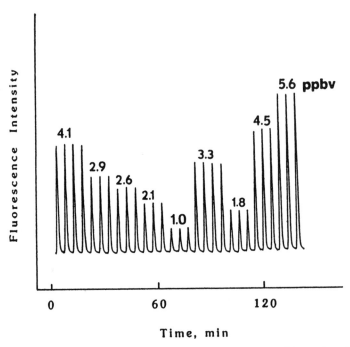

Figure 5. A calibration sequence for the H_2O_2 instrument. (Reproduced with permission from reference 55. Copyright 1988 Pergamon Press.)

comparison studies at different stages in their development and have generally fared well (88–91).

The lessons of the field studies dealing with the sampling of ambient air all have to do with the deposition of particulate matter and how such deposition affects collection efficiencies in the porous membrane DS devices and analyte losses in the inlet lines and the valves. A strategy for the facile field determination of collection efficiencies using two serial DS devices has been developed (55). However, although periodic recalibration, a part of the routine protocol, can to a large extent correct for any analyte loss or loss of collection efficiency in the scrubber itself, losses prior to the scrubber cannot be corrected for.

A unique DS with two Celgard tubular membranes was used by Sigg (92) to collect two independent liquid effluents for the simultaneous measurement of organic peroxides and hydrogen peroxide; Sigg used a differential analytical scheme similar to that reported in reference 90. Such devices have been used successfully in a number of balloon flights (92).

Recently, the annular geometry porous membrane DS coupled to a fluorometric detection system was shown to be applicable to the sensitive measurement of ambient ammonia (59). The detection scheme is based on

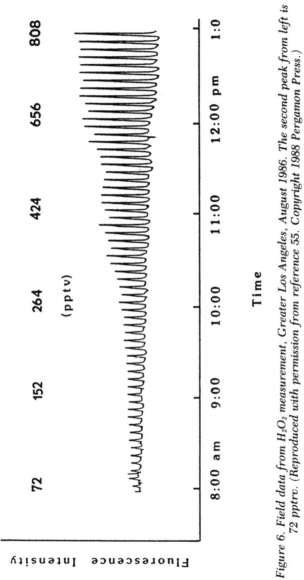

Figure 6. Field data from H₂O₂ measurement, Greater Los Angeles, August 1986. The second peak from left is 72 pptv. (Reproduced with permission from reference 55. Copyright 1988 Pergamon Press.)

the reaction of ammonia with *o*-phthaldialdehyde and sulfite in a heated in-line reactor to produce 1-sulfonatoisoindole (*93*). Getting an analyte into the liquid phase often permits a greater latitude in manipulations than is otherwise possible. For example, the selective destruction of H_2O_2 over organic peroxides with a MnO_2 catalyst has been demonstrated (*91, 94*). The work on ammonia also illustrated a facile in-line preconcentration procedure generally applicable to DS-based analyzers. The liquid-phase portion of the analytical system is shown in Figure 7. Carrier water (W) is deionized by a small in-line ion exchanger (C) that flows through the DS. The *o*-phthaldialdehyde reagent (O) and buffered sulfite reagent (B) are added to the DS effluent. The stream flows through mixing coil (M) and then through reaction coil (R) in heated block (H). The stream, debubbled by tubular porous membrane (T), is detected by fluorescence detector (F). Inset a shows the conventional mode; b shows the preconcentration mode. The DS is connected between the points A and A'. In the preconcentration mode, a six-port loop valve (V) is installed between A and A' and the DS is contained within its loop. During the concentration step, the flow (W) bypasses the DS and establishes the system liquid-phase blank. Meanwhile, air is sampled through the DS, any evaporative loss being made up by the liquid in the reservoir (W). During the measurement mode, the valve is switched and the contents of the DS are flushed to the system. For a 3-min–2-min load–inject cycle, the LOD is 45 pptrv of NH_3, although accuracy at these

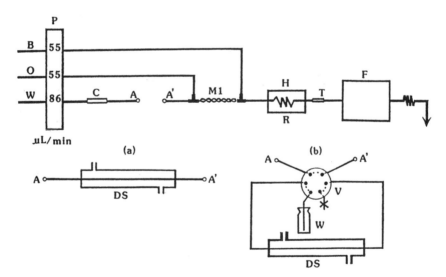

Figure 7. The DS-based ammonia analyzer liquid-phase schematic. Insets a and b show in-line or in-loop (preconcentration) configurations, respectively. (Reproduced from reference 59. Copyright 1989 American Chemical Society.)

levels can be compromised by major variations in relative humidity unless corrections are made. Instrument performance at low levels is shown in Figure 8.

A DS-based NH_3 instrument, similar to that just described, was among several others tested in a European intercomparison study in Rome in 1989 and was found to perform well (95).

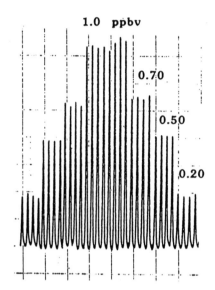

1.0 ppbv

0.70

0.50

0.20

Figure 8. Performance of the DS-based ammonia instrument at low levels of NH_3 (g). (Reproduced from reference 59. Copyright 1989 American Chemical Society.)

Many specific and highly sensitive fluorometric and electrochemical detection methods for various analytes are available. The combination of such detection schemes with a DS-based collection system provides a combination of sensitive and affordable instrumentation for atmospheric measurements. A step-by-step construction and operation manual for a DS-based fluorometric H_2O_2 analyzer is available (94). With a change in the reagents, the calibration source, and the conditions of the fluorometric measurement, such an instrument is readily reconfigured for a different analyte. The 1992 fabrication cost of a complete DS instrument that utilizes fluorometric detection and includes a thermostated calibration source from commercially available components is approximately $12,000.

Diffusion Scrubber Coupled-Ion Chromatography.   Over the past decade, ion chromatography (IC) has become the technique of choice for the determination of anions. Even with conventional sorbent-coated denuders for acid gases, the actual determination is most commonly conducted by IC. The recognition of this fact led to the coupled DS–IC system (96). The system is shown schematically in Figure 9. Items A through H constitute

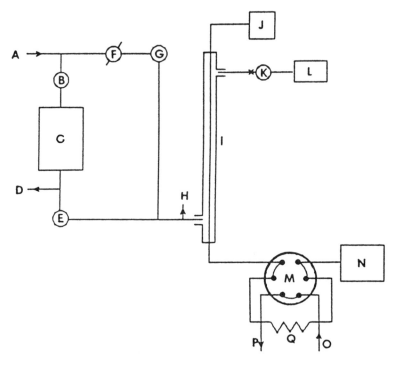

Figure 9. Configuration of the DS–IC system: A, clean air input; B, mass-flow controller; C, permeation device chamber; D and H, vents; E, needle valve–rotameter; F, needle valve; G, mass-flow meter; I, diffusion scrubber; J, scrubber liquid reservoir; K, needle valve–rotameter; L, suction pump; M, injection valve; N, peristaltic pump; O, eluent flow; P, downstream chromatographic components; and Q, sample loop. (Reproduced from reference 96. Copyright 1989 American Chemical Society.)

the calibration source. Sample–zero–calibrant gas is drawn through the DS (Celgard membrane and annular geometry). A mildly pneumatically pressurized (1–2 psi) reservoir (J) contains 1 mM H_2O_2, the scrubber liquid. It is connected to the DS membrane, and the effluent liquid is aspirated by peristaltic pump (N) through the 100-μL loop (Q) of chromatographic injection valve (M) at a rate of 16 μL/min. This arrangement is preferred over pumping the liquid through the DS because it maintains a constant effluent flow rate from the DS even when the extent of the scrubber liquid evaporation varies because of variable inlet air humidity. Every 6 min (the approximate time required to perform the chromatography), valve M switches for a long enough period (30–60 s) to completely inject the collected sample into the chromatographic system. Thus a new chromatogram, representing the preceding collection period, is obtained every 6 min. There is ~10% liquid-phase carryover from one sample to the next. The actual integration time can be effectively larger than the nominal time resolution for sticky analyte gases like HNO$_3$ because of adsorption–desorption on inlet surfaces. In the original work (96), the system performance was primarily illustrated with SO$_2$ as the test gas, and the LOD was reported to be better than 20 pptrv of SO$_2$. Much work has been done with this system in the analysis of ambient air. The use of dilute H$_2$O$_2$ as a scrubber liquid ensures the complete oxidation of collected S(IV) to S(VI) and thereby improves the LOD for SO$_2$. At the same time this scrubber liquid does not significantly oxidize collected nitrite to nitrate. A representative ion chromatogram of ambient air is shown in Figure 10—identities of peaks a through g are due to unresolved organic acids, HCl, HONO, CO$_2$, unknown, SO$_2$ (320 pptrv), and HNO$_3$ (~700 pptrv), respectively. Clearly, the system is applicable for the simultaneous

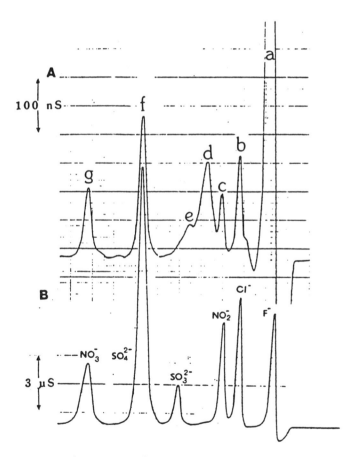

Figure 10. An ambient air ion chromatogram (A) is shown in comparison with a standard liquid-phase ion chromatogram (B). (Reproduced from reference 96. Copyright 1989 American Chemical Society.)

determination of a number of acid gases. In airborne applications where significant changes in the inlet pressure relative to the calibration conditions are involved and sampling is conducted at a constant mass flow rate, pressure corrections are necessary; the method has been reported (97). The DS–IC instrument has undergone airborne intercomparison studies for the determination of SO_2 with satisfactory performance (98).

More recently, Lindgren (99) utilized a similar conductometric DS–IC instrument for the measurement of atmospheric HCl; an LOD of 20 pptrv was reported. The rapid response time of the instrument could readily demonstrate the washout of HCl by precipitation. The detection of certain analytes (e.g., NO_2^-) is more sensitively accomplished with UV detection than with conductometry. Tandem conductivity and UV detection were used for the determination of HONO (100); particulate nitrite does not commonly occur, and this work utilized a 1-m-long Celgard membrane coiled into a 0.3- × 40-cm helix suspended inside a 6-mm-diameter jacket tube. The longer membrane length permits a greater collection efficiency, and its helical form maintains the compactness, although it does increase the extent of particle deposition. The LOD for HONO is 24 to 16 pptrv at $Q = 0.5$ to 2.1 L/min, respectively. The interference from NO_2 is $\sim 0.025\%$ on a molar basis, with or without added NO or SO_2. The instrument has been used extensively for ambient measurements. HONO maxima at the measurement location were clearly associated with traffic. Persistent daytime concentrations of 100–500 pptrv were also observed, much higher than could be accounted for by interference from NO_2. Both the HCl and HONO instruments used integral calibration sources based on the equilibrium vapor pressure of the acid over the corresponding ammonium salts.

Miscellaneous Diffusion Scrubber Applications. Dasgupta et al. (101) compared two DS-based schemes for determining SO_2, both involving reactions in which the collected analyte is converted by appropriate chemistry into some other more detectable compound. In the first system, the Celgard-membrane-based annular geometry DS is used with a water scrubber. The collected S(IV) in the scrubber effluent is oxidized to sulfate by the enzyme sulfite oxidase with the simultaneous production of an equivalent amount of H_2O_2. This H_2O_2 is then detected by the previously mentioned fluorometric procedure involving the oxidation of 4-hydroxyphenylacetate. Obviously, the H_2O_2 must be removed from the sample gas for this technique to work; this was successfully accomplished with an internally silvered glass tube (catalytic decomposition predenuder for H_2O_2) without removing the SO_2. The second system used a simple tubular DS with a 2-μm-pore PTFE membrane. This scheme is unique in that the measured analyte remains in the gas phase and the jacket volume through which the liquid flows is not constrained; leakage problems are obviated for these reasons. A 2.5 μM solution of $Hg_2(NO_3)_2$ flows in the jacket, and SO_2 reacts with Hg_2^{2+} at the gas–liquid interface to produce Hg(II) and Hg(0). The latter is present in the exit gas from the DS and is collected on a gold coil. After collection for any desired period of time, the coil is flash-heated electrically, and the liberated Hg is measured by a commercial conductometric gold film Hg sensor. LODs for the two systems were comparable (80–170 pptrv for system 1 and 60–150 pptrv for system 2), and no major interferences were found for either (as long as H_2O_2 was preremoved in system 1). Parallel ambient measurements in the 400- to 1000-pptrv range correlated with an r^2 of 0.88 (r, linear correlation coefficient). If used as a solution, the enzyme is relatively expensive, and the H_2O_2-translation system can be more practical if an immobilized sulfite oxidase enzyme is used. With the Hg-translation system, diffusion of the Hg(0) in and out of the liquid phase increases the instrument response time to ~5 min, even though only 1 min of sampling time was typically used.

The DS represents an attractive collection interface to devise continuous-flow analytical systems even when discrimination between gas and particles is not a particularly important issue. In unpublished work, a Celgard-membrane-based annular geometry DS was used with an absorbance detector (based on a green-light-emitting diode emitting at 555 nm) to determine NO_2; the well-known Griess–Saltzman chemistry was used (*102*). The Griess–Saltzman reagent was directly pumped at 13 μL/min through the DS, and the absorbance of the DS effluent was monitored. Normally it would not be permissible to pump a reagent with a significant amount of dissolved solids because of evaporative deposition as mentioned. However, the uptake of NO_2 by pure water is essentially negligible; thus, complete humidification of the inlet sample–zero air can be achieved by passing it through a porous membrane immersed in water without significant loss of NO_2. At the same time, HONO is removed. The humidity of the air stream precludes any evaporation of the scrubber liquid and permits the very low flow rate. Otherwise, evaporative losses and consequent variation in blank values would greatly compromise the reliability of the data. The low flow rate and the intrinsic sensitivity of this colorimetric method allows, with inexpensive equipment, an LOD of 250 pptrv of NO_2 with 10-min time resolution.

Diffusion-Scrubber-Based Analyzers: Limitations and Future Considerations. As may be evident from the foregoing discussion, Celgard-based porous membrane annular geometry DS devices have thus far been the mainstay of DS-based analyzers. Collection efficiency of porous membrane devices does decrease with particle deposition. In the short term, the lumen of the fiber needs to be flushed with a wetting solvent and dried thoroughly before reuse every 24–48 h, depending on the particle loading of the air sampled. Over a longer term, the inside of the jacket needs to be washed as well to remove deposited material. The DS construction described in reference 96 uses a demountable membrane that can be replaced in minutes. Some, however, maintain that periodic cleaning is sufficient and that membrane replacement is not necessary even in year-long field use (*103*). In any case, the need for periodic cleaning does deter from long-term unattended use. All ambient air studies carried out thus far with DS-based analyzers have been carried out without any attempts to remove coarse particles from the sample, for example, with a cyclone or an impactor, prior to entry to the DS. Experience indicates that the large majority of the particles deposited in the DS are coarse particles, and particle deposition experiments (*40*) confirm this trend. Preremoval of coarse particles may greatly reduce the frequency of cleaning necessary. The effect of the inlet geometry was also studied for 0.1- to 3-μm particles in these experiments— as may be intuitively guessed, the extent of loss is larger with a T-type than a Y-type air inlet geometry. Much of this deposition occurs in the entrance region and not on the membrane itself. The only commercial version of the DS presently available (Analytek, Umeå, Sweden) utilizes PTFE end fittings in which air flows in–out through ports at an angle of 45° relative to the main body of the DS. This geometry should have deposition losses equal to or less than those of the Y-inlets. However, minimum deposition was observed with a straight inlet system where the tubes connecting to the termini enter and exit through the jacket walls (Figure 11). This design has become the one of choice in my laboratory.

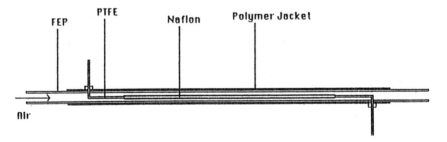

Figure 11. The straight inlet DS.

Frequently, a major limitation of DS-based collection systems is that they operate at substantially subquantitative collection efficiencies at the typical sampling rates used. This situation increases the probability of error because of large thermal variations that affect diffusive transport. For these reasons, should wet denuders (vide infra) prove to be viable continuous collection devices, they may well replace DS-based systems. Their ability to more quantitatively remove gases may also spur the development of combined gas–particle analyzer systems in which, for example, the acid gases are removed by the denuder and analyzed; the particles are then collected by the impactor equivalent of a wet denuder, and the acidity associated with the aerosol is determined. Despite the bounties promised by the wet denuder, in certain cases, the DS may remain superior. Nafion-membrane-based DS systems of the geometry shown in Figure 11 can collect gases like H_2O_2 or HCHO with efficiencies comparable to those attainable by a single-tube wet denuder of same length, display no change in collection efficiency over time (no pores to be blocked), and are more easily interfaced to the liquid-phase analyzer system. Using slightly larger Nafion membrane tubes and a geometry where the air is sampled inside the membrane and the scrubber liquid is flowing outside, Lind (*80*) showed excellent collection efficiency for HNO_3 (g), coupled such a DS to an automated IC system, and has used it successfully in field experiments.

Wet Denuders

As the name implies, wet denuders are devices in which the denuder active surfaces are continuously wetted by the scrubber liquid, and in a true continuous analyzer the effluent is continuously removed. In many ways, the wet denuder represents all of the desirable features—a continuously renewed surface, the possibility of using any scrubber liquid, and the best collection efficiency that a given scrubber liquid will allow for a given geometry. Further, if the liquid film is indeed always present, the denuder wall for all practical purposes is composed of the liquid, and relatively few restrictions are placed on the construction material of the denuder. With a conductive scrubber liquid, electrostatic particle losses are also expected to be minimum. In its simplest form, a wet denuder is a tube with the scrubber liquid flowing down as a uniform film along its inner wall and the liquid being collected at the bottom without a major disruption of the boundary layer of the air flow, which proceeds from the bottom up. Water or an aqueous solution is most commonly the scrubber liquid of choice for determining a variety of atmospheric gases of interest. Water has a relatively large surface tension, and it is particularly difficult therefore to maintain a uniform film. Further, unlike the liquid flow with a DS, the liquid flow is not in a closed system; there may be attendant difficulties in maintaining an inflow–outflow balance in the face of a variable extent of evaporative loss.

Significant progress has already been made toward a continuous-flow wet denuder. The wetted wall column (WWC) of Fendinger and Glotfelty (*104*), based on the original work of Emmert and Pigford (*105*), utilizes a mechanically etched glass tube and U-shaped copper wire pieces along the upper interior walls to disrupt the surface tension of the water pumped from the top and make a uniform film. The air inlet–outlet geometry, in regard to potential particle losses, was far from that desired in a wet denuder (although it was of little consequence to these authors, who were solely interested in achieving gas–liquid equilibration). Most importantly, the air–liquid flow-rate ratio in the Fendinger and Glotfelty WWC was of the order of 100 (air flow rate 50–200 mL/min and water flow rate 1.4–2.0 mL/min), whereas in a wet denuder air–liquid flow rate ratios 2 orders of magnitude greater (air flow rate 1000–2000 mL/min and liquid flow rate 100–200 μL/min) would be desirable to achieve meaningful preconcentration factors. I have attempted to adapt a chemically or mechanically etched glass tube to air inlet–outlet geometries that have low probabilities of particle loss. With desirable air–liquid flow-rate ratios as stated, the integrity of a continuous film covering the surface, if formed in the first place, was short-lived. Invariably, within a few hours dry areas would be apparent, and in the worst case a single narrow stream of water would flow down one wall.

If the deposition of particles was not a factor or was in fact desired, the design would be much simpler. Most likely, such a device would operate in the turbulent flow mode; the theoretical aspects have been discussed by Lucero (106) in exemplary detail for a hypothetical device in which gaseous sample molecules, particles, or both from a turbulent gas stream are dissolved into a laminar film liquid stream that flows in turn to a chromatograph sampling valve.

The first reported wet denuder operated in discrete cycles and was of annular geometry; analysis of the scrubber liquid was carried out off-line (107). The denuder (d_o = 4.5 cm, d_i = 4.2 cm, and L = 30 cm) is shown in Figure 12, and the complete setup is shown in Figure 13. The two concentric tubes are held together by PTFE spacers, and during operation the annular space is charged with 25 mL of scrubber liquid and the outer tube is rotated at 40 rpm by a motor-driven belt. This operation results in an aqueous film, ~0.5 mm thick, on the surfaces defining the annulus. A pair of denuders is used in parallel in the field instrument, each with Q = 32 L/min. One scrubber liquid is a 0.5 mM HCOOH–HCOONa buffer of pH 3.7, and the other is a phosphate-buffered (pH 7) solution containing 4-hydroxyphenylacetate, peroxidase, and formaldehyde. The first solution is used for the collection of HCl, HNO_3, and NH_3, and the second is used for the collection of SO_2 and H_2O_2. At the completion of the sampling cycle (40 min), the denuder pair is tilted by an electric jack, and the denuder effluents are pumped out to individual autosampler tubes by a peristaltic pump. The denuders are washed with 2 mL of the appropriate scrubber liquids before they are charged with the new scrubber solutions, and then the sampling cycle is repeated. All of these operations are fully automated. The first denuder inlet is equipped with a cyclone to remove coarse particles. The second denuder inlet is equipped with provisions for premixing NO with the sampled air to eliminate O_3 interference in H_2O_2 measurement (53). At the stated flow rate, all of these analyte gases are collected with 90%+ efficiency, and the attainable LODs are stated to be 290, 80, 130, 190, and 7 pptrv for NH_3, HNO_3, HCl, SO_2, and H_2O_2, respectively. In field intercomparison with other techniques, no major interferences for any of the analytes of interest were apparent.

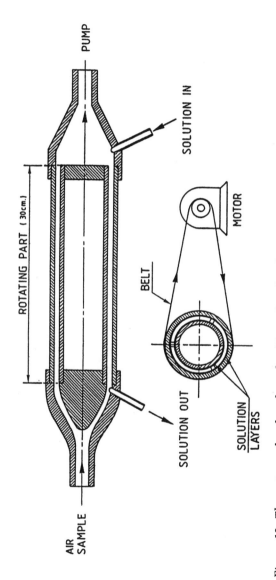

Figure 12. The wet annular denuder tube. (Reproduced with permission from reference 107. Copyright 1988 Pergamon Press.)

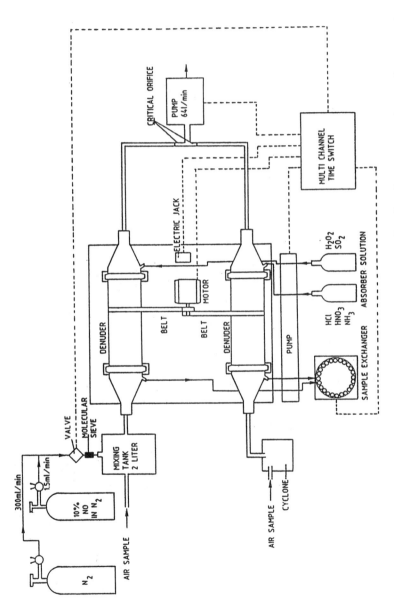

Figure 13. Wet annular denuder experimental arrangement. (Reproduced with permission from reference 107. Copyright 1988 Pergamon Press.)

Experiments have been conducted in my laboratory on various single-tube wet denuder designs (108). These devices have been coupled to an IC for semicontinuous analysis. The basic system design is largely the same as that described in reference 96, except that an eight-port dual stack slider valve is used for chromatographic injection. Such a valve is configured to have two injection loops; while one is in the load mode, the other is in the inject mode. Each of the loops contains a short preconcentration column. The effluent from the wet denuder is pumped by a peristaltic pump through the preconcentration column(s). The typical input liquid flow rate is 100–700 μL/min. (Unlike in reference 96, the flow resistance of the loop containing the preconcentrator column is too high for the collected effluent to be aspirated through it. The column must be pumped, and some losses of labile analytes like NO_2^- on passage through poly(vinyl chloride) (PVC) pump tubing have been observed.) In the face of variable evaporative losses (up to 30 μL/min for dry air at $Q = 2$ L/min), the choice is either to let some of the effluent liquid be wasted and pump an air-free stream to the preconcentrator or to pump the entire effluent liquid and some air through the preconcentrator. The latter alternative was chosen because it was found that as long as the IC eluent was degassed, the small amounts of air on the precolumn are dissolved under pressure in the eluent and do not cause any quantitation or reproducibility problems if sufficient restriction is added to the detector cell exit to prevent bubble formation. The use of the preconcentrator columns also allows relatively large liquid flow rates through the denuder, and this design facilitates complete wetting of the denuder surface. Although the use of preconcentrators is generally applicable for ionic analytes, this setup is not applicable to analytes like H_2O_2 and HCHO.

The wet denuder designs are shown in Figure 14. Figure 14a shows an internally threaded (7 mm, one thread per millimeter) denuder that was fabricated from a glass-filled PTFE tube (1/4-inch i.d., 1/2-inch o.d.). Liquid pumped in through the wall at the top follows the thread and is collected by aspiration at the bottom. It is not known if the uneven wall hinders laminar flow development. The second denuder, shown in Figure 14b, contains a porous polypropylene membrane (5.5-mm i.d., 1.5-mm wall) jacketed by a PVC tube. Liquid is forced through the membrane walls, collects at a well at the bottom, and is aspirated from there. The design shown in Figure 14c contains a rolled sheet of a thin polycarbonate membrane (Nuclepore) within a 9.4-mm-i.d. PVC or PFA Teflon tube. The liquid is introduced simultaneously through four symmetrically placed needle apertures at the top periphery of the tube and collected at the bottom in a fashion similar to that in Figure 14b. None of these designs are wetted with pure water reliably over long periods. However, a solution of a nonionic fluorocarbon surfactant (0.3% Zonyl FSN) passed through a mixed-bed ion exchanger is satisfactory. The LODs attainable by the IC system suffer by a factor of 2 to 3 with continued injection of the surfactant-containing samples, relative

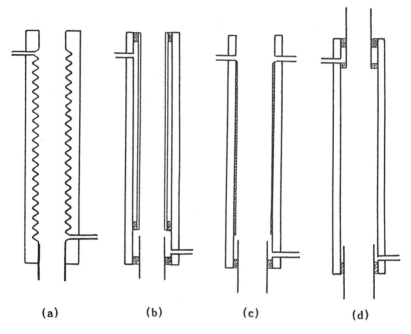

(a)   (b)   (c)   (d)

*Figure 14. Experimental wet denuder designs. (Reproduced from reference
108. Copyright 1991 American Chemical Society.)*

to purely aqueous samples, because of increased baseline noise. It was possible to use pure water in the denuder shown in Figure 14c with the polycarbonate film replaced by a tissue paper. However, this setup was not
studied in depth because of concerns about the integrity of collected samples.
The response time from denuder c was excellent (<2 min) even at low liquid
flow rates. In comparison, denuder a and especially denuder b showed
substantially longer response times. The holdup volume in device a presented by the threads is not insignificant. In device b, diffusion in and out
of the pores of the relatively thick wall makes, in effect, a very thick film.
At a flow rate of 100 μL/min, up to 30 min is required to obtain a plateau
response.

The design shown in Figure 14d exhibited the best performance. It
consisted of a glass tube (8-mm i.d.) first treated with triethylorthosilicate,
which bonds to the glass. Before this coating is completely cured, the tube
is filled with fine-mesh, thin layer chromatographic grade silica gel that bonds
in turn to the surface. After overnight curing, the excess silica gel is removed,
and then the interior is cleaned with compressed air and repeatedly washed
with water. The resulting silica surface is readily wetted by water. Water is
pumped into a space between the air inlet tube and the glass tube and is
sealed with a ring of filter paper at the bottom periphery and with adhesive

at the top. The effluent is collected at the bottom as in denuders b and c. A fitting was designed subsequently (*108*) to facilitate the introduction and aspiration of the scrubber liquid. The system was extensively tested with SO_2 as the test gas and dilute H_2O_2 as the scrubber liquid. Ambient air measurements were also made. With a length of 40 cm, the device exhibited collection efficiencies that agreed within experimental error with equation 1. With $Q = 1$ L/min and a liquid flow rate of 700 μL/min, a 7-min sampling cycle provided an LOD of 4 to 5 pptrv for SO_2.

Vecera and Dasgupta (*109*) described a different wet denuder, also of the simple tubular design. Male halves of a fitting are permanently glued to the denuder tube and threaded into appropriate female halves for introduction and removal of the scrubber liquid. A wettable, highly porous soft glass layer is created within a glass tube by coating the tube with sodium silicate solution and excess silica and then heating it to high temperatures. The denuder was used in a system expressly designed to measure HONO and HNO_3. Most of the denuder effluent was aspirated and preconcentrated on a 1.5- × 3-mm ion-exchange column for 12 min and then eluted onto a 1.5- × 10-mm column that separated nitrate and nitrite. The column effluent proceeded through a cadmium reductor that reduced nitrate to nitrite. Griess–Saltzman reagent was added continuously to the column effluent, and the resulting azo dye formed was detected colorimetrically at 555 nm. This inexpensive low-pressure (peristaltically pumped) system provided an LOD of 20 and 200 pptrv for HONO and HNO_3, respectively. In field experiments, persistent daytime presence of HONO was again noted.

More recently, Simon and Dasgupta (*110*) used a parallel-plate wet denuder in which the scrubber liquid is made to flow down two closely spaced silica-coated parallel plates facing each other with the air sample flowing upward in the gap. Excellent collection efficiency (90% +) was established for SO_2 even at flow rates of 15 L/min. The denuder effluent was preconcentrated and chromatographed as in reference 108. On the basis of signal-to-noise data observed at SO_2 test concentrations of 19 pptrv, the calculated LOD of the system is on the order of 0.5 pptrv.

Summary

It is tempting but admittedly unwise to make predictions in an area where, much as the red queen informed Alice, "it takes all the running one can do to keep in the same place. If you want to get somewhere else, you must run at least twice as fast as that!" Regardless, not only are diffusion-based collection systems coupled to automated analyzers expected to make increasing contributions to meeting the measurement challenges in atmospheric chemistry, wet denuder technology will likely mature before the millennium ends. If an equivalent "wet impactor" can be designed in which impaction rather than diffusion vectors are used for collection, then aerosol composition, at least in terms of its readily soluble constituents, can be determined by using the same technology. It is conceivable that chromatographic instrumentation can be time-shared by sequential gas and aerosol collection devices. The recent work of Lind (80) in combining membrane-based diffusion scrubbers to collect HNO_3 (g) and in examining a bead-packed continuously wetted glass coil to collect particulate nitrate has clearly established the feasibility and attractiveness of such an approach. There is relatively little extant data on the occurrence of many organic compounds in the ambient atmosphere and on their distribution in the aerosol versus the gas phase. Wet denuders are capable of using any scrubber liquid, and at a minimum the effluent can be semicontinuously analyzed by liquid chromatographic analyzers. The technology for true high-resolution, high-sensitivity analysis exists today; the extraordinarily high separating power obtained by subjecting each effluent liquid chromatographic peak to capillary zone electrophoresis (111) and the ability to sensitively identify both inorganic and organic eluates from a capillary system by electrospray ionization mass spectrometry (112) have already been demonstrated. Should even a fraction of this technology come to be deployed in a systematic study of atmospheric chemistry, it may reveal much about trace atmospheric constituents.

Acknowledgment

This work would have been impossible without the cheerful help of D. P. Y. Chang, University of California, Davis. This work was supported by the State of Texas Advanced Research program and the Electric Power Research Institute (RP 1630–55).

References

1. Schiff, H. I.; Harris, G. W.; Mackay, G. I. In *The Chemistry of Acid Rain: Sources and Atmospheric Processes*; Johnson, R. W.; Gordon, G. E., Eds.; ACS Symposium Series 349; American Chemical Society: Washington, DC, 1988; pp 274–288.
2. Platt, U.; Perner, D.; Harris, G. W.; Winer, A. M.; Pitts, J. N., Jr. *Nature (London)* **1980**, *285*, 312–314.
3. Hanst, P. L. *Adv. Environ. Sci. Technol.* **1971**, *2*, 91–210.
4. U.S. Environmental Protection Agency. *Fed. Regist.* **1978**, *43*(121), 26962–26986.
5. Okabe, H.; Splitstone, P. L.; Ball, J. J. *J. Air Pollut. Control Assoc.* **1973**, *23*, 514–516.
6. Steffenson, D. M.; Stedman, D. H. *Anal. Chem.* **1974**, *46*, 1704–1708.
7. Warren, G. J.; Babcock, G. *Rev. Sci. Instrum.* **1970**, *41*, 280–282.
8. Yunghans, R. S.; Monroe, W. A. *Technicon Symposia, 1965: Automation in Analytical Chemistry*; Mediad Inc.: New York, 1966; pp 279–284.
9. Lazrus, A. L.; Kok, G. L.; Lind, J. A.; Gitlin, S. N.; Heikes, B. G.; Shetter, R. E. *Anal. Chem.* **1986**, *58*, 594–597.

10. Lazrus, A. L.; Fong, K. L.; Lind, J. A. *Anal. Chem.* **1988**, *60*, 1074–1078.
11. Abbas, R. A.; Tanner, R. L. *Atmos. Environ.* **1984**, *15*, 277–281.
12. Vecera, Z.; Janak, J. *Anal. Chem.* **1987**, *59*, 1494–1498.
13. Cofer, W. R., III; Edahl, R. A., Jr. *Atmos. Environ.* **1986**, *20*, 979–984.
14. Ingham, D. B. *Aerosol Sci.* **1975**, *6*, 125–132.
15. Lee, K. W.; Gieseke, J. A. *Atmos. Environ.* **1980**, *14*, 1089–1094.
16. Townsend, J. S. *Philos. Trans. R. Soc. London, A* **1900**, *193*, 129–158.
17. Gormley, P. G.; Kennedy, M. *Proc. R. Ir. Acad., Sect. A* **1949**, *52*, 163–169.
18. Bowen, B. D.; Levine, S.; Epstein, N. *J. Colloid Interface Sci.* **1976**, *54*, 375–390.
19. Cheng, Y.-S. In *Air Sampling Instruments*, 7th ed.; Hering, S. V., Ed.; American Conference of Governmental Industrial Hygienists: Cincinnati, 1989; pp 405–419.
20. Thomas, J. W. *J. Colloid Sci.* **1955**, *10*, 246–255.
21. Fish, B. R.; Durham, J. L. *Environ. Lett.* **1971**, *2*, 13–21.
22. Pack, M. R.; Hill, A. C.; Thomas, M. D.; Transtrum, L. G. *ASTM Spec. Tech. Publ.* **1959**, *281*, 27–44.
23. Crider, W. L.; Barkley, N. O.; Knott, M. L.; Slater, R. *Anal. Chim. Acta* **1969**, *47*, 237–241.
24. Stevens, R. K.; Dzubay, T. G.; Russwurm, G.; Rickel, D. *Atmos. Environ.* **1978**, *12*, 55–68.
25. Possanzini, M.; Febo, A.; Liberti, A. *Atmos. Environ.* **1983**, *17*, 2605–2610.
26. Coutant, R. W.; Callahan, P. J.; Kuhlman, M. R.; Lewis, R. G. *Atmos. Environ.* **1989**, *23*, 2205–2211.
27. Stevens, R. K.; Pardue, L. J.; Barnes, H. M.; Ward, R. P.; Baugh, J. P.; Bell, J. P.; Sauren, H.; Sickles, J. E.; Hodson, L. L. In *Transactions, TR–17; Visibility and Fine Particles*; Mathai, C. V., Ed.; Air and Waste Management Association: Pittsburgh, PA, 1990; pp 122–130; AWMA Transaction Series.
28. Koutrakis, P.; Fasano, M.; Slater, J. L.; Spengler, J. D.; McCarthy, J. F.; Leaderer, B. P. *Atmos. Environ.* **1989**, *23*, 2767–2773.
29. Durham, J. L.; Ellestad, T. G.; Stockburger, L.; Knapp, K. T.; Spiller, L. L. *J. Air Pollut. Control Assoc.* **1986**, *36*, 1228–1232.
30. Ellestad, T. G.; Knapp, K. T. *Atmos. Environ.* **1988**, *22*, 1595–1600.
31. Ferm, M. Ph.D. Dissertation, University of Göteborg, Sweden, 1986.
32. Ali, Z.; Paul Thomas, C. L.; Alder, J. F. *Analyst (London)* **1989**, *114*, 759–769.
33. Hering, S. V.; Lawson, D. R.; Allegrini, I.; Febo, A.; Possanzini, M.; Sickles, J. E., II; Anlauf, K. G.; Wiebe, A.; Appel, B. R.; John, W.; Ondo, J.; Wall, S.; Braman, R. S.; Sutton, R.; Cass, G. R.; Solomon, P. A.; Eatough, D. J.; Eatough, N. L.; Ellis, E. C.; Grosjean, D.; Hicks, B. B.; Womack, J. D.; Horrocks, J.; Knapp, K. T.; Ellestad, T. G.; Paur, R. J.; Mitchell, W. J.; Pleasant, M.; Peake, E.; MacLean, A.; Pierson, W. R.; Brachaczek, W.; Schiff, H. I.; Mackay, G. I.; Spicer, C. W.; Stedman, D. H.; Winer, A. M.; Biermann, H. W.; Tuazon, E. C. *Atmos. Environ.* **1988**, *22*, 1519–1539.
34. McKays, W.; Crawford, M. E. *Convective Heat and Mass Transfer*, 2nd ed.; McGraw-Hill: New York, 1980; pp 66–68.
35. Schlichting, H. *Boundary-Layer Theory*, 6th ed.; McGraw-Hill, New York, 1963; pp 176–178.
36. Chang, D. P. Y. University of California—Davis, personal communication, 1990.
37. Mercer, T. T. *Aerosol Technology in Hazard Evaluation*; Academic: New York, 1973.
38. Friedlander, S. K. *Smoke, Dust and Haze*; Wiley: New York, 1977.

39. Ye, Y.; Tsai, C.-J.; Pui, D. Y. H.; Lewis, C. W. *Aerosol Sci. Technol.* 1991, *14*, 102–111.
40. Zhang, G.; Dasgupta, P. K.; Cheng, Y.-S. *Atmos. Environ.* 1991, *25A*, 2717–2729.
41. Appel, B. R. California Department of Health Services, Berkeley, personal communication, 1990.
42. McMurry, P. H.; Stolzenburg, M. R. *Atmos. Environ.* 1987, *21*, 1231–1234.
43. Murphy, D. M.; Fahey, D. W. *Anal. Chem.* 1987, *59*, 2753–2759.
44. Cooney, D. O.; Kim, S.; Davis, E. J. *Chem. Eng. Sci.* 1974, *29*, 1731–1738.
45. Corsi, R. L.; Chang, D. P. Y.; Larock, B. *Environ. Sci. Technol.* 1988, *22*, 561–565.
46. Dasgupta, P. K.; McDowell, W. L.; Rhee, J.-S. *Analyst (London)* 1986, *111*, 87–90.
47. Braman, R. S.; Shelley, T. J.; McClenny, W. A. *Anal. Chem.* 1982, *54*, 358–364.
48. Possanzini, M.; Febo, A.; Cecchini, F. *Anal. Lett.* 1985, *18*, 681–693.
49. Gormley, P. G. *Proc. R. Ir. Acad., Sect. A* 1938, *45*, 59–63.
50. Winiwarter, W. *Atmos. Environ.* 1989, *23*, 1997–2002.
51. Stephan, K. *Chem.-Ing. Tech.* 1972, *34*, 207–212.
52. Rosenberg, C.; Winiwarter, W.; Gregori, M.; Pech, G.; Casenkey, V.; Puxbaum, H. *Fresenius Z. Anal. Chem.* 1988, *331*, 1–7.
53. Tanner, R. L.; Markovits, G. Y.; Ferreri, E. M.; Kelly, T. J. *Anal. Chem.* 1986, *58*, 1857–1865.
54. Lundberg, R. E.; Reynolds, W. C.; Kays, W. M. NASA Technical Note D-1972, National Aeronautics and Space Administration: Washington, DC, 1963.
55. Dasgupta, P. K.; Dong, S.; Hwang, H.; Yang, H.-C.; Genfa, Z. *Atmos. Environ.* 1988, *22*, 949–964.
56. Vossler, T. L.; Stevens, R. K.; Paur, R. J.; Baumgardner, R. E.; Bell, J. P. *Atmos. Environ.* 1988, *22*, 1729–1733.
57. Koutrakis, P.; Wolfson, J. M.; Brauer, M.; Spengler, J. D. *Aerosol Sci. Technol.* 1990, *12*, 607–612.
58. Ström, L. *Atmos. Environ.* 1972, *6*, 133–142.
59. Genfa, Z.; Dasgupta, P. K.; Dong, S. *Environ. Sci. Technol.* 1989, *23*, 1467–1474.
60. McClenny, W. A.; Galley, P. C.; Braman, R. S.; Shelley, T. J. *Anal. Chem.* 1982, *54*, 365–369.
61. Fox, D. L.; Stockburger, L.; Weathers, N.; Spicer, C. W.; Mackay, G. I.; Schiff, H. I.; Eatough, D. J.; Mortensen, F.; Hansen, L. D.; Shepson, P. B.; Kleindienst, T. E.; Edney, E. O. *Atmos. Environ.* 1988, *22*, 575–586.
62. Roberts, J. M.; Norton, R. B.; Goldan, P. D.; Fehsenfeld, F. C. *J. Atmos. Chem.* 1987, *5*, 217–238.
63. Appel, B. R.; Tokiwa, Y.; Kothny, E. L.; Wu, R.; Povard, V. *Atmos. Environ.* 1988, *22*, 1566–1573.
64. Roberts, J. M.; Langford, A.; Goldan, P. D.; Fehsenfeld, F. C. *J. Atmos. Chem.* 1988, *7*, 137–152.
65. Langford, A. O.; Goldan, P. D.; Fehsenfeld, F. C. *J. Atmos. Chem.* 1989, *8*, 359–376.
66. Lindqvist, F. *Atmos. Environ.* 1985, *19*, 1671–1680.
67. Lindqvist, F. *J. Air Pollut. Control Assoc.* 1985, *35*, 19–23.
68. Tanner, R. L.; Kelly, T. J.; Dezaro, D. A.; Forrest, J. *Atmos. Environ.* 1989, *23*, 2213–2222.

69. Slanina, J.; Lamoen-Doornenbal, L. V.; Lingerak, W. A.; Meilof, W.; Klockow, D.; Niessner, R. *Int. J. Environ. Anal. Chem.* **1981**, *9*, 59–70.
70. Slanina, J.; Schoonebeek, C. A. M.; Klockow, D.; Niessner, R. *Anal. Chem.* **1985**, *57*, 1955–1960.
71. Dasgupta, P. K.; Lundquist, G. L.; West, P. W. *Atmos. Environ.* **1979**, *13*, 767–774.
72. Forrest, J.; Tanner, R. L.; Dasgupta, P. K.; Lundquist, G. L.; West, P. W. *Atmos. Environ.* **1979**, *13*, 1604–1605.
73. Slanina, J.; Keuken, M. P.; Schoonebeek, C. A. M. *Anal. Chem.* **1987**, *59*, 2764–2766.
74. Keuken, M. P.; Wayers-Ijpelaan, A.; Mols, J. J.; Otjes, R. P.; Slanina, J. *Atmos. Environ.* **1989**, *23*, 2177–2185.
75. Klockow, D.; Niessner, R.; Malejczyk, M.; Kiendl, H.; vom Berg, B.; Keuken, M. P.; Wayers-Ypellan, A.; Slanina, J. *Atmos. Environ.* **1989**, *23*, 1131–1138.
76. Bos, R. *J. Air Pollut. Control Assoc.* **1980**, *30*, 1222–1224.
77. Dasgupta, P. K. *Atmos. Environ.* **1984**, *18*, 1593–1599.
78. Phillips, D. A.; Dasgupta, P. K. *Sep. Sci. Technol.* **1987**, *22*, 1255–1267.
79. Dasgupta, P. K. In *Ion Chromatography*; Tarter, J. G., Ed.; Marcel-Dekker: New York, 1987; pp 220–224.
80. Lind, J. A. National Center for Atmospheric Research, Boulder, personal communication, 1991.
81. Dong, S.; Dasgupta, P. K. *Environ. Sci. Technol.* **1987**, *21*, 581–588.
82. Hwang, H.; Dasgupta, P. K. *Anal. Chem.* **1986**, *58*, 1521–1524.
83. Dasgupta, P. K.; Yang, H.-C. *Anal. Chem.* **1986**, *58*, 2839–2844.
84. Hwang, H.; Dasgupta, P. K. *Environ. Sci. Technol.* **1985**, *19*, 255–258.
85. Dasgupta, P. K.; Dong, S. *Atmos. Environ.* **1986**, *20*, 565–570.
86. Dong, S.; Dasgupta, P. K. *Environ. Sci. Technol.* **1986**, *20*, 637–640.
87. Lindgren, P. F. *Anal. Chem.* **1991**, *63*, 1008–1011.
88. Kleindienst, T. E.; Shepson, P. B.; Hodges, D. N.; Nero, C. M.; Arnts, R. R.; Dasgupta, P. K.; Hwang, H.; Kok, G. L.; Lind, J. A.; Lazrus, A. L.; Mackay, G. I.; Mayne, L. K.; Schiff, H. I. *Environ. Sci. Technol.* **1988**, *22*, 53–61.
89. Kleindienst, T. E.; Shepson, P. B.; Nero, C. M.; Arnts, R. R.; Tejada, S. B.; Mackay, G. I.; Mayne, L. K.; Schiff, H. I.; Lind, J. A.; Kok, G. L.; Lazrus, A. L.; Dasgupta, P. K.; Dong, S. *Atmos. Environ.* **1988**, *22*, 1931–1939.
90. Lawson, D. R.; Winer, A. M.; Biermann, H. W.; Tuazon, E. C.; Mackay, G. I.; Schiff, H. I.; Kok, G. L.; Dasgupta, P. K.; Fung, K. *Aerosol Sci. Technol.* **1990**, *12*, 64–76.
91. Dasgupta, P. K.; Dong, S.; Hwang, H. *Aerosol Sci. Technol.* **1990**, *12*, 98–104.
92. Sigg, A. Physics Institute, University of Bern, Switzerland, personal communication, 1990.
93. Genfa, Z.; Dasgupta, P. K. *Anal. Chem.* **1989**, *61*, 408–412.
94. Dasgupta, P. K. "An Improved Analytical Technique for Gas and Aqueous Phase Hydrogen Peroxide—Instrument Manual", National Technical Information Services: Springfield, VA; PB 88–239 025/AS.
95. Neftel, A. Physics Institute, University of Bern, Switzerland, personal communication, 1990.
96. Lindgren, P. F.; Dasgupta, P. K. *Anal. Chem.* **1989**, *61*, 19–24.
97. Dasgupta, P. K.; Lindgren, P. F. *Environ. Sci. Technol.* **1989**, *23*, 895–897.
98. Kok, G. L.; Schanot, A. J.; Lindgren, P. F.; Dasgupta, P. K.; Hegg, D. A.; Hobbs, P. V.; Boatman, J. F. *Atmos. Environ.* **1990**, *24A*, 1903–1908.

99. Lindgren, P. F. *Atmos. Environ.* **1992**, *26A*, 43–49.
100. Vecera, Z.; Dasgupta, P. K. *Environ. Sci. Technol.* **1991**, *25*, 255–260.
101. Dasgupta, P. K.; Ping, L.; Zhang, G.; Hwang, H. In *Biogenic Sulfur in the Environment*; Saltzman, E. S.; Cooper, W. J., Eds.; ACS Symposium Series 393; American Chemical Society: Washington, DC, 1989; pp 380–401.
102. Mir, K. A.; Dasgupta, P. K. Texas Tech University, unpublished studies, 1988.
103. Sexton, K. Department of Public Health, University of North Carolina, personal communication, 1990.
104. Fendinger, N. J.; Glotfelty, D. E. *Environ. Sci Technol.* **1988**, *22*, 1289–1293.
105. Emmert, R. E.; Pigford, R. L. *Chem. Eng. Prog.* **1954**, *50*, 78–95.
106. Lucero, D. P. *J. Chromatogr. Sci.* **1985**, *23*, 293–303.
107. Keuken, M. P.; Schoonebeek, C. A. M.; Wensveen-Louter, A.; Slanina, J. *Atmos. Environ.* **1988**, *22*, 2541–2548.
108. Simon, P. K.; Dasgupta, P. K.; Vecera, Z. *Anal. Chem.* **1991**, *63*, 1237–1242.
109. Vecera, Z.; Dasgupta, P. K. *Anal. Chem.* **1991**, *63*, 2210–2216.
110. Simon, P. K.; Dasgupta, P. K., submitted for publication in *Anal. Chem.*
111. Bushey, M. M.; Jorgenson, J. W. *Anal. Chem.* **1990**, *62*, 978–984.
112. Chapman, E. G.; Barinaga, C. J.; Udseth, H. R.; Smith, R. D. *Atmos. Environ.* **1990**, *24A*, 2951–2957.

RECEIVED for review March 20, 1991. ACCEPTED revised manuscript May 14, 1992.

Fast-Response Chemical Sensors Used for Eddy Correlation Flux Measurements

A. C. Delany

National Center for Atmospheric Research, P.O. Box 3000, Boulder, CO 80307

The requirements for chemical sensors suitable for use in eddy correlation direct measurements of surface fluxes are examined. The resolution of chemical sensors is examined and defined in terms of surface flux and commonly measured micrometeorological parameters. Aspects of the design and operation of sensor systems are considered. In particular, the effects of the inlet ducting, the sensing volume, and the signal processing on the ability to measure surface fluxes were analyzed.

"A MAJOR LIMITATION TO RESEARCH on surface-exchange and flux measurements is the lack of sensitive, reliable, and fast-response chemical species sensors that can be used for eddy correlation flux measurement. Therefore we recommend that continued effort and resources be expended in developing chemical species sensors with the responsiveness and sensitivity required for direct eddy correlation flux measurements." This recommendation (1) was assigned the first priority in the report of the recent Global Tropospheric Chemistry workshop jointly convened by the National Science Foundation, the National Aeronautics and Space Administration, and the National Oceanic and Atmospheric Administration. The authors of the report recognized that the limited availability of fast, accurate chemical sensors is a major measurement challenge in the field of atmospheric chemistry.

One of the greatest uncertainties in the understanding of the mechanisms that control the chemical composition of the atmosphere concerns the exchange of trace species between the atmosphere and the surface. These surface exchanges include both emission and deposition and are intimately

0065–2393/93/0232–0091$06.00/0

connected to processes involving the biosphere. The dry deposition of atmospheric acids and oxidants, the natural emission of biogenic hydrocarbons, the anthropogenically perturbed cycles of nutrients, and many other important aspects of surface exchange are involved in these processes. To investigate surface exchange, measurements of emission and deposition fluxes must be made over selected representative sites. Several techniques to measure these chemical fluxes have been developed. However, the most direct technique for measuring the emission or deposition is eddy correlation. This method is a fundamentally direct technique that has the added advantage of not disturbing the nature of the surface.

The concepts involved in the eddy correlation technique and the factors involved in the design of appropriate chemical sensors are briefly examined in this chapter. The discussion focuses on surface-layer measurements because it is in this layer that the atmosphere–biosphere interaction is most readily examined.

Eddy Correlation

The eddy correlation technique directly determines the flux of an atmospheric trace constituent through a plane parallel to the surface. For the determination of surface emission and deposition fluxes, the method is rigorous when specific criteria are met. Ideally, the meteorological conditions controlling the state of turbulence should not vary over the course of the measurements. The surface viewed by the sensors should be horizontally uniform, both in its physical and chemical–biological aspects, and should stretch for a distance much greater than the height at which the measurements are made. This height should be much larger than the scale of the surface roughness and the intrinsic scale of the sensors. The extent to which these criteria can be relaxed with the method remaining valid is a subject of ongoing debate (2). The eddy correlation technique has been examined thoroughly, and a considerable literature exists that deals with potential errors that can result because of flow distortion (3), inappropriate signal processing (4), failure to make the necessary corrections for density effect (5), and the chemical reactivity of the measured species (6). Stratagems have been designed to make the most advantageous compromise between the need to accrue the best statistics and the desire to operate within periods without significant changes in meteorological conditions (7). Approaches dealing with the problems of correlated and uncorrelated noise have been explored (8). However, the basic requirement for the measurement of the surface flux of atmospheric chemical species involves the ability to make the appropriate chemical measurements.

Because the eddy correlation method may be considered as defining the instantaneous upward or downward transport of the constituent and then averaging contributions to give the net flux, it must take into account the frequency range of the turbulence responsible for vertically transporting the constituent in the atmosphere. Kaimal et al. (9) found that the scalar flux cospectra scaled with a nondimensional frequency, $n = fz/\overline{u}$, where f is the measurement frequency, z is the height at which the measurement is made, and $\overline{u}$ is the mean wind speed. Under neutral or unstable conditions, such as occur during most daytime circumstances, 90% of the flux is measured with $n \leq 1$. Thus, for measurements on a tower at 10 m with a 5-m/s wind, $f_{max} \leq (1)(5)/10 = 0.5$ Hz. The highest required frequency is only 0.5 Hz. Because of the Nyquist requirement the required sampling rate is 1 Hz. For stable nighttime conditions or if fluxes must be measured with greater than 10% accuracy, then higher frequencies need to be measured.

Kaimal et al. also determined that the lowest frequency that needs to be included is $n \approx 0.01$. Thus, for the tower measurements mentioned the lowest frequency that needs to be included is $f_{min} \approx (0.01)(5)/10 \approx 0.005$ Hz. This calculation indicates an averaging time of 200 s. Unfortunately, this time would not allow a sufficient sampling of the lowest frequency contribution. For such sampling to occur a period an order of magnitude longer is required, and hence a period of 30 min is generally required.

The technique requires simultaneous fast and accurate measurements of both the vertical velocity and the trace species in question. Fortunately the technology for the measurement of turbulence with the necessary resolution is available. Sonic anemometers can readily yield air motion data with the required resolution (10). Likewise, the ability to handle the air motion and chemical concentration data with modern computer data systems is well in hand (11). Thus these aspects can be ignored, and the major limitation can be dealt with: the availability of appropriate chemical sensors with sufficient time and chemical resolution.

Chemical Resolution Required

The effect of a surface flux on the concentration of an atmospheric constituent measured at some height is to impose a variance upon that concentration as turbulence intermittently transports air up from the surface, where the constituent is either enhanced or depleted, to the level of the sensor. The chemical resolution of the sensor defines the extent to which this fluctuation in concentration, c', can be assessed. The required resolution for a sensor to be used for eddy correlation measurements of fluxes thus depends on the intensity of the flux to be measured. A greater flux gives a greater c' and a more relaxed requirement for chemical resolution. A smaller flux gives a smaller c' and a more stringent requirement for chemical resolution. The relationship between the micrometeorological environment prevailing at the specific location and the chemical resolution required is somewhat more complex. In general, atmospheric conditions that tend to dissipate or mix out chemical variability will impose a greater stricture on the chemical resolution, and conditions that allow chemical variability to persist will ease the requirement of chemical resolution.

Businger and Delany (12) examined this problem and derived a relationship that defines the chemical resolution required for sensors used for eddy correlation measurements. Their approach was to specify the standard deviation of the chemical concentration σ_c (the root mean square of the chemical fluctuation c') in terms of the surface chemical flux, F_c, and readily measured micrometeorological parameters.

$$\sigma_c = |F_c| \frac{\sigma_\theta}{u_*|\theta_*|} \tag{1}$$

where $u_* = (-\overline{u'w'})^{1/2}$ is the friction velocity, $\theta_* = -\overline{w'\theta'}/u_*$ is the characteristic potential temperature, $\sigma_\theta = (\overline{\theta'\theta'})^{1/2}$ is the standard deviation of the potential temperature, u is the horizontal velocity, w is the vertical velocity, and θ is the potential temperature.

It was then argued that if the flux is to be known to $\pm 10\%$ there is a resolution requirement that $R_c \simeq 0.1\sigma_c$ for the worst-case scenario.

$$R_c = 0.1\,\sigma_c = 0.1\,|F_c| \frac{\sigma_\theta}{u_*|\theta_*|} \tag{2}$$

The value of $\sigma_\theta/u_*|\theta_*|$, which Businger and Delany termed the atmospheric parameter, is plotted in Figure 1 as a function of atmospheric stability, z/L, and friction velocity, u_*.

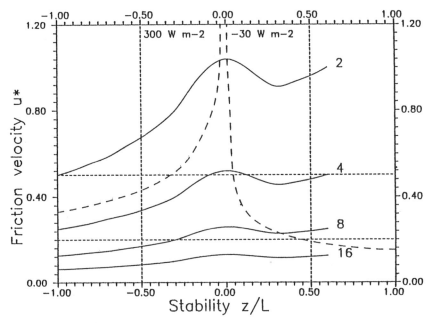

Figure 1. Isopleths of R_c, a measure of the chemical resolution of the instrument with the eddy correlation technique.

This atmospheric stability versus friction velocity plot is capable of representing any atmospheric condition except that when the stability is neutral. The information contained in Figure 1 and equation 2 allows the relationship between the value of the surface flux and the required chemical sensor resolution to be estimated for a full range of atmospheric conditions.

Chemical Sensors

The factors that influence the chemical resolution of sensors are well understood and are not discussed here. This section reviews the factors that control the temporal resolution of sensors to be used for eddy correlation. In the analysis of the design of chemical sensors to be used for eddy correlation it is instructive to consider the different components of chemical sensor systems separately to determine the influences that they have on the temporal response to variations in the atmospheric concentration of a trace constituent. Of course this analysis is an oversimplification because the total systems operate in a more complex fashion, but it is a useful exercise.

Inlet System. The simplest inlet system is that of the open-path in situ sensor. This system is the ideal; only subtle effects are expected. Thus, flow distortion can produce air density changes that interfere with the measurement of molecular density by optical absorption techniques, and size fractionation of large aerosol particles can result because of airflow around the body of the sensor. For many chemical sensors this open path cannot be achieved, and atmospheric air must be ducted to the sensor. Even when optical absorption techniques are used, the need to reduce the pressure broadening of absorption lines can necessitate the atmospheric sample being drawn into a reduced-pressure cell for analysis. For analysis techniques involving chemiluminescence, flame photometry, electron capture, and other techniques, it is often imperative to enclose the sensing volume, hence the air must be ducted to the sensor. The distance that the sample must be ducted depends on several factors. If the physical size of the chemical sensor system is large or if it requires special environmental housing, then it cannot be placed near the turbulence sensor or the flow distortion would be too severe. However, too great a separation can impose line loss and other penalties. Considerations of chemistry and turbulence can lead to a compromise balancing the disadvantages.

The most obvious result of ducting the atmospheric sample from the vicinity of the sonic anemometer to the chemical sensor is the introduction of a time delay. This time lag must be eliminated before the correlation between chemical concentration and the vertical air motion variances is made to yield the covariance. Several different approaches have been taken to determine the length of the delay. One simple method involves spiking a balloon with the compound involved, and then inflating and bursting it in such a manner that the sonic anemometer path is interrupted at the same instant the sample is released at the chemical sensor intake. A fast solenoid valve may be used to enable the rapid injection of the trace compound into the sample intake. A more sophisticated approach utilizes the fact that the chemical concentration and the temperature are correlated in the atmosphere. By adjusting the time lag of the chemical sensor data stream, one can determine the maximum possible correlation (or anticorrelation), and the correct value can be obtained. A more serious difficulty arises if the chemical compound interacts with the material of the ducting. Although an appropriate calibration procedure, one that involves the entire inlet system, will allow the effect on the mean concentration of the trace constituent to be corrected for, the calibration may not compensate for the effect on the fluctuation. For the effect of chemisorption of a compound on the wall of the duct, for example, elevated concentrations would cause adsorption of molecules that would be released when the concentration decreased. The net effect of this smoothing would be to diminish the fluctuation and hence the perceived flux.

Even without chemical interaction the effect of a velocity profile across the section of the duct leads to a similar effect. The result, again, is an attenuation of variance. Lenschow and Raupach (13) investigated this aspect and presented their results in terms of the half-power frequency, $f_{0.5}$, the frequency at which the variance has been diminished by 50%. This half-power frequency is given by the expression

$$f_{0.5} = n_{0.5} \times \text{velocity}/(\text{radius} \times \text{length})^{1/2} \tag{3}$$

where $n_{0.5}$ is a dimensionless frequency with a value of $n_{0.5} = 0.92/(Re \times Sc)^{1/2}$ for laminar flow and a value of $n_{0.5} = 0.066Re^{1/16}$ for turbulent flow, Re is the Reynolds number (Re = diameter × velocity × density/viscosity), and Sc is the Schmidt number (Sc = molecular viscosity/molecular diffusivity; Sc is approximately 0.5 for common gases).

The best single system to use to duct atmospheric air to chemical sensors used cannot be defined for eddy correlation measurements because there are many parameters capable of adjustment and other considerations of cost, available electrical power, and so forth. However, the problem can be illustrated by an analysis of a realistic example.

A fast chemical sensor that operates at a reduced pressure of 50 torr (6700 Pa) and with a flow of 1 standard liter per minute must be maintained in an instrument shelter. Considerations of flow distortion require that 10 m separate the sensor from its intake on the tower near the sonic anemometer. Three ducting arrangements can be considered. The first would involve drawing air along 10 m of 1/4-inch tubing (0.2 cm internal radius) and controlling the flow and pressure at the sensor itself. The second option would place the pressure–flow controller at the inlet and allow the inlet intake to flow at the reduced pressure. The third course would be to use a high-flow manifold of 2-inch pipe with a flow of 10 cubic feet per minute (2.5-cm internal radius and 4.7 liters per second) to bring the air to the sensor, and then sampling would be done from the manifold. Table I gives the relevant dimensions and calculated parameters. The results of using each of the three options are shown schematically in Figure 2.

Table I. Dimensions and Calculated Parameters for the Three Different Inlet Configurations

| | Inlet Configuration | | |
| | Ambient | Reduced | |
Parameter	Pressure	Pressure	Manifold
Length (cm)	1.0×10^3	1.0×10^3	1.0×10^3
Radius (cm)	2.0×10^{-1}	2.0×10^{-1}	2.5
Density (g/cm³)	1.2×10^{-3}	7.9×10^{-5}	1.2×10^{-3}
Ambient flow (cm^{-3} s^{-1})	1.7×10^1	2.6×10^2	4.7×10^3
Time delay (s)	7.4	4.8×10^{-1}	4.2
Re	3.5×10^2	3.5×10^2	7.8×10^3
$f_{0.5}$ (s^{-1})	7.0×10^{-2}	7.0×10^{-2}	1.2×10^{-1}
$n_{0.5}$ (s^{-1})	6.6×10^{-1}	1.0×10^1	5.5×10^{-1}

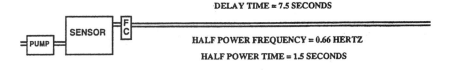

DELAY TIME ≈ 7.5 SECONDS

HALF POWER FREQUENCY = 0.66 HERTZ

HALF POWER TIME = 1.5 SECONDS

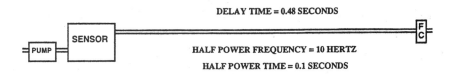

DELAY TIME = 0.48 SECONDS

HALF POWER FREQUENCY = 10 HERTZ

HALF POWER TIME = 0.1 SECONDS

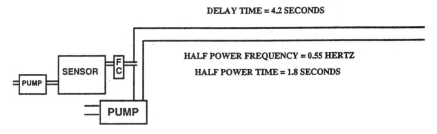

DELAY TIME = 4.2 SECONDS

HALF POWER FREQUENCY = 0.55 HERTZ

HALF POWER TIME = 1.8 SECONDS

Figure 2. Three possible options for ducting atmospheric air from the inlet on the tower to the sensor in the instrument shelter.

Sensing Volume.   The sensing volume of a sensor is the volume where the air is actually monitored. The sensing volume is the reaction chamber of a flame photometric detector or a chemiluminescence device, the field of view of an open-path sensor, or the White cell of a reduced-pressure optical system. The residence time of the sample within the sensing volume ultimately limits the temporal resolution of most chemical sensors.

For an in situ sensor the residence time for the chemical constituent is defined by the sensing path of the particular technique. In the best possible case this path is the same as for the sonic anemometer, which measures air motion, and hence there is a correspondence when the two factors, c and w, are correlated.

For most chemical or optical sensors the size and flow rate of the reactor or optical cell defines the residence time. The time resolution cannot be better than the residence time τ, although it can be worse.

$$\tau = \text{volume/ambient flow} \tag{4}$$

Thus, for a nitric oxide sensor dependent on ozone chemiluminescence, a 1.5-liter reaction chamber operating at 60 torr with a flow of 1.0 standard liter per second has a residence time of approximately $\tau = (1.5/1.0) \times (60/760) = 0.12$ s.

Signal Processing.   Although the temporal resolution of a sensor is ultimately limited by the residence time of the sensing volume, resolution can be further degraded by the necessity to integrate the signal for some longer period in order to accumulate sufficient data for adequate statistics. This scenario can best be illustrated by considering a sensor that involves a statistically characterized output such as photon counting. For such a sensor the smallest change in concentration that can be reliably detected is one that generates a change of output greater than or equal to the statistical uncertainty associated with the total number of counts. This value may be taken as the square root of the total number. Because the number of counts depends on the integration time, the statistical uncertainty depends on some specific time interval. For a sensor in which S is the chemical sensitivity of a sensor, the count rate generated by some concentration of the specific chemical species, and B is the background of a sensor, the count rate that is independent of the chemical concentration, the chemical resolution of this sensor, R_τ, is the smallest change in chemical concentration that can be seen against the mean chemical concentration, $\bar{c}$, when the measurement is performed for time τ.

$$R_\tau = (B\tau + S\tau\bar{c})^{1/2}(S\tau)^{-1} \tag{5}$$

For the case when the background is negligible

$$R_\tau = (S\tau\bar{c})^{1/2}(S\tau)^{-1} = (\bar{c}/S\tau)^{1/2} \tag{6}$$

The detection limit of a sensor, D_τ, is a special case of chemical resolution. This limit is the smallest increase of concentration that can be seen against the background when the measurement is performed for time τ. Both the chemical resolution and the detection limit are dependent upon measurement time. The longer the integration period, the smaller the chemical resolution and the detection limit. Thus

$$D_\tau = D_{(1\ s)}(1.0/\tau)^{1/2} \tag{7}$$

where $D_{(1\ s)}$ is the 1-s integration detection limit. However, the detection limit (and the chemical resolution) do not decrease indefinitely. A limit is reached when nonstatistical variation, including drift and artifact response, become dominant. Likewise τ (and the temporal resolution) cannot be decreased indefinitely. A detector has a minimum response time that is associated with the clearance of the sample of air from the sensing volume.

Conclusion

More information is needed about the surface emission and deposition of trace atmospheric species. These fluxes can often be best measured by the eddy correlation technique with fast chemical sensors in conjunction with micrometeorological instrumentation. As analytical techniques for trace species progress, fast and sensitive sensors are becoming available for field research. Consideration must be given to matching the chemical sensors to the eddy correlation technique.

Acknowledgment

The National Center for Atmospheric Research is sponsored by the National Science Foundation.

References

1. *Global Tropospheric Chemistry: Chemical Fluxes in the Global Atmosphere*; Lenschow, D. H.; Hicks, B. B., Eds.; National Center for Atmospheric Research: Boulder, CO, 1989.
2. Businger, J. A. *J. Clim. Appl. Meteorol.* **1986**, *25*, 1100–1124.
3. Wyngaard, J. C. *Boundary Layer Meteorol.* **1988**, *42*, 19–26.
4. Shaw, W. J.; Tillman, J. E. *J. Appl. Meteorol.* **1980**, *19*, 90–97.
5. Webb, E. K.; Pearman, G. I.; Leuning, R. *Q. J. R. Meteorol. Soc.* **1980**, *106*, 85–100.
6. Lenschow, D. H.; Delany, A. C. *J. Atmos. Chem.* **1987**, *5*, 301–309.
7. Arya, S. P. *Introduction to Micrometeorology*; Academic: San Diego, 1988; 307 pp; International Geophysics Series; Vol. 42.
8. Lenschow, D. H.; Kristensen, L. *J. Atmos. Oceanic Technol.* **1985**, *2*, 68–81.

9. Kaimal, J. C.; Wyngaard, J. C.; Izumi, Y.; Cote, O. R. *Q. J. R. Meteorol. Soc.* **1972**, *98*, 563–589.
10. Wyngaard, J. C. *Annu. Rev. Fluid Mech.* **1981**, *13*, 399–423.
11. Businger, J. A.; Dabberdt, D. W.; Delany, A. C.; Horst, T. W.; Martin, C. L.; Oncley, S. P.; Semmer, S. R. *Bull. Am. Meteorol. Soc.* **1990**, *71*, 1006–1011.
12. Businger, J. A.; Delany, A. C. *J. Atmos. Chem.* **1990**, *10*, 399–410.
13. Lenschow, D. H.; Raupach, M. M. *J. Geophys. Res.* **1991**, *96*, 15259–15268.

RECEIVED for review March 20, 1991. ACCEPTED revised manuscript September 15, 1992.

4

Tropospheric Sampling with Aircraft

Peter H. Daum and Stephen R. Springston

Environmental Chemistry Division, Department of Applied Science, Brookhaven National Laboratory, Upton, NY 11973

The use of aircraft in atmospheric sampling places stringent requirements on the instruments used to measure the concentrations of atmospheric trace gases and aerosols. Some of these requirements, such as minimization of size, weight, and power consumption, are general; others are specific to individual techniques. This review presents the basic principles and considerations governing the deployment of trace gas and aerosol instrumentation on an aircraft. An overview of common instruments illustrates these points and provides guidelines for designing and using instruments on aircraft-based measurement programs.

IN SITU MEASUREMENTS have always been an important component of atmospheric chemical studies because of the enormous complexity of the atmosphere and the impossibility of realistically simulating the interactions between all of the relevant species in the laboratory. Surface measurements are extremely useful in identifying important species and characterizing the chemical processes in which they are involved. However, atmospheric composition and chemical reactivity are spatially varying quantities, and surface measurements are not capable of providing data for the study of many important chemical phenomena. Furthermore, these features of the atmosphere are frequently linked to static and dynamic meteorological phenomena that occur over a vast range of scales and cannot be understood by sampling at a fixed location.

In theory, spatially resolved measurements of concentrations of atmospheric trace substances could be made by remote sensing from either surface-based or satellite platforms, but remote sensing capabilities have not yet been developed for many species. Furthermore, both surface- and satellite-based remote sensing methods have distinct limitations in resolving

0065–2393/93/0232–0101$09.00/0
© 1993 American Chemical Society

horizontal and vertical variations in concentrations on scales that may be important to understanding various atmospheric chemical properties. For these reasons a need for in situ sampling of atmospheric regimes inaccessible to surface- or satellite-based instruments (by altitude, geographic, or technical constraints) will remain for the foreseeable future.

Platforms capable of conducting such measurements include tethered balloons, free balloons such as those deployed for meteorological soundings or for stratospheric measurements, and aircraft. Although each of these platforms has been used in atmospheric chemical studies, only aircraft have been used by a relatively broad community because payload and electrical power limitations or cost (the latter an especially important consideration for stratospheric measurements) limits the use of balloons in the study of phenomena not readily accessible by other means. More importantly, however, balloons sample the atmosphere in fundamentally different ways than aircraft. Tethered balloons, because they only have access to a narrow cylinder above their mooring, essentially sample in a Eulerian framework. Free balloons essentially sample in a Lagrangian framework. Although sampling in such frameworks is important for the study of specific classes of problems, such measurement strategies do not provide the information necessary to understand many of the important chemical properties of the atmosphere.

As the atmospheric chemical community focuses more intently on issues of global concern, a focus that requires the survey of the chemical composition of the remote atmosphere and the study of chemical processes occurring well above the surface, a larger number of programs will use aircraft either as the principal sampling platform or to augment measurements made by other means. Because aircraft measurements impose different and frequently more stringent sets of requirements on techniques used to measure trace atmospheric species than surface measurements, there is a need to review these requirements and to show how various measurement methods have been designed or altered to meet the needs of aircraft use. This chapter focuses on measurement of trace chemical species; this definition includes trace gases as well as aerosol particles and cloud droplets. Measurements of winds, atmospheric state parameters, solar radiation, and similar parameters are not treated. Furthermore, this review is restricted to the use of aircraft and instruments for tropospheric measurements. This chapter is divided into two main sections: the first section emphasizes general issues involved in sampling from aircraft; the second section surveys specific techniques that have been used for aircraft measurement of species that have been identified to be of common interest to a large segment of the chemical community.

General Considerations

Aircraft measurement of trace atmospheric species emphasizes several problems in the composite sampling–analytical process that are not routinely encountered to the same degree during measurements at a stationary surface site. The first problem is distortion of sample composition during the sampling process. For aerosol species this distortion is an especially significant problem because sampling is accomplished at high relative velocity (50–200 m/s) and fractionation of the aerosol by size due to nonisokinetic sampling, turbulence, and inertial losses at bends in the sampling lines can occur. For trace gases special attention must be paid to design and implementation of inlet and distribution systems to avoid adsorptive losses or contamination of the sampling system. Second, there are often important engineering problems associated with the use of instruments in an aircraft that are not usually encountered in surface sampling applications. Weight, size, and electrical power consumption are frequently severe constraints on aircraft. Furthermore, instruments may be subjected to extreme variations in temperature and humidity and to high levels of vibration. Thus, it is frequently not possible to deploy conventional state-of-the-art analytical instruments on aircraft without substantial modifications to minimize the influence of these factors on instrument performance. Third, aircraft measurements are frequently made over a wide range of altitudes, and instruments will be exposed to a range of pressures more extreme than would ever be encountered during operation at the surface (depending on whether the aircraft is pressurized and how the sample inlet system is designed). Because instruments frequently exhibit large changes in sensitivity with pressure, using a sensor on an aircraft requires that the pressure response of the composite instrument–sampling system be well understood. No instrument, whether commercial or custom-made, can be deployed in an aircraft without a consideration of the effects of this environment. This section explores these and related issues and gives some guidelines on how these problems can be resolved.

Concentration Units.   Any discussion of aircraft measurements must begin with a review of various ways in which abundances of trace atmospheric constituents are specified. Such nomenclature is frequently a point of some confusion. In addition, because instruments used on aircraft are operated over a range of pressures, the fundamental way that an instrument senses a species and generates an output signal that is then converted to a measure of abundance is an important issue. The latter concern is discussed in the next section.

Several methods are commonly used to specify the abundance of substances in the atmosphere. For gaseous constituents common practice is to specify abundances as mixing ratios, or equivalently as mole fractions of the species in air. This quantity is simply the ratio of the partial pressure of a substance to the total pressure. The advantage of this unit is that it is independent of pressure and temperature, and for an atmospheric component that is well mixed, the mixing ratio will be constant as the pressure or temperature changes. Common units for specifying mixing ratios are parts per million (ppmv), parts per billion (ppbv), and parts per trillion (pptrv), meaning ratios of the partial pressure to the total pressure (or molar ratios) of 1 in 10^6, 1 in 10^9, and 1 in 10^{12}, respectively. To distinguish from common usage in specifying solution-phase concentrations, where ppb and so forth indicate the mass mixing ratio, the suffix v is added to indicate mixing ratio by volume.

Other common ways of expressing abundances, particularly of solid or liquid particles, is to express them as concentrations in units of micrograms per cubic meter or nanomoles per cubic meter. For purposes of consistency, concentrations expressed in these units should be normalized to standard conditions of temperature and pressure. Because there is some confusion as to what constitutes standard conditions in atmospheric chemistry (273 K and 1.013 bar are commonly used in chemistry and physics and 293 K and 1.013 bar are used in engineering), it is important to define the standard conditions that are assumed when reporting data. This explicit definition is frequently not done. Concentrations expressed in these units can be easily converted to mixing ratios by use of the ideal gas law:

$$MR \text{ (ppbv)} = C \text{ } (\mu g/m^3) \times \frac{22.4}{M} = 0.0224 \text{ } C \text{ (nmol/m}^3) \tag{1}$$

where MR is the mixing ratio, M is the molecular weight, C is the concentration expressed in the specified units, and standard conditions are assumed to be 273 K and 1.013 bar.

Altitude Response.   Pressure response is an issue that needs to be addressed for every instrument deployed on an aircraft. First, it must be decided how chemical abundances are to be reported. If standard practice is followed and they are reported as mixing ratios, then it must be determined whether the instrument is fundamentally a mass- or a concentration-dependent sensor, because this definition determines the first-order means by which instrument response is converted to mixing ratios as a function of pressure. In this context, a mass-sensitive detector is a device with an output signal that is a function of the mass flow of analyte molecules; a concentration-sensitive detector is one in which the response is proportional to the absolute concentration, that is, molecules per cubic centimeter.

For sensors that are truly mass sensitive and for which the mass flow of sample through the sensing element is held constant as a function of pressure (for example, by use of electronic mass-flow controllers), instrument response is proportional to the mixing ratio independent of the pressure. For concentration-sensitive detectors, such as simple spectrophotometric instruments measuring absorbance or fluorescence, instrument response is a function of the absolute concentration, and the response will decrease for a constant mixing ratio as the pressure decreases. For example, the response of a pulsed fluorescence SO_2 instrument sampling air containing a fixed mixing ratio of SO_2 will be less at 850 mbar than at 1000 mbar because there will be 15% fewer SO_2 molecules present in the active volume of the detector. The correct mixing ratio can be obtained by multiplying the instrument response by the ratio P_o/P_i, where P_o is the pressure at which the calibration was performed and P_i is the pressure at which the measurement is made. Similar first-order corrections are adequate for most instruments based on concentration-sensitive detectors, such as the UV photometric O_3, the gas filter correlation (GFC) CO, and similar instruments. The response of concentration-sensitive detectors having sensing chambers that operate at constant absolute pressure should not need to be adjusted for altitude if concentrations are reported as mixing ratios.

The need for more complicated considerations in conversion of instrument response to mixing ratios usually arises when instruments that are based on mass-sensitive detectors are used. Common reasons are either that the mass flow is not held constant or that the process whereby the flow of analyte molecules is converted to an electrical signal changes as the pressure changes. These effects are illustrated by a discussion of the pressure response of two instruments commonly used to measure atmospheric trace gases, both based on detection schemes that are inherently mass sensitive.

The first example is the O_3–ethylene chemiluminescent detector for measurement of O_3. Such detectors are inherently mass sensitive because the photomultiplier current is a function of the flow of O_3 molecules through the detector, provided other factors controlling response remain constant. Thus, as long as the mass flow of sample is fixed, instrument response will be proportional to the O_3 mixing ratio and independent of pressure. However, all of the commercially available O_3–ethylene detectors use flow restrictors to control sample flow. If these restrictors operate under choked flow conditions, the mass flow will decrease as the pressure decreases, and the response of the instrument to a fixed mixing ratio of O_3 will decrease with pressure. If the decrease in mass flow with pressure is linear, the O_3 detector will exhibit characteristics of a concentration-sensitive detector (response to fixed mixing ratio decreasing with pressure) even though it is inherently a mass-sensitive detector, and mixing ratios can be calculated from instrument response in the same way as with concentration-sensitive detectors. (This is the case because the process generating the chemiluminescent signal is not a strong function of pressure.) Behavior of a commercial O_3 chemiluminescent detector using choked flow restrictors to control sample flow was evaluated (reference 1; see also references 2 and 3). Under operating conditions of the instrument, response to a fixed O_3 mixing ratio decreased with pressure in a manner that could be predicted from ideal gas law considerations; that is, the change in response was virtually identical to the change in response of a UV O_3 sensor in parallel measurements. If the response had been assumed to be proportional to the mixing ratio (mass sensitive), then O_3 concentrations would have been reported incorrectly. It was also demonstrated that maintaining constant sample mass flow caused the O_3 sensor response to be proportional to the mixing ratio independent of pressure, that is, to respond as a mass-sensitive detector.

The second example is the commonly used flame photometric detector for measurement of SO_2 and aerosol sulfur. In this detector, sulfur compounds are determined by measuring the light emitted by excited-state S_2 molecules formed when sulfur compounds are introduced into a hydrogen-rich flame. Because two sulfur atoms combine to form the emitting species, detector response is approximately proportional to the square of the sulfur concentration. The detector is mass sensitive in the sense that the signal is proportional to the mass flow of sulfur molecules passing through the detector; it is concentration sensitive in the sense that the signal is proportional (approximately) to the square of the sulfur concentration in the active region of the detector. In the instrument's commercial form, sample flow is determined by a flow restrictor, and thus mass flow decreases with pressure; it would be predicted that the response to a given sulfur mixing ratio would decrease with altitude. However, because of the way in which the hydrogen flow is controlled, the H_2–O_2 ratio in the flame also changes with pressure. This situation causes not only changes in sensitivity larger than would be predicted on the basis of mass-flow reduction with pressure but also large shifts in the background emission of the flame. The dependence of both the baseline response and sulfur sensitivity on pressure can be minimized (4) by controlling the mass flow of sample and hydrogen. The small residual dependence on pressure is presumably a manifestation of the concentration-sensitive characteristics of the detector due to the reduction in burner pressure with altitude.

This discussion points out several issues that need to be addressed when deployment of an instrument on an aircraft is being considered. First, it cannot be assumed that an instrument based on a mass-sensitive detector will give a response proportional to the mixing ratio independent of pressure. Response may vary, because the flow control system does not maintain constant mass flow over a relevant range of pressures. Similarly, it cannot be assumed that a flow control system based on restrictors will operate under choked flow conditions over the required range of pressures. Pumping efficiency changes with pressure, and at higher altitudes pumps may not be able to maintain the pressure drop required for choked flow. Under these conditions, the flow will not be a simple function of the ambient pressure, and instrument response cannot easily be converted to either mixing ratios or absolute concentrations. For these reasons, it is good practice to use electronic mass-flow controllers to control sample flow either directly or by difference on any mass-sensitive detector used on an aircraft.

However, even the use of mass-flow controllers may not be sufficient to stabilize instrument response as the pressure changes. For many detectors, chamber pressure varies with altitude even if the sample mass flow is maintained at a constant value. This variance may cause a change in the efficiency of the processes converting analyte flow to an electrical or optical signal proportional to concentration. As demonstrated by the examples mentioned, it is difficult to predict, a priori, the pressure response of a composite sampling system–detector. It is thus necessary to characterize instruments, particularly mass-sensitive instruments based on chemiluminescent processes, before using them in aircraft measurements to assure that the pressure response is well understood.

Instrument Time Response.   Instrument response time is always a consideration in aircraft measurements because aircraft velocities can be very high, upwards of 200 m/s. Because instruments have a finite response time, spatial resolution is often an important consideration when samples are being taken from an aircraft. About the slowest speed that can be maintained on a conventional aircraft with any realistic payload capability is 50 m/s. Thus, if an instrument has a 10-s response time, the spatial resolution of the measurement is on the order of 0.5 km; for an aircraft sampling at 200 m/s, the spatial resolution is on the order of 2 km. Slow instrument response distorts the signal in regions of strong spatial gradients, decreasing peak heights and smoothing the signal. Reconstruction of the actual signal is not trivial and will generally require extensive laboratory characterizations of instrument response characteristics in order to provide an adequate differential transform. A more complete discussion of this topic is given in reference 5 and the references contained therein.

Whether response time is a significant issue in an aircraft sampling program depends on the objective of the measurements. For sampling programs designed to characterize spatially compact regions of the atmosphere, such as power plant plumes or single cumulus clouds, fast instrument response time is of paramount importance because these features may only be 0.5 to several kilometers in extent. Features of interest in some cloud studies may have scales of only a few meters. For measurement programs in which the objective is characterization of remote regional air masses where concentrations are relatively homogeneous, instrument time response may not be important.

For most measurement techniques a trade-off exists between time response and sensitivity. Instruments can be made to respond more rapidly, but then noise levels and detection limits increase. To fulfill certain measurement objectives for which sensitivity is not a problem, it may be useful to enhance the time response of the instrument and accept the decrease in sensitivity. In any event, careful consideration must be made in the design of any sampling program to ascertain what level of concentrations needs to be measured and what time–space resolution is required. These issues will determine the characteristics of the airborne platform and the instruments that will be used and will dictate how they will be deployed.

Sample Inlets.   The design and construction of any system for aircraft sampling involves implementing an inlet that will not alter the composition of the sample air. For trace gases an inlet is required that will not change the concentration of trace gases as they transit the inlet system; for aerosol sampling an inlet is required that preserves the size distribution of the aerosol particles during the sampling process. For trace gases, the type of inlet that is required depends strongly on the characteristics of the gases that are being sampled. For substances that are relatively inert, such as CH_4, CO, NO, and so forth, common materials such as glass, Teflon, and stainless steel will suffice; for reactive gases or gases that have a tendency to adsorb on surfaces (e.g., HNO_3), the choice of materials for sample inlet lines is much more critical. Surface adsorption on the tubing material is only part of the problem. Clean surfaces, made of even the most inert material, quickly become coated with a layer of moisture and fine particles. Short residence times and laminar flow conditions minimize the influence of the wall and aid in preserving the integrity of sample components.

Sampling aerosols from an aircraft is inherently a difficult process, and no generally accepted method for sampling over the entire size range of interest (0.01–10 μm) has been developed. The problem is difficult because aircraft sample at a high velocity and the aerosol particles must be brought to that velocity during the sampling process. Because of this requirement, distortion of the size distribution (by size-selective sampling at the probe tip or by evaporative losses of volatile components due to adiabatic heating) and losses at the inlet or in subsequent turbulent deposition or inertial impaction are of great concern.

The first requirement for sampling aerosols is to preserve their size distribution. To accomplish this, sampling must be conducted isokinetically; that is, the sampling rate must be such that the velocity of air relative to the probe tip is the same just inside the probe as it is outside. If these velocities are different, sampling is anisokinetic and will lead to a distortion in the size distribution of the sampled aerosol. If air is sampled subisokinetically, large particles will be oversampled relative to smaller particles; the smaller particles will follow the streamlines around the probe tip, while the large particles will still penetrate the inlet. If sampling is superisokinetic, smaller particles will be oversampled. This situation is illustrated in Figure 1. A second requirement is to move the aerosol particles from the probe tip to the sample collection point or to the inlet of an instrument without losses either by turbulent processes or inertial deposition at bends in the inlet tubing. These losses can be minimized by keeping the inlet lines as straight as possible and, where bends are required, by using a large radius of curvature.

A typical aircraft sampling probe consists of a conically shaped tip with sharp leading edges (Figure 2). The probe is shaped in this way to minimize distortion of the airstream by the nozzle at the point of sample entry into

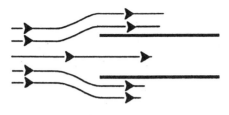

SUBISOKINETIC

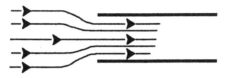

SUPERISOKINETIC

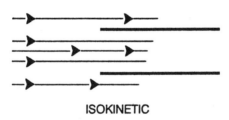

ISOKINETIC

Figure 1. Representation of an aircraft aerosol sampling inlet showing flow streamlines for subisokinetic, superisokinetic, and isokinetic sampling. Under subisokinetic conditions large particles will be oversampled; under superiso-kinetic conditions small particles will be oversampled; and under isokinetic conditions, the particle size distribution will be preserved.

the system. The diameter of the probe increases behind the nozzle (usually at an angle of about 7°) to decrease the linear flow rate, minimize losses at bends in the tubing, and bring the sample into the interior of the aircraft. The radius of curvature of required bends can be quite large, particularly if coarse-mode aerosol particles are to be sampled. For example, Sheridan and Zoller (6) used an inlet with a radius of curvature of ~75 cm to minimize losses of large particles by inertial deposition.

An alternative probe design using a rounded profile on the leading edge that followed criteria developed for aircraft engine intakes at low Mach

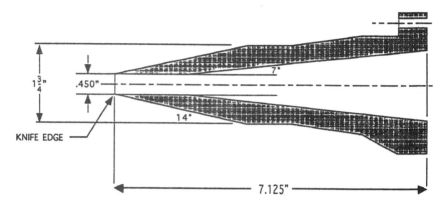

Figure 2. Diagram of a typical sharp-edged probe for isokinetic sampling of aerosol particles from an aircraft. The probe is sized to provide ~500 L/m at a sampling velocity of 50 m/s. The 7° expansion of the probe behind the probe tip is to slow the air velocity. The interior of the probe is polished to a mirror surface.

numbers has been applied to aerosol sampling from aircraft (7). This intake is reported to avoid distortion of the pressure field at the nozzle tip and the resulting problems associated with flow separation and turbulence. This design has been reported to be much more efficient at collecting coarse aerosol particles than isokinetic inlets with sharp leading edges, and presumably because of its shape the design is more forgiving of slight misalignment with the local airflow.

The ability of currently used aircraft probes to accurately sample aerosols has been questioned. Huebert et al. (8) conducted a comparative study of several different types of aerosol probes, all mounted on the same aircraft. The results suggested that substantial losses of particles occurred in all of the inlet systems. Because of the limited nature of the study, however, the causes of the aerosol losses could not be identified. The results of the Huebert study prompted a workshop to reexamine the entire issue of aerosol sampling from aircraft (9). An important conclusion of the workshop was that currently there is insufficient knowledge to "adequately describe important characteristics of airflow and particle trajectories at flight speeds" of aerosol sampling probes used on aircraft.

A second conclusion was that coarse-mode particles are typically undersampled by currently used aircraft probes and that although the sampling efficiencies for accumulation-mode particles seem higher, there is insufficient experimental data to assure that currently used sampling methods are adequate. It was also suggested that inlet performance is highly sensitive to both orientation with respect to the local airflow and to location on the airframe and that aerosol sampling problems are more severe at higher aircraft speeds.

The workshop report suggested that use of shrouds around sharp-edged inlets to align streamlines along the probe axis may be an effective means of reducing turbulent losses of aerosols at the probe tip. Such inlet systems have been used on conventional aircraft and helicopters (*9, 10*) but have not been adequately tested at typical aircraft sampling speeds. Double diffuser inlets in which the sample air velocity is decreased in two stages also seem promising.

Several types of inlets can be used to provide air for trace gas sampling. In many instances a forward facing "ram air" inlet is used (so called because the forward motion of the aircraft "rams" air into the tube). Ram air inlets do not sample isokinetically and are not suitable for aerosols, but this situation is generally not a problem for trace gases. Ram air inlets are usually vented to the aircraft cabin to avoid overpressurization of the instruments and consequent errors in instrument response. Venting also maintains high air velocities through the sample lines, minimizing sample losses and improving response speed. Vented inlet lines can only be used in unpressurized aircraft because pressurization may cause the airflow to reverse or to be insufficient to meet the needs of the instrumentation; the result would be sampling of cabin air. Reverse flow inlets, in which the opening of the inlet faces away from the airflow, are also common. Reverse flow inlets are frequently used to avoid intake of cloud droplets and precipitation into the sampling line. These lines cannot be vented because the venturi effect will cause sufficient negative pressure in the line to extract cabin air. The negative pressure generated by such lines can render certain instruments such as the sulfur flame photometric detector inoperable, because the reduced pressure extinguishes the flame.

Other more sophisticated sampling lines have also been used. Heikes et al. (*11*) used a constant pressure sampling line to eliminate the pressure-dependent response of a flame photometric detector used for SO_2 measurements. In this system the sampling line was connected to a high-capacity pump that evacuated the sampling line to a pressure equivalent to the highest altitude at which measurements were made. A feedback system consisting of a pressure sensor and fast-acting valve maintained the pressure at the set value as the altitude changed.

Regardless of which inlet system is chosen for trace gas sampling, minimization of wall losses during the sampling process is an important consideration. As a general rule, losses can be minimized by keeping the inlet volume flow as high as possible. A further consideration is constructing the inlet of materials that do not have a high affinity for the species that is being sampled. A common practice is to use various types of Teflon, either in the form of a tube that is inserted into a metal shell, together forming the sampling system, or by coating the Teflon directly on the tube. Whatever material is chosen for the sampling line should be rigorously tested for possible losses in the laboratory before installation on an aircraft.

An additional issue that seems obvious but that is frequently overlooked in installations is to ensure that the sampling line is not in a position to sample the aircraft engine exhaust or heater fumes and is not directly behind the propeller envelope. The sampling inlet should also be placed sufficiently far from the fuselage so that it is in the free airstream and not within the boundary layer of the aircraft.

Weight, Power, and Size Considerations. One of the axioms of aircraft sampling is that if an instrument has sufficient sensitivity and time response for the intended measurement, then it either weighs too much or draws too much power to be used on the available aircraft. Such considerations are important when designing an instrument for aircraft use or when preparing a commercial instrument for use on aircraft. One of the typical problems that is encountered in aircraft sampling is a limitation on the amount of power available. This problem is especially significant on small aircraft (where the power available may be as little as 3–4 kV • A) but is also a consideration on large aircraft. Normally, aircraft electrical systems provide 28-V DC and 400-Hz AC power. Unfortunately, most laboratory instruments operate on 110- or 220-V 60-Hz power. Thus, most aircraft set up for atmospheric research use 60-Hz 120-V AC inverters to meet the electrical requirements of research equipment. Because some power loss is involved in the inversion process, it is important when planning any aircraft installation to minimize the amount of 60-Hz power that is required. Because many measurement techniques (such as filter pack sampling systems or chemiluminescence detectors for oxides of nitrogen) use high-capacity pumps requiring large amounts of power, one way of maximizing efficiency is to drive these pumps with 28-V DC motors to avoid losses from the inversion process. This is common practice in aircraft installations. Another source of high power consumption is instrument heaters (such as those typically used on gas chromatographs), and it is important when using these systems to minimize the capacity of the heater that is required or to redesign the method or instrument so that heaters are not needed.

The electrical power available to equipment is subject to interruptions. These outages may be predictable, such as for takeoffs and landings in some aircraft, or random because of circuit overloads or pilot discretion. Instrument systems must often operate without lengthy warm-up periods, both at the beginning of a flight and after in-flight power interruptions. Data systems must be sufficiently robust to recover data taken before unexpected power outages. Battery backup for the data system is one alternative but greatly adds to the weight. A better solution is to write all data to disk or tape backup as they are recorded. The data processing program must be able to recover data streams that terminate unexpectedly.

Weight and size of instruments are also important considerations. The first (and most obvious) consideration is that the sampling system cannot exceed the payload of the available aircraft. For many aircraft, particularly small twin-engine aircraft, the payload can be quite small (<250 kg), and when designing a sampling system it is important to minimize weight wherever possible. Even for large aircraft, weight is a consideration because there is a trade-off between instrument payload and range. While this may not be an important issue for local sampling of short duration, the payload–fuel trade-off may be the limiting factor for long-range flights for sampling in remote regions.

Equipment installations must meet the center of gravity (CG) requirements of the aircraft. Every aircraft has a range of CGs that can be tolerated, thus an instrument, particularly a large one, cannot be installed at an arbitrary position within the aircraft. As a first-order approximation, installation of the heaviest equipment (e.g., large pumps) should be planned to be near the normal CG of the aircraft. Center of gravity considerations can be quite critical for small aircraft and somewhat less so for larger aircraft.

The influence of vibration on instrument operation must also be considered in aircraft installation. If problems arise, they can usually be solved by using standard vibrational analysis techniques to select appropriate devices (such as dashpots) to isolate instruments from the airframe. Vibrational problems with instruments incorporating precisely aligned optical components may require more complex solutions, including redesign or remounting of some of the critical components in the instrument. Electrical noise, arising from aircraft radio transmissions, is also a frequent problem. It is important to identify this problem early in an installation so that sensitive instruments can be mounted at a location where the potential for these interferences is minimal (away from radios and antennas). It is also good practice to shield all data transmission leads. Installation of instruments on aircraft may also require modification to strengthen the equipment to the point where it can withstand the load requirements imposed by applicable Federal Aviation Administration regulations.

Review of Techniques for Measurement of Trace Gases and Aerosols from Aircraft

Techniques used in aircraft sampling can be divided into categories: those providing concentration information on a continuous basis and those providing such information on a sporadic basis. The latter techniques can be further divided into techniques wherein the sample is collected by some means and is returned to the laboratory for subsequent analysis and techniques wherein the sample is collected and analyzed on the aircraft.

It is important to measure continuously as many species as possible to achieve the maximum possible time–space resolution. However, the complexity of instrumentation required for measurement of certain classes of compounds may be such that aircraft installation is not feasible. Alternatively, the analytical procedure may be so time consuming that it is not worthwhile to install the equipment on the aircraft. In these cases it will be necessary to collect samples for subsequent analysis in the laboratory. There are two fundamental problems with batch sampling techniques: time–space resolution is limited and batch sampling requires that the composition of sample not change between the time of collection and the time of analysis. Analysis of collected samples in a laboratory setting, however, allows application of a broad array of techniques that would be impossible to operate in an aircraft environment.

In selecting instruments for installation on an aircraft it should be noted that commercially available instrumentation should not be used for aircraft sampling until the operation of the instrument is understood in that environment. An equally important consideration is that many commercial instruments have been designed for the purposes of determining whether pollutant concentrations exceed environmental standards. Because these concentrations are frequently quite high, such instruments may not be able to quantify abundances of species at the low levels significant for an understanding of atmospheric processes, particularly those occurring in the remote or upper troposphere. For these reasons, significant effort has been devoted by atmospheric chemists over the last several decades to developing techniques that can measure low concentrations, and to a lesser extent to adapting these techniques for aircraft use.

The following section is a review of instruments and techniques that have been used and/or specifically developed for aircraft measurement of trace species. To limit the scope, focus is restricted to species that we have identified to be of interest to a large segment of the atmospheric chemical community. The objective is to give the reader a starting point from which to either choose or develop an instrument or technique for aircraft sampling. This section is divided into two parts: the first covers techniques that are used for batch sampling or analysis of atmospheric constituents; the second is a review of continuous analysis methods.

Systems for Batch Sampling. *Filter Packs.* One of the most reliable and adaptable methods for measuring concentrations of atmospheric trace gases and aerosols is the filter pack sampling system. Indeed, collection by filter is one of the few methods available for obtaining comprehensive information on aerosol composition. In filter methods, sample air is passed through a sequence of filter elements, each of which is designed to collect a different species. The various collection elements are designed and positioned in the filter pack assembly so that each element selectively collects the desired components at high efficiency and passes components to be collected on downstream filters with equally high efficiency. The exposed filters are analyzed in the laboratory to determine the mass of the collected substances. Atmospheric concentrations are then computed from the mass of material collected on the filter and the sample air volume.

A number of sequential multistage filter pack sampling systems have been used in aircraft. Because many of these systems are similar, it is useful to provide a general description pointing out different approaches for measurement of various species. The first stage in all filter pack systems is a particle filter for collecting aerosol. Several types of materials are used; the most common are made of treated quartz (12) or Teflon (6, 13, 14). Size-segregated sampling of the aerosol has been accomplished by use of two filters of different pore size arranged sequentially. John et al. (15) described a system in which coarse (>1.0 μm) and fine (<1.0 μm) fractions are collected with an 8-μm (pore size) Nuclepore filter, followed by a 2-μm (pore size) Zefluor Teflon filter. This system has been used in several aircraft sampling programs (e.g., references 7 and 16). Subsequent to exposure, the filters are returned to the laboratory for analysis of the collected materials. Usually, aerosol filters are extracted with water and the extract is analyzed for various soluble materials, commonly strong acid, Na^+, NH_4^+, Ca^{2+}, Mg^{2+}, NO_3^-, SO_4^{2-}, or Cl^-. Representative examples of the various extraction and analytical procedures have been published (13, 17).

Frequently, the second element in a sequential system is used to collect gaseous HNO_3. Both NaCl-impregnated filters (18) and nylon filters (13, 19) have been used for this purpose. Both of these filters have been shown to exhibit high collection efficiency (>90%) for nitric acid. Nylon filters have an advantage in that they generally require no pretreatment or impregnation before use, in contrast to the NaCl cellulose filters, which require both. However, nylon filters can exhibit high and variable blanks that may compromise their use in sampling regimes where low HNO_3 concentrations are expected. The third element in a sequential filter system may be a base-treated cellulose filter for collection of SO_2 or an impregnated filter for collection of gaseous NH_3. SO_2 filters are usually pre-washed cellulose filters impregnated with an aqueous solution of K_2CO_3 with glycerol added as a humectant. Daum and Leahy (17) and Anlauf et al. (13) give preparation procedures that call for an impregnating solution containing 250 g of K_2CO_3 and 100 mL of glycerol per liter. It has been reported that use of a less concentrated solution leads to lower detection limits for SO_2 when ion chromatography is used as the method for analyzing the filter extract (20). Filters for collecting gaseous ammonia have been fabricated from either cellulose filters impregnated with an aqueous solution of citric acid and glycerol (21) or with oxalic acid in ethanol (22).

The detection limit for measuring a species by filter collection and subsequent analysis is determined either by the uncertainty in the filter blank, that is, the variability in the amount of a species in an unexposed filter, or by the limit of detection imposed by the analytical method. Most frequently, the limit of detection is determined by blank variability, thus care in preparation and handling of filters is important when using such systems for aircraft sampling, particularly if sampling is conducted in remote areas where concentrations are low. In this regard, it is crucial to establish the blank level or residual material (and its variability) of unexposed filters. The blank level may be established by carrying extra filters on each sampling mission and exposing them to all of the deployment procedures with the exception of actual air sampling. It has been our practice to carry at least one such filter on every flight.

The time–space resolution that may be achieved with filter sampling techniques is dependent on the collection rate, limit of detection, and ambient concentrations. In aircraft applications, filters are typically operated at high flows (100–500 L/min) to maximize the mass accumulation rate. At these flow rates, sampling times on the order of 20 to 30 min are generally sufficient for measurement of substances in the urban troposphere. For sampling in the upper troposphere or in areas remote from pollutant sources, collection times of several hours may be necessary to obtain measurable quantities of material.

One of the important concerns in the application of filter methodology to aircraft sampling is generation of artifacts, particularly as they pertain to measurement of aerosol nitrate, nitric acid, and ammonia. These species are in dynamic equilibrium, and under conditions where HNO_3 and NH_3 concentrations are sufficiently high to form the solid NH_4NO_3, the possibility of evaporation of this material followed by collection of artifact HNO_3 and NH_3 should be considered (for a recent discussion of these artifacts, see the articles in reference 23). Similarly, artifact gaseous HCl may be formed by interaction of HNO_3 or H_2SO_4 with NaCl particles previously collected on an aerosol filter. Such artifacts can be minimized in the first instance by restricting collection of samples to a single air mass so that the collected sample is not exposed to air that has significantly different concentrations of HNO_3 or NH_3. In the second instance, artifact formation can be minimized by keeping filter loading as small as possible, consistent with the limit of detection requirements.

Cloud Water and Precipitation Collectors. Several methods have been developed for collecting cloud water samples (*24–26*). Probably the device most commonly used in warm clouds is the slotted rod collector developed by the Atmospheric Science Research Center at the State University of New York (SUNY) at Albany. Commonly known as the ASRC collector (*25*), this collector consists of an array of rods constructed from Delrin (a form of nylon). Each rod is hollow and has a slot located at its forward stagnation line. The rod radius determines the collection efficiency as a function of particle size, the rods are sized to collect cloud droplets but not submicrometer aerosol particles, and the 50% cutoff is calculated to be at about 3 μm.

The original design of the collector was refined several times to enhance collection efficiency and to more firmly establish the operational characteristics of the collector. Major modifications have included lengthening the collector rods to give a larger collection area, sheathing the rods in steel to prevent bending and the consequent dropping of collection efficiency at higher airspeeds, and constructing the rods of Teflon rather than Delrin to reduce the possibility of losses of sample constituents (particularly nitrate) during the collection process (27). Further modifications include mounting the rods in a single row, not the staggered multirow configuration of the original collector, to prevent interaction of the aerodynamic flow around the rods (28). With this modified collector, collection efficiencies upwards of 80% have been measured on the basis of the collection area of the rods, the cloud liquid water content, and the mass of collected water.

Mounting of the cloud water collectors on the aircraft is a critical issue because flow-field effects can easily distort the size distribution of drops. If at all possible, the collector should be mounted on a pylon so that the collector is in the free airstream. Substantially greater efficiencies can be achieved if the collector is mounted with a forward inclination of about 12° to 15° relative to a perpendicular from the aircraft longitudinal centerline. This kind of mounting accounts for the nose-up attitude at which most aircraft fly under cruise conditions and also provides a component of the airstream to drive impacted cloud droplets down the rod into the collection vessel, minimizing losses due to blow-off (28).

Supercooled cloud droplets can be collected as ice through the riming process. Isaac and Daum (29) report the use of several different types of collectors. Collection of warm precipitation from aircraft has generally been accomplished with a SUNY collector consisting of a single large-diameter (~8 cm) slotted rod (~2-cm slot width) (V. Mohnen, private communication). This collector has never been subjected to realistic laboratory or field testing, so neither actual size distributions collected nor collection efficiency has been determined. Collection of snow from aircraft has been attempted, but with limited success. The most promising approach is the use of a cyclone wherein the air is required to follow a circular path and the snow particles are centrifuged out of the air in which they are suspended. The snow particles settle to the bottom of the cyclone and can be drawn off into a collection bottle for analysis. Several different types of cyclones have been used to collect snow from aircraft (29).

Gas Samples for Subsequent Laboratory Analysis.   Collection of air samples for later analysis in the laboratory is a common technique used for aircraft sampling. Whole air sampling for stable compounds (CO, CO_2, halocarbons, and low-molecular-weight hydrocarbons) is usually accomplished by filling a container to a pressure of about 2 atm (203 kPa) with a metal bellows pump. Alternatively, containers may be evacuated in the laboratory to a low pressure and filled during flight by simply opening a valve at the appropriate time. Containers are typically constructed of stainless steel that has been electropolished or treated in some way to reduce surface activity

(30–34). Samples have also been collected in glass flasks (35) and in bags fabricated from inert materials (36, 37). An alternative whole air sampling method involves cryogenic collection by submerging flasks in liquid nitrogen while pressurizing the container with sample air (38).

An important concern in applying any of these techniques is losses from wall reactions or other processes either during pumping or in the container during the time between sampling and analysis. Apparently, stability is not altogether predictable. Oliver et al. (33) reported sample composition to be unaffected by storage in steel containers for up to 30 days in the presence of water, CO_2, NO_x, and O_3. Similarly, Ehhalt et al. (32) reported no measurable loss of light hydrocarbons over periods of months. However, other studies have indicated the possibility of substantial losses (30). Materials that have been collected by cryosampling techniques may experience losses during the analysis when the collected sample is heated. The basic problems of collecting gaseous samples for laboratory analysis have been outlined elsewhere (39).

Other Batch Sampling Techniques. Solid adsorbents, such as Tenax or Poropak–Q, have been used to sample higher molecular weight hydrocarbons (more than eight carbons) (40–42). Tenax does not trap water or permanent gases, so the subsequent gas chromatographic analysis is simplified. Thermal desorption of these traps can produce artifacts arising from degradation of the sample or the trap. The presence of O_3 can further reduce sample integrity (43).

Samples of OCS and CS_2 have been collected by passing air through coils of Teflon tubing submerged in liquid argon (44). Collecting the sample in this way concentrated the species of interest. However, blockage of the coil by ice formed by condensation of ambient water vapor limited the sampled air volume to ~600 mL. The coils were kept refrigerated by the cryogen prior to analysis. Gold wool has been used to trap dimethyl sulfoxide after SO_2 is first removed (45). The sample is later thermally desorbed and analyzed by gas chromatography with flame photometric detection.

Chromatographic Techniques. These techniques have long been applied to the problems of separation and analysis of trace atmospheric species. For stable species, batch samples are usually collected as described in the preceding section and transported to the laboratory for subsequent analysis. However, some compounds are not sufficiently stable to survive transport intact. In situ chromatographic analyses have been used for these samples. Usually, chromatography is used on aircraft in a batch mode: samples are collected, preconcentrated, and separated on a column, and the individual species are detected as they elute; the process is then repeated for the next sample. Thus, as with other batch techniques, time resolution is limited.

Packed column technology has been used in airborne gas chromatographs for the separation and quantitation of sulfur species (*46, 47*) and peroxyacetic nitric anhydride (*48*). The combination of sample preconcentration and sensitive detectors has yielded detection limits that are superior to corresponding continuous sensors. For SO_2, a detection limit of 25 pptrv was claimed, and for peroxyacetic nitric anhydride the detection limit was roughly 60 pptrv for an ~50-cm³ air sample. Analysis times for samples were on the order of 10 min.

Several engineering factors have discouraged more frequent utilization of gas chromatographs aboard aircraft. Laboratory research instruments are large, heavy, and have power requirements that exceed the capabilities of most aircraft. However, smaller, high-quality gas chromatographs are now commercially available and can be easily modified to reduce power consumption.

Some altitude effects on the operation of chromatographic instruments are anticipated. To achieve reproducible retention times for identifying compounds, mobile-phase flows need to be controlled so that they are independent of ambient pressure. Detectors may also respond to changes in pressure. For example, the electron capture detector is a concentration-sensitive sensor and exhibits diminished signal as the pressure decreases. Other detectors, such as the flame ionization detector, respond to the mass of the sample and are insensitive to altitude as long as the mass flow is controlled.

Although the modification of existing gas chromatographs for flight is not trivial, no fundamental problems prevent the use of this separation technique on aircraft. The presence of a chromatographic system aboard the Viking mission to Mars confirms that chromatographic instruments can operate within stringent weight, power, and size limitations in an inhospitable environment (*49*). As the atmospheric community expands its interest to include a larger number of species, chromatographic techniques should become more widely used on aircraft. Analysis times can, in principle, be decreased to 1 or 2 min (*50*), a time response that is comparable to some of the slower continuous sensors. Furthermore, implementation of capillary columns, such as those used in laboratory analyses of air samples (*31*), will greatly enhance the scope of measurements that can be made during an aircraft sampling program.

Continuous Measurement Methods for Trace Gases and Aerosols.   *Ozone.*   Three basic types of ozone instruments have been used in aircraft: the ultraviolet photometric method and two chemiluminescent techniques measuring, respectively, light emitted from the reaction of O_3 with ethylene and light emitted from the reaction of O_3 with NO. Ultraviolet absorption photometry is one of the preferred methods for measuring O_3 from aircraft because of the stability and reliability of commercially available instruments. The method is specific for O_3 provided there are no immediate sources of aromatic hydrocarbons (51). However, because of the small absorption coefficient of O_3 at the 254-nm measurement wavelength, commercially available detectors have time responses between 30 and 60 s, and fine features of the O_3 distribution cannot be resolved.

The O_3–ethylene chemiluminescent instrument measures the luminescent signal at 435 nm resulting from the reaction of a large excess of ethylene with O_3. The chemiluminescent reaction is second order, and, to achieve a response that is rapid and linear with changes in the O_3 mixing ratio, the ethylene concentration is maintained at a high value, typically by adjusting the pressure across a flow restrictor. Sample flow is controlled by a flow restrictor downstream of the reaction chamber. Because the detector is inherently a mass-sensitive detector, commercially available detectors can be modified to respond to the O_3 mixing ratio independent of altitude by use of mass-flow controllers (see also the preceding discussion in Altitude Response). Such modifications are described by Gregory et al. (3). Ethylene chemiluminescent detectors exhibit response times that are considerably less (typically, $1/e$ response times of less than 3 s; $1/e$ response time is the time required for a 63% response to a step change in concentration) than those of UV photometric instruments, and they are thus more broadly applicable to aircraft measurements.

The final O_3 measurement technique discussed here is based on the O_3–NO chemiluminescent reaction. Essentially, this instrument is an NO–O_3 chemiluminescent detector for determination of NO operated in reverse. That is, a high concentration of NO is used to measure O_3 as opposed to a high concentration of O_3 being used to measure NO. Because of the kinetics of the ozone–NO reaction, this detector responds faster than the ethylene-based detector, and $1/e$ response times less than 0.1 s have been reported. Design and operating characteristics of an instrument capable of providing eddy correlation measurements of O_3 flux from an aircraft are given by Pearson and Stedman (52). A revised version of this instrument that uses considerably less power and thus places fewer constraints on its use in aircraft was reported (53). The instrument uses a smaller reaction chamber than the Pearson–Stedman instrument, requires less power, and generates less heat. The detection limit of this instrument is reported to be 1–2 ppbv, and it has a $1/e$ response time of 0.45 s and a precision of about 2% or 2 ppbv. This instrument has been used for airborne O_3 flux measurements in the Amazon (54).

Sulfur Dioxide. Both flame photometric and pulsed fluorescence methods have been applied to the continuous measurement of SO_2 from aircraft. In the flame photometric detector (FPD), sulfur compounds are reduced in a hydrogen-rich flame to the S_2 dimer. The emission resulting from the transition of the thermally excited dimer to its ground state at 394 nm is measured by using a narrow band-pass filter and a photomultiplier tube.

The response of the instrument is roughly proportional to the square of the sulfur concentration. Two major problems are associated with the use of commercially available flame photometric detectors on aircraft. First, major changes in background current and more modest changes in detector sensitivity occur as the inlet pressure changes. Second, commercial detectors exhibit detection limits (greater than 1 ppbv, depending on make and model) that are too high to allow the measurement of ambient concentrations away from SO_2 source regions.

These problems have been largely eliminated by modifying commercial detectors with the addition of mass-flow controllers to maintain constant flows of sample air and hydrogen as the altitude changes and by using SF_6-doped hydrogen to enhance sensitivity (4, 55, 56). Controlling the mass flow holds the burner H_2-to-O_2 ratio at a value independent of altitude and fixes the sulfur mass flow for a given atmospheric sulfur mixing ratio, greatly stabilizing instrument response. The use of H_2 doped with 60-ppbv SF_6 decreases the limit of detection. Because the atmospheric signal is measured on top of a 60-ppbv background, the response is linear up to ~15-ppbv concentrations of atmospheric SO_2, even though the FPD is inherently nonlinear. The limit of detection for aircraft measurements with this modified detector have been reported to be as low as ~0.1 ppbv and up to ~0.3 ppbv under turbulent atmospheric conditions (57).

Because the FPD responds to both aerosol and gaseous sulfur species, it has also been possible to modify these instruments to continuously measure aerosol sulfur by selectively removing gaseous sulfur compounds with a lead(II) oxide–glycerol coated denuder (55). Use of such an instrument for airborne measurements of aerosol sulfur in and around broken clouds has been reported (57). In principle, speciation between aerosol sulfate, disulfate, and sulfuric acid by selective thermal decomposition (58, 59) can also be achieved. Flame photometric detectors have also been used as selective detectors for gas chromatography. Thornton and Bandy (60) reported the use of a chromatographic system with a flame photometric detector for airborne measurement of SO_2 and OCS with a detection limit of 25 pptrv.

The pulsed fluorescence method operates by optically stimulating SO_2 molecules with a UV source (190–230 nm) and measuring the resulting fluorescence. The excitation source is pulsed to achieve high excitation intensity and to encode the concentration information in the form of a chopped signal. By chopping the signal, dark current drift of the photomultiplier tube is eliminated. A scrubber is used to remove hydrocarbon interferences that fluoresce at similar wavelengths. A commercially available version of this detector (Thermo Environmental 43S, Hopkington, Massachusetts) has a detection limit better than 0.3 ppbv and a time constant of roughly 2 min. This system has several advantages over the FPD for aircraft use: there are no consumable gases as with the FPD, the detector is simple to operate, it exhibits nominal zero shift with altitude, and the output is easily converted to mixing ratios. For these reasons, the pulsed fluorescence detector has largely supplanted the flame photometric detector for aircraft measurement of SO_2 in the relatively polluted lower troposphere. The relatively slow response of commercial instruments is not optimal for aircraft sampling but can be improved at the expense of sensitivity. The instrument does not, however, have sufficient sensitivity for measurements in remote regions or in the upper troposphere.

A third measurement technique for SO_2 that may have future application in aircraft measurements was described (61). In this method gaseous SO_2 is removed from the airstream with a diffusion scrubber containing 1 mM H_2O_2. The scrubber eluant is analyzed in flight for SO_4^{2-} by ion chromatography. Because H_2O_2 quantitatively converts SO_2 to sulfate in the scrubber, the SO_4^{2-} concentration is proportional to the gaseous SO_2 concentration for a given set of sampling conditions. SO_2 was measured at concentration levels of 20 pptrv with 6-min time resolution. The diffusion scrubber method was compared with the pulsed fluorescence and filter techniques in airborne tests (20). All three techniques operated with no difficulty in both pressurized and unpressurized aircraft and were shown to give equivalent response to ambient SO_2 mixing ratios, although there was significant scatter below 200 pptrv.

Oxides of Nitrogen. Two highly sensitive techniques for determining odd-nitrogen compounds have been developed that are based on chemiluminescence reactions with either NO or NO_2. The first group of detectors measures the chemiluminescence resulting from the reaction of NO with O_3. Three classes of instruments based on this reaction can be considered for aircraft use. The first class is commercial instruments that typically have detection limits on the order of 1–10 ppbv. These instruments are not particularly suitable for aircraft-based monitoring of ambient air outside of urban areas because of their limited sensitivity. Relatively minor modifications were described that can improve the sensitivity of these instruments by ten- to a hundredfold, thus they are suitable for monitoring continental air (62–64). This intermediate class of instruments is capable of detection limits on the order of 0.1 ppbv. A third class of instruments has been developed for measuring concentrations in the low parts per trillion range. These instruments typically use high-intensity ozone sources, large vacuum pumps for high sample throughput, cryogenically cooled photomultiplier tubes to reduce background noise, and photon counting electronics to improve measurement statistics. These detectors are custom-made (for example, see reference 65) and have been extensively compared in flight to sensors based on other techniques (48, 66, 67).

Various approaches have been described to convert other nitrogen species to NO before detection by the ozone chemiluminescence reaction. The most common converter is hot molybdenum, which converts all of the higher oxides of nitrogen, including NO_2, HNO_3, peroxyacetic nitric anhydride, N_2O_5, NO_3, and many organic nitrogen species, to NO (collectively these species are referred to as NO_y). Heated gold catalysts (68) have also been used for determinations of NO_y. The conversion efficiency of these hot converters is usually close to 1 but must be periodically checked. Selective photolytic conversion of NO_2 to NO by irradiation with UV light has been used to measure NO_x (the sum of NO_2 and NO). NO_2 is then determined by the difference (69). The conversion efficiency of NO_2 to NO is less than one and must be measured by performing a gas-phase titration of standard NO mixtures with O_3. NO chemiluminescence detectors generally do not use expendables but require large pumps to achieve high sample throughput. This requirement imposes considerable weight and power restrictions on an aircraft.

NO_2 has also been selectively determined based on its chemiluminescent reaction with luminol (LMA-3, Scintrex/Unisearch, Concord, Ontario). Incoming air impinges on a wick containing a proprietary luminol solution. Emitted light is then detected by a photomultiplier tube. The instrument is mechanically simple, lightweight, and compact, and it consumes little power so it is practical for use in aircraft. The luminol reaction is specific for NO_2; however, NO can be determined by oxidizing it to NO_2. Drawbacks associated with the luminol chemiluminescent detector include nonlinear response below a few parts per billion, a minor interference from ozone, and partial response from peroxyacetic nitric anhydride. These problems can be significant when low concentrations (<3 ppbv) are measured. However, correction factors can be applied to account for their effect. Evaluation of the use of this instrument for aircraft sampling, including modifications to eliminate pressure dependence, was reported (70). A commercial instrument (LPA-4, Scintrex/Unisearch, Concord, Ontario) is also available to determine peroxyacetic nitric anhydride by catalytic conversion to NO_2 and detection by means of luminol chemiluminescence. In general, though, the low concentrations and wide assortment of individual organic odd-nitrogen species require sample preconcentration and subsequent analysis by gas chromatography. Some of these approaches are discussed in the previous section Chromatographic Techniques.

Tunable diode laser absorption spectroscopy (TDLAS) has been used to measure oxides of nitrogen during flight (71). By tuning the laser to specific infrared absorption bands, the technique can selectively measure each compound. Detection limits are higher (25–100 pptrv for a 3-min response time) than the best chemiluminescent methods, and the instrumentation is less amenable to aircraft operations than the chemiluminescence techniques because of weight and size.

A sensor based on two-photon laser-induced fluorescence has been described for detection of $NO-NO_x-NO_2$ (72). NO is detected directly on the basis of its fluorescent properties. NO_2 is first converted to NO by photofragmentation at 353 nm with a XeF excimer laser. Detection limits approach the low parts per trillion level for 2-min integration times. This technique is still in the developmental stage but holds promise for future aircraft measurements.

Carbon Monoxide. Methods for determining carbon monoxide include detection by conversion to mercury vapor, gas filter correlation spectrometry, TDLAS, and grab sampling followed by gas chromatograph (GC) analysis. The quantitative liberation of mercury vapor from mercury oxide by CO has been used to measure CO (73). The mercury vapor concentration is then measured by flameless atomic absorption spectrometry. A detection limit of 0.1 ppbv was reported for a 30-s response time. Accuracy was reported to be ±3% at tropospheric mixing ratios. A commercial instrument providing similar performance is available.

GFC spectrometry is based on measurement of the absorption of broadband infrared radiation by carbon monoxide. Specificity for CO is achieved by alternating the infrared beam between wavelengths that are absorbed and transmitted by CO. In practice, this process is done by chopping the light with a gas filter containing CO. When the filter is in place, all of the absorption bands are saturated and the beam cannot be further attenuated by CO in the sample cell. This signal serves as a reference. Absorption of the unfiltered light by CO in the cell is measured relative to the reference beam to determine the CO concentration. A multipass White cell increases sensitivity. Detectors based on this technology are commercially available but lack sufficient sensitivity for aircraft measurements. Modifications of a commercial GFC detector to improve its sensitivity were described (74). This instrument was used in several aircraft sampling programs (75, 76).

TDLAS measures the absorption of monochromatic light by CO in a multipass flow cell (77). The time response is controlled by the flow of air through the cell. In theory, the instrument is fast enough for application to aircraft measurement of CO flux by eddy correlation. A reported precision of ~1 ppbv or ~1% is superior to other techniques. The instrument is not commercially available.

Grab samples obtained during flight can be subsequently analyzed by GC. Because CO is stable in passivated containers, this approach is straightforward. Discrete sampling limits the resolution of rapid changes in concentration. The TDLAS and two different GC trapping methods have been evaluated during an airborne intercomparison (78).

Gaseous Hydrogen Peroxide. Methods for determination of gas-phase hydrogen peroxide have been reviewed (79–81). Hydrogen peroxide is determined by either scrubbing air with an aqueous solution and measuring the resultant liquid-phase peroxide or by measuring the peroxide directly with a spectroscopic technique. The method that has been most commonly used for aircraft sampling is the POHPAA (p-hydroxyphenylacetic acid) technique developed by Lazrus and co-workers (82, 83). In the POHPAA technique, peroxides are removed from the air with a scrubber coil and subsequently undergo the horseradish peroxidase catalyzed reaction with POHPAA to form a fluorescent dimer species. The fluorescent intensity of the dimer is measured and is directly proportional the peroxide concentration in the air. Because the method is sensitive to all peroxides, a second channel is used in which H_2O_2 is selectively destroyed by catalase; H_2O_2 is computed from the difference. The accuracy and precision with which H_2O_2 can be determined with this technique is critically dependent on the catalase activity. Thus, the catalase activity must be periodically monitored and adjusted to assure that it is at an appropriate level to minimize errors in the determination of H_2O_2. The method is reported by various aircraft users (reference 84 and references therein) to have a detection limit of less than 0.1 ppbv and a time delay of ~3 min, although the time response is considerably faster, ~1 min. A nonenzymatic method has been developed that avoids the catalase activity problem in distinguishing H_2O_2 from total peroxide (85).

Particle Measurements. A variety of instruments is available for measuring the number density and size distribution of particles sampled from airborne platforms. This discussion is restricted to instruments that measure particles smaller than 50 μm (cloud droplets and aerosol particles) because these particles are of most interest to atmospheric chemists.

Particle Measurement Systems (PMS; Boulder, Colorado) has developed a series of probes for aircraft measurement of atmospheric particles. Several of these probes are discussed here. The airborne Active Scattering Aerosol Spectrometer Probe (ASASP–100X) measures the number concentration and size distribution of particles in the (approximate) size range 0.12 to 3.12 μm by measuring the light scattered when the particles pass through the active cavity of a laser. In this probe, particles are forced into a conical deceleration chamber by the motion of the aircraft. A pump pulls about 1 cm³/s of this air into the detection region through a narrow nozzle that aerodynamically focuses the particle stream to a diameter (150 μm) that is small with respect to the diameter of the laser beam. During transit of the laser beam, particles scatter radiant energy, which is sensed by a photodetector. The output of the photodetector is a series of pulses (one for each particle) with sizes proportional to the scattering intensity. A pulse height analyzer groups the pulses into counts by size; the ASASP uses a pulse height analyzer with 15 channels. Because scattering intensity is proportional to particle size, the count in each channel represents the number of particles within a specified size range. A newer version of this probe, the Passive Cavity Aerosol Spectrometer Probe (PCASP–100X) has been introduced. It utilizes the same sampling techniques as the ASASP, but the sensing area has been moved external to the laser and the laser has been changed to improve sensitivity and stability.

Scattering intensity measured by the pulse height analyzer is related to particle size by calibration with monodisperse latex spheres or nearly monodisperse NaCl particles. Calibration uncertainties have been studied and discussed (86–91). These studies show that the smallest particles that can be sensed by the ASASP probe are somewhat larger than the 0.12 μm stated by the manufacturer. Similarly, it is reported that detection of particles larger than about 2 μm is unreliable because of attenuation of the laser power.

Distortion of the particle size during the sampling process is a concern in the use of this probe on an aircraft. Compressional heating due to deceleration of the particles may distort the size distribution, because evaporation of water from aerosol particles reduces their diameters. Likewise, particle sizes can be reduced by use of a heater, incorporated into some models of this probe, to prevent icing when supercooled clouds are being flown through. One study (88) indicated that the probe heater removes most of the water from aerosol particles sampled at relative humidities of 95%. Thus, size distributions of aerosol particles measured with the probe heater on correspond to that of the dehydrated aerosol. These results were confirmed by a later study (90) in which size distributions of aerosols measured with a nonintrusive probe were compared to size distributions measured with a de-iced PCASP probe. Measurement of the aerosol size distribution with the probe heater on may be an advantage in certain studies.

The PMS Forward Scattering Cloud Droplet Spectrometer Probe (FSSP–100) determines particle size by measurement of the scattering intensity over a prescribed range of angles during particle interaction with a focused laser beam. In a way similar to the operation of the ASASP probe, scattered light pulses are measured by a photosensor and sized with a pulse height analyzer; the height of the pulses is proportional to the particle size. The FSSP can measure particles with diameters between 0.5 and 47 μm in four selectable size ranges. Each range divides particles into 15 equally spaced size classes. The response of the FSSP is a function of particle shape, refractive index, and absorption as well as size. Furthermore, the response function is multivalued because scattered intensity is not a smooth function of the particle diameter. This situation can lead to errors in determination of the size distribution of both aerosol particles and cloud droplets (86, 91). Extensive information regarding calibration and operation of this probe has been published (91–97).

A newer version of the FSSP–100, the FSSP–300, is currently available. The probe is similar to the FSSP–100 except that the electronics are faster and it uses a different laser and slit arrangement. The major difference is that it is configured to measure particles in the size range ∼0.3 to 20 μm in 31 separate size channels (compared to the FSSP–100's 15 size channels).

The additional channels in the lower portion of the size spectrum overlap the sizes measured by the ASASP and the PCASP probes and have allowed intercomparisons between aerosol size distributions measured by these two probes (90). Dye et al. have evaluated the capabilities of this probe for measuring the number density and size distribution of atmospheric aerosols (98).

The integrating nephelometer (99, 100) measures the extinction coefficient, b_{scat}, due to light scattering from gases and aerosols. A sample cell is illuminated with a diffuse source, and the scattered light is measured in both the forward- and backscattered directions with a photodetector. The geometry of the system physically integrates the intensity of scattered light and provides a close approximation to the scattering component of extinction. Correction for Rayleigh scattering by gases is done by filtering particles from the incoming airstream. Calibrations are performed by measuring the extinction of pure gases such as carbon dioxide and dichlorodifluoromethane (Freon).

The integrating nephelometer has been used on aircraft for several purposes. The most common uses are to measure visual range in atmospheric haze studies or to locate atmospheric boundaries such as temperature inversions by noting discontinuities in b_{scat}. However, submicrometer aerosol mass can also be estimated from b_{scat} measurements because the ratio of light scattering coefficient to aerosol particle volume is approximately independent of the size distribution of the aerosol for particle radii comparable to the wavelength of light used (101–103). It has also been noted that b_{scat} can serve as a surrogate measurement of sulfate in areas where sulfate and associated species constitute a fairly uniform fraction by volume of the submicrometer aerosol (reference 56 and references cited therein). In these applications, the relationship between b_{scat} and the mass of an aerosol constituent holds only to the extent that the amount of water associated with the aerosol particle is constant. Thus, it is common practice to use a preheater in the b_{scat} instrument to bring particles to a reference relative humidity. The application of b_{scat} for airborne study of the distribution of sulfate in broken clouds has been described (56).

Condensation nuclei counters (CNCs) measure the concentration of particles in the size range 0.001 to 0.1 μm. They operate by exposing a sample of ambient air to a high supersaturation by humidifying the incoming air and then abruptly reducing the pressure to cause adiabatic cooling. Supersaturations on the order of 400% are achieved and are sufficient to cause even the smallest particles to activate and grow. These larger particles are then usually detected optically by light scattering (104). Because all droplets are, in theory, approximately the same size at a given time following the expansion, the scattered light intensity is directly proportional to the concentration of condensation nuclei.

Adaptation of a CNC for aircraft use is straightforward. Unlike larger particles, condensation nuclei are small enough that they are not easily deposited to walls by inertial processes, so the physical shape of the inlet is not a significant issue. However, metal rather than plastic tubing should be used for sampling lines to avoid electrostatic precipitation of particles (*105*). Particle concentrations ranging from ~1 to 300,000 cm^{-3} can be measured with a CNC; however, to assure reliable measurements, calibrations should be performed over the range of particle concentrations expected (*106*). CNC counting efficiency decreases rapidly for particles smaller than 0.01 μm (*107*).

Cloud Liquid Water Content. Three different methods have been used to measure cloud liquid water content from aircraft. The most commonly used instrument is the Johnson–Williams detector (Cloud Technology Inc., Palo Alto, California). This device uses an electrically heated resistance wire mounted perpendicular to the airstream to sense the water content. As cloud droplets strike the wire, they evaporate, cooling the wire and causing an imbalance in a bridge circuit, of which the wire is one arm. The magnitude of the imbalance is a function of the cloud liquid water content. Another wire, mounted parallel to the airstream so that it is not subject to water drop impingement, serves to compensate for variations in airspeed, altitude, and air temperature. Although the operating principal of the Johnson–Williams detector is simple, the device requires calibration to yield accurate results (*108*). The device begins to underestimate the mass of droplets with diameters larger than 30 μm. A carefully calibrated probe has a response time of less than 0.1 s, a detection limit of about 0.03 g/m^3, and an accuracy of about 20%.

The operating principle of the CSIRO (Australian Commonwealth Scientific and Industrial Research Organization) King probe (Particle Measuring Systems Inc., Boulder, Colorado) is similar in concept to that of the Johnson–Williams probe. The King probe measures the amount of power necessary to maintain a heated wire at a constant temperature, whereas the Johnson–Williams probe measures the change in resistance due to cooling of the wire by water evaporation. The probe consists of a heated coil of wire that is maintained at a constant temperature. The amount of excess power required to maintain the wire at this temperature when it is impacted by water droplets is measured and is proportional to the cloud liquid water content. The nominal response time of the instrument is 0.05 s, and it has an accuracy of 20%. This instrument uses less power than a Johnson–Williams probe, an important consideration in aircraft applications.

A third commonly used method for determining cloud liquid water content is integration of the droplet size spectrum as measured by a PMS FSSP probe. Estimates of cloud liquid water content using this technique are subject to large errors due to uncertainties in determining the number concentrations of droplets in the largest size ranges.

Acknowledgment

We gratefully acknowledge the Department of Energy Atmospheric Chemistry Program for support of this work. This work was performed under the auspices of the United States Department of Energy under contract DE–AC02–76CH00016.

References

1. Gregory, G. L.; Hudgins, C. H.; Edahl, R. A., Jr. *Environ. Sci. Technol.* **1983**, *17*, 100–103.
2. O'Brien, R. J.; Hard, T. M.; Mehrabzadeh, A. A. *Environ. Sci. Technol.* **1983**, *17*, 560–562.
3. Gregory, G. L.; Hudgins, C. H.; Edahl, R. A., Jr. *Environ. Sci. Technol.* **1983**, *17*, 562–564.
4. D'Ottavio, T.; Garber, R. W.; Tanner, R. L.; Newman, L. *Atmos. Environ.* **1981**, *15*, 197–203.
5. Gregory, G. L. In *Proceedings of the In-Situ Air Quality Monitoring from Moving Platforms Specialty Conference*; American Pollution Control Association: San Diego, 1982; pp 6–25.
6. Sheridan, P. J.; Zoller, W. H. *J. Atmos. Chem.* **1989**, *9*, 363–381.
7. Andreae, M. O.; Berresheim, H.; Andreae, T. W.; Krita, M. A.; Bates, T. S.; Merrill, J. T. *J. Atmos. Chem.* **1988**, *6*, 149–173.
8. Huebert, B. J.; Lee, G.; Warren, W. L. *J. Geophys. Res.* **1990**, *95*, 16369–16381.
9. Baumgardner, D.; Huebert, B. J.; Wilson, C. *Meeting Review: Airborne Aerosol Inlet Workshop*; National Center for Atmospheric Research: Boulder, CO, 1991; Technical Note NCAR/TN–362 + 1A, pp 1–285.
10. Torgeson, W. L.; Stern, S. C. *J. Appl. Meteorol.* **1966**, *5*, 205–212.
11. Heikes, B. G.; Kok, G. L.; Walega, J. G.; Lazrus, A. L. *J. Geophys. Res.* **1987**, *92*, 915–931.
12. Leahy, D. F.; Phillips, M. F.; Garber, R. W.; Tanner, R. L. *Anal. Chem.* **1980**, *52*, 1779–1780.
13. Anlauf, K. G.; Wiebe, H. A.; Fellin, P. *J. Air Pollut. Control Assoc.* **1986**, *36*, 715–723.
14. Huebert, B. J.; Robert, C. H. *J. Geophys. Res.* **1985**, *90*, 2985–2990.
15. John, W. S.; Hering, S.; Reischl, G.; Sasaki, G.; Goren, S. *Atmos. Environ.* **1983**, *17*, 373–382.
16. Talbot, R. W.; Andreae, M. O.; Berresheim, H.; Artaxo, P.; Garstung, M.; Harriss, R. C.; Beecher, K. M.; Li, S. M. *J. Geophys. Res.* **1990**, *95*, 16,955–16,969.
17. Daum, P. H.; Leahy, D. F. *The Brookhaven National Laboratory Filter Pack System for Collection and Determination of Air Pollutants*; Brookhaven National Laboratory: Upton, NY, 1985; Report BNL 31381R; pp 1–43.
18. Forrest, J.; Tanner, R. L.; Spandau, D.; D'Ottavio, T.; Newman, L. *Atmos. Environ.* **1980**, *14*, 137–144.
19. Norton, R. B.; Huebert, B. J. *Atmos. Environ.* **1983**, *17*, 1355–1364.
20. Kok, G. L.; Schanot, A. J.; Lindgren, P. F.; Dasgupta, P. K.; Hegg, D. A.; Hobbs, P. V.; Boatman, J. F. *Atmos. Environ.* **1990**, *24A*, 1903–1908.
21. Anlauf, K. G.; MacTavish, D. C.; Wiebe, H. A.; Schiff, H. I.; Mackay, G. I. *Atmos. Environ.* **1988**, *22*, 1579–1586.

22. Solomon, P. A.; Larson, S. M.; Fall, T.; Cass, G. R. *Atmos. Environ.* **1988,** *22,* 1587–1594.
23. *Atmos. Environ.* **1988,** *22A,* 1517–1669.
24. Scott, W. D. *Atmos. Environ.* **1978,** *12,* 917–924.
25. Winters, W.; Hogan, A.; Mohnen, V.; Barnard, S. *ASRC Airborne Cloudwater Collection System*; Atmospheric Science Research Center: Albany, NY, 1979; ASRC–SUNYA Publication No. 728.
26. Walters, P. T.; Moore, M. J.; Webb, A. H. *Atmos. Environ.* **1983,** *17,* 1083–1091.
27. Huebert, B. J.; Baumgardner, D. *Atmos. Environ.* **1985,** *19,* 843–846.
28. Huebert, B. J.; VanBramer, S.; Tschudy, K. L. *J. Atmos. Chem.* **1988,** *6,* 251–263.
29. Isaac, G. A.; Daum, P. H. *Atmos. Environ.* **1987,** *21,* 1587–1600.
30. Greenberg, J. P.; Zimmerman, P. R. *J. Geophys. Res.* **1984,** *88,* 4767–4778.
31. Westberg, H.; Lonneman, W.; Holdren, M. *Identification and Analysis of Organic Pollutants in Air*; Keith, L. H., Ed.; Butterworth: Boston, 1984; pp 323–337.
32. Ehhalt, D. H.; Rudolph, J.; Meixner, F.; Schmidt, U. *J. Atmos. Chem.* **1985,** *3,* 21–29.
33. Oliver, K. D.; Pleil, J. D.; McClenny, W. A. *Atmos. Environ.* **1986,** *20,* 1403–1412.
34. Singh, H. B.; Viezee, W.; Salas, L. J. *J. Geophys. Res.* **1988,** *93,* 15861–15878.
35. Conway, T. J.; Steele, L. P. *J. Atmos. Chem.* **1989,** *9,* 81–100.
36. Evans, L. F.; Weeks, I. A.; Eccleston, A. *J. Clean Air, Melbourne* **1985,** *19,* 21–29.
37. Zimmerman, P. R.; Greenberg, J. P.; Westberg, C. E. *J. Geophys. Res.* **1988,** *93,* 1407–1416.
38. Rasmussen, R. A.; Khalil, M. A. K. *J. Geophys. Res.* **1988,** *93,* 1417–1421.
39. Huebert, B. J. *Atmos. Technol.* **1980,** *12,* 30–34.
40. Bertoni, G.; Bruner, F.; Liberti, A.; Perrino, C. *J. Chromatogr.* **1983,** *203,* 263–270.
41. Billings, W. N.; Bidleman, T. F. *Atmos. Environ.* **1983,** *17,* 383–391.
42. Brown, R. H.; Purnell, C. J. *J. Chromatogr.* **1979,** *178,* 79–90.
43. Walling, J. F.; Bumgarner, J. E.; Driscoll, D. J.; Morris, C. M.; Riley, A. E.; Wright, L. H. *Atmos. Environ.* **1986,** *20,* 51–57.
44. Carroll, M. A. *J. Geophys. Res.* **1985,** *90,* 10483–10486.
45. Berresheim, H.; Andreae, M. O.; Ayers, G. P.; Gillett, R. W.; Merrill, J. T.; Davis, V. J.; Chameides, W. L. *J. Atmos. Chem.* **1990,** *10,* 341–370.
46. Thornton, D. C.; Drieger, A. R., III; Bandy, A. R. *Anal. Chem.* **1986,** *58,* 2688–2691.
47. Thornton, D. C.; Bandy, A. R.; Drieger, A. R., III *J. Atmos. Chem.* **1989,** *9,* 331–346.
48. Gregory, G. L.; Hoell, J. M., Jr.; Ridley, B. A.; Singh, H. B.; Gandrud, B. W.; Salas, L. J.; Shetter, J. *J. Geophys. Res.* **1990,** *95,* 10077–10087.
49. Novotny, M.; Hayes, J. M.; Bruner, F.; Simmonds, P. G. *Science (Washington, D.C.)* **1975,** *189,* 215–217.
50. Desty, D. H.; Goldup, A. *Coated Capillary Columns—An Investigation of Operating Conditions, Gas Chromatography 1960*; Scott, R. P. W., Ed.; Butterworths: London, 1960; pp 162–183.
51. Grosjean, D.; Harrison, J. *Environ. Sci. Technol.* **1985,** *19,* 862–875.
52. Pearson, R., Jr.; Stedman, D. *Atmos. Technol.* **1980,** *12,* 51–55.
53. Gregory, G. L.; Hudgins, C. H.; Ritter, J. A.; Lawrence, M. *Proceedings of the 6th Symposium of Meteorological Instruments*; American Meteorological Society: Boston, 1987; pp 136–139.

54. Ritter, J. A.; Lenschow, D. H.; Barrick, J. D. W.; Gregory, G. L.; Sachse, G. W.; Hill, G. F.; Woerner, M. A. *J. Geophys. Res.* **1990**, *95*, 16875–16886.
55. Garber, R. W.; Daum, P. H.; Doering, R. F.; D'Ottavio, T.; Tanner, R. L. *Atmos. Environ.* **1983**, *17*, 1381–1385.
56. Tanner, R. L.; Daum, P. H.; Kelly, T. J. *Int. J. Environ. Anal. Chem.* **1983**, *13*, 323–335.
57. ten Brink, H. M.; Schwartz, S. E.; Daum, P. H. *Atmos. Environ.* **1987**, *21*, 2035–2052.
58. Huntzicker, J. J.; Hoffman, R. S.; Ling, C-S. *Atmos. Environ.* **1978**, *12*, 83–88.
59. Cobourn, W. G.; Husar, R. B.; Husar, J. D. *Atmos. Environ.* **1978**, *12*, 89–98.
60. Thornton, D. C.; Bandy, A. R. *Global Biogeochem. Cycles* **1987**, *4*, 317–328.
61. Lindgren, P. F.; Dasgupta, P. K. *Anal. Chem.* **1989**, *61*, 19–24.
62. Delany, A. C.; Dickerson, R. R.; Melchior, F. L., Jr.; Wartburg, A. F. *Rev. Sci. Instrum.* **1982**, *53*, 1899–1902.
63. Dickerson, R. R.; Delany, A. C.; Wartburg, A. F. *Rev. Sci. Instrum.* **1984**, *55*, 1995–1998.
64. Kelly, T. J. *Modifications of Commercial Oxides of Nitrogen Detectors for Improved Response*; Brookhaven National Laboratory: Upton, NY, 1986; BNL Informal Report 38000; pp 1–31.
65. Drummond, J. W.; Volz, A.; Ehhalt, D. H. *J. Atmos. Chem.* **1985**, *2*, 287–306.
66. Gregory, G. L.; Hoell, J. M., Jr.; Carroll, M. A.; Ridley, B. A.; Davis, D. D.; Bradshaw, J.; Rodgers, M. O.; Sandholm, S. T.; Schiff, H. I.; Hastie, D. R.; Karecki, D. R.; Mackay, G. I.; Harris, G. W.; Torres, A. L.; Fried, A. *J. Geophys. Res.* **1990**, *95*, 10103–10127.
67. Gregory, G. L.; Hoell, J. M., Jr.; Torres, A. L.; Carroll, M. A.; Ridley, B. A.; Rodgers, M. O.; Bradshaw, J.; Sandholm, S. T.; Davis, D. D. *J. Geophys. Res.* **1990**, *95*, 10129–10138.
68. Bollinger, M. J.; Sievers, R. E.; Fahey, D. W.; Fehsenfeld, F. C. *Anal. Chem.* **1983**, *55*, 1980–1986.
69. Kley, D.; McFarland, M. *Atmos. Technol.* **1980**, *12*, 63–69.
70. Kelly, T. J.; Spicer, C. W.; Ward, G. F. *Atmos. Environ.* **1990**, *24A*, 2397–2403.
71. Schiff, H. I.; Karecki, D. R.; Harris, G. W.; Hastie, D. R.; Mackay, G. I. *J. Geophys. Res.* **1990**, *95*, 10147–10153.
72. Sandholm, S. T.; Bradshaw, J. D.; Dorris, K. S.; Rodgers, M. O.; Davis, D. D. *J. Geophys. Res.* **1990**, *95*, 10155–10161.
73. Seiler, W.; Giehl, H.; Roggendorf, P. *Atmos. Technol.* **1980**, *12*, 40–45.
74. Dickerson, R. R.; Delany, A. C. *J. Atmos. Technol.* **1988**, *5*, 424–431.
75. Dickerson, R. R.; Huffman, G. J.; Luke, W. T.; Nunnermacker, L. J.; Pickering, K. E.; Leslie, A. C. D.; Lindsey, C. G.; Slinn, W. G. N.; Kelly, T. J.; Daum, P. H.; Delany, A. C.; Greenberg, J. P.; Zimmerman, P. R.; Boatman, J. F.; Ray, J. D.; Stedman, D. H. *Science (Washington, D.C.)* **1987**, *235*, 460–466.
76. Kleinman, L. I.; Daum, P. H. *J. Geophys. Res.* **1991**, *96*, 991–1005.
77. Sachse, G. W.; Hill, G. F.; Wade, L. O.; Perry, M. G. *J. Geophys. Res.* **1987**, *92*, 2071–2081.
78. Hoell, J. M., Jr.; Gregory, G. L.; McDougal, D. S.; Sachse, G. W.; Hill, G. F.; Condon, E. P.; Rasmussen, R. A. *J. Geophys. Res.* **1987**, *92*, 2009–2019.
79. Kleindienst, T. E.; Shepson, P. B.; Hodges, D. N.; Nero, C. M.; Arnts, R. A.; Dasgupta, P. K.; Hwang, H.; Kok, G. L.; Lind, J. A.; Lazrus, A. L.; Mackay, G. I.; Mayne, L. K.; Schiff, H. I. *Environ. Sci. Technol.* **1988**, *22*, 53–61.

80. Sakugawa, H.; Kaplan, I. R.; Tsai, W.; Cohen, Y. *Environ. Sci. Technol.* **1990**, *24*, 1452–1462.
81. Kok, G. L.; Heikes, B. G.; Lind, J. A.; Lazrus, A. L. *Atmos. Environ.* **1989**, *23*, 283.
82. Lazrus, A. L.; Kok, G. L.; Gitlin, S. N.; Lind, J. A.; McLaren, S. E. *Anal. Chem.* **1985**, *57*, 917–922.
83. Lazrus, A. L.; Kok, G. L.; Lind, J. A.; Gitlin, S. N.; Heikes, B. G.; Shetter, R. E. *Anal. Chem.* **1986**, *58*, 594–597.
84. Daum, P. H.; Kleinman, L. I.; Hills, A. J.; Lazrus, A. L.; Leslie, A. C. D.; Busness, K.; Boatman, J. *J. Geophys. Res.* **1990**, *95*, 9857–9871.
85. Lee, J. H.; Tang, I. N.; Weinstein-Lloyd, J. B. *Anal. Chem.* **1990**, *62*, 2381–2384.
86. Pinnick, R. G.; Auvermann, H. J. *J. Aerosol Sci.* **1979**, *10*, 55–74.
87. Garvey, D. M.; Pinnick, R. G. *J. Aerosol Sci. Technol.* **1983**, *2*, 477–488.
88. Isaac, G. A.; Leaitch, W. R.; Strapp, J. W. *Atmos. Environ.* **1990** *24A*, 3033–3046.
89. Kim, Y. J.; Boatman, J. F. *Aerosol Sci. Technol.* **1990**, *12*, 665–672.
90. Strapp, J. W.; Leaitch, W. R.; Liu, P. S. K. In *Meeting Review: Airborne Aerosol Inlet Workshop*; National Center for Atmospheric Research: Boulder, CO, 1991; Technical Note NCAR/TN–362 + 1A; p 109.
91. Tsonis, A. A.; Leaitch, W. R.; Couture, M. D. *J. Atmos. Technol.* **1987**, *4*, 518–526.
92. Pinnick, R. G.; Rosen, J. M. *J. Aerosol Sci.* **1979**, *10*, 533–538.
93. Pinnick, R. G.; Garvey, D. M.; Duncan, L. D. *J. Appl. Meteorol.* **1981**, *20*, 1049–1057.
94. Dye, J. E.; Baumgardner, D. *J. Atmos. Technol.* **1984**, *1*, 329–344.
95. Baumgardner, D.; Strapp, J. W.; Dye, J. E. *J. Atmos. Technol.* **1985**, *2*, 626–632.
96. Baumgardner, D.; Spowart, M. *J. Atmos. Technol.* **1990**, *7*, 666–672.
97. Kim, Y. J.; Boatman, J. F. *J. Atmos. Technol.* **1990**, *7*, 673–680.
98. Dye, J. E.; Gandrud, B. W.; Baumgardner, D.; Sanford, L. A. *Geophys. Res. Lett.* **1991**, *17*, 409–412.
99. Butcher, S. S.; Charlson, R. J. *An Introduction to Air Chemistry*; Academic: New York, 1972; p 241.
100. Charlson, R. J. *Atmos. Technol.* **1980**, *12*, 10–14.
101. Pinnick, R. G.; Jennings, S. G.; Chylek, P. *J. Geophys. Res.* **1980**, *85*, 4059–4066.
102. Waggoner, A. P.; Weiss, R. E.; Alquist, N. C.; Covert, D. S.; Charson, T. J. *Atmos. Environ.* **1981**, *14*, 623–626.
103. Lewis, C. W. *Atmos. Environ.* **1981**, *15*, 2639–2646.
104. Skala, G. F. *Anal. Chem.* **1963**, *35*, 702–706.
105. Cooper, G.; Langer, G. *J. Aerosol Sci.* **1978**, *9*, 65–75.
106. Bodhaine, B. A.; Murphy, M. E. *J. Aerosol Sci.* **1980**, *11*, 305–312.
107. Liu, B. Y. H.; Kim, C. S. *Atmos. Environ.* **1977**, *11*, 1097–1100.
108. Strapp, J. W.; Schemenauer, R. S. *J. Appl. Meteorol.* **1982**, *21*, 98–108.

RECEIVED for review March 20, 1991. ACCEPTED revised manuscript September 23, 1992.

5

In Situ Measurements of Stratospheric Reactive Trace Gases

William H. Brune[1] and Richard M. Stimpfle[2]

[1]Department of Meteorology, Pennsylvania State University, University Park, PA 16802
[2]Department of Earth and Planetary Sciences, Harvard University, Cambridge, MA 02138

In situ measurements of the abundances of reactive trace gases have been essential to the understanding of stratospheric photochemistry. The measurement of those gases that directly affect the abundance of ozone—NO, NO$_2$, OH, HO$_2$, ClO, Cl, BrO, Br, and O—are of particular interest. The stratospheric environment, with its low temperatures, large range in pressure, and solar ultraviolet light, offers many measurement challenges. Simultaneous measurements of a number of trace gas species are required to develop an understanding of their distributions, and some of these measurements have been made from instruments mounted on helium-filled balloons and high-altitude aircraft. Although much has been learned about the workings of the stratosphere and, in particular, the mechanisms affecting the distribution of ozone, a truly predictive understanding has yet to be developed.

T HE IMPACT OF ANTHROPOGENIC ACTIVITY on stratospheric photochemistry has been an important motivation for the measurements of stratospheric trace gases for the last 20 years. Although the stratosphere is remote, the changes induced in the ozone layer by these changing reactive trace gases and the resultant increase in ultraviolet radiation are of concern for all living things on the surface of the earth. Thus, much stratospheric research has been focused on ozone depletion. Those reactive trace gases that have direct impact on ozone—NO, NO$_2$, OH, HO$_2$, ClO, Cl, BrO, Br, and O—are of particular interest. The photochemical systems are highly interactive, how-

0065–2393/93/0232–0133$13.80/0
© 1993 American Chemical Society

ever, and many components not directly related to ozone loss are nonetheless important for understanding what that loss is now and might be in the future. The strategy for research in the stratosphere has been to develop computer simulations to predict trends in photochemistry and ozone change. Incorporated in these simulations are laboratory data on chemical kinetics and photolytic processes and a theoretical understanding of atmospheric motions. An important aspect of this approach is knowing if the computer models represent the conditions of the stratosphere accurately enough that their predictions are valid. These models are made credible by comparisons with stratospheric observations.

Measurements either from the ground or from satellites have been a major contribution to this effort, and satellite instruments such as LIMS (Limb Infrared Monitor of the Stratosphere) on the Nimbus 7 satellite (1) in 1979 and ATMOS (Atmospheric Trace Molecular Spectroscopy instrument), a Fourier transform infrared spectrometer aboard Spacelab 3 (2) in 1987, have produced valuable data sets that still challenge our models. But these remote techniques are not always adequate for resolving photochemistry on the small scale, particularly in the lower stratosphere. In some cases, the altitude resolution provided by remote techniques has been insufficient to provide unambiguous concentrations of trace gas species at specific altitudes. Insufficient altitude resolution is a handicap particularly for those trace species with large gradients in either altitude or latitude. Often only the most abundant species can be measured. Many of the reactive trace gases, the key species in most chemical transformations, have small abundances that are difficult to detect accurately from remote platforms.

In situ measurements of stratospheric reactive trace gas abundances provide an opportunity to test the fundamental photochemical mechanisms (3). The advantage of such measurements is that they are local, so the simultaneous measurements of trace gases place a true constraint on the possible photochemical mechanisms. These measurements are also able to resolve small-scale spatial and temporal structure in the trace constituent fields. The disadvantage of in situ measurements is that they do not capture the global or perhaps even seasonal view of photochemical transformations because they are seldom done frequently enough or in enough places to provide that information. Another disadvantage of in situ measurements is that they must be made from platforms in the stratosphere, and these remote observational outposts have their liabilities.

One of the success stories of in situ measurements in the stratosphere is the confirmation that the rapid loss of ozone over Antarctica each October is indeed caused by photochemistry related to the release of chlorofluorocarbons at the surface of the earth. Ground-based measurements of the primary chlorine culprit, ClO, and O_3 have given a similar picture (4), but not with the fine detail possible from the in situ techniques, as shown in

Figure 1. This graph shows the rapid variation of ClO and O_3 as the edge of the chemically perturbed region in the Antarctic polar vortex is penetrated by the National Aeronautics and Space Administration (NASA) ER–2 high-altitude aircraft over the Palmer Peninsula of Antarctica on September 16, 1987 (5). It is one of a series of 12 snapshots, or individual flights, during the Airborne Antarctic Ozone Experiment (AAOE) that show the development of an anticorrelation between ClO and O_3 that began as a correlation in mid-August. When these two measurements are combined with all the others from the ER–2 aircraft, the total data set provides a provocative picture of how such chemistry occurs and what it is capable of doing to ozone.

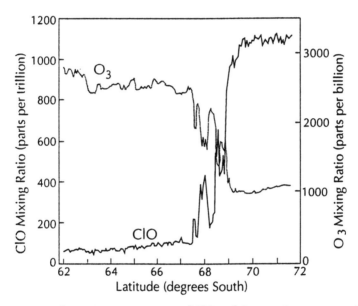

Figure 1. Simultaneous measurements of ClO and O_3 over Antarctica on September 16, 1987, during the AAOE mission. The boundary of the chemically perturbed region at 69°S is clearly shown by the rapid increase in the ClO mixing ratio and the rapid decrease in the O_3 mixing ratio. There is an anticorrelation between ClO and O_3 near the boundary.

The measurements in the midlatitude stratosphere during the last four years have been equally successful because more related species are being measured simultaneously, and these data sets are placing serious constraints on photochemical models (2). This situation is dramatically different from that in the mid-1970s through the mid-1980s, when confirmation that certain trace gas species were present in the stratosphere in approximately the correct abundances was still an issue (6). Analyses of more recent measurements indicate that the current understanding is not complete, however. But efforts to measure the major components of all the trace gas families, especially during diurnal or seasonal variations, are solidifying the understanding of stratospheric photochemistry at the midlatitudes.

The challenges that face the scientist studying the stratosphere are the same as those that face scientists in other fields. First is the challenge of knowing what needs to be measured to better understand the natural system being studied. In the stratosphere, this understanding includes not only the trace gas abundances and their spatial and temporal variations, but also the transport of those species on both short and long time scales. Distributions of and correlations among long-lived trace gases have provided the best clues for how stratospheric transport works. Second is the challenge of how to make these measurements. This skill includes knowing not only how to measure the abundances of individual trace gas species, but how to combine several instruments or measurements together and how to deploy those instruments throughout an experiment.

The goal of this chapter is to instill an understanding of in situ measurements of stratospheric reactive trace gases and of how the challenges of knowing what measurements to make and how to make them are being met. Because most of the other chapters in this book concern measurements in the troposphere, a brief overview of the characteristics of the stratosphere—its physical state and its trace gas composition and photochemistry—is first presented. This discussion contains a general statement of the current understanding of stratospheric photochemistry and the areas where the lack of knowledge is critical. Some general guidelines are presented of what the challenges in stratospheric measurements are, and examples of these challenges are given from instruments that have been flown either on helium-filled balloons or on high-altitude aircraft.

The Stratospheric Environment and Trace Gas Distribution

The stratosphere, beginning at roughly 10 km above the earth and extending up to 50 km, contains 10% of the air in the atmosphere and 90% of the ozone. It and the mesosphere are the regions where much of the solar ultraviolet light deposits its energy as heat; this situation results in a positive temperature gradient that maintains vertical dynamic stability. The stratosphere thus evolves in relative isolation from the troposphere, where turbulent mixing occurs, and only a slow exchange occurs between the two regions. Although the processes of exchange between the troposphere and stratosphere are not completely understood, it is known that air and a few trace constituents that are not collected and precipitated out of clouds enter the stratosphere predominantly in the tropics, can be chemically transformed, and exit a few years later in middle to high latitudes. The admission of only a relatively few trace gases into the stratosphere, the exposure of gases to the ultraviolet light that is screened from the troposphere by ozone and to the resultant photochemistry, the mixing, and the poleward transport produce a trace gas composition that is significantly different from that in the troposphere.

The temperatures in the stratosphere range from a low of 200 K at 14 km to 250 K at 50 km for typical midlatitude conditions (7). In the polar regions, particularly over the South Pole, temperatures can fall to as low as 185 K over the entire height of the stratosphere (8). The average temperature profile for middle latitudes has higher temperatures than those observed at 70°S during the Airborne Antarctic Ozone Expedition in August 1987 or at 75°N during the Airborne Arctic Stratospheric Experiment in January 1989 (Figure 2). For comparable altitudes and seasons, the Arctic is slightly warmer than the Antarctic (9). In general, the coldest regions are the lower equatorial stratosphere just above the tropopause and the wintertime lower polar stratosphere, and these regions exhibit temperatures at or below 200 K.

The air pressure in the stratosphere ranges from 100 mbar near 15 km to 0.1 mbar near 50 km, falling off exponentially with altitude as shown in Figure 2 (9). As in the troposphere, altitude and pressure are used as vertical

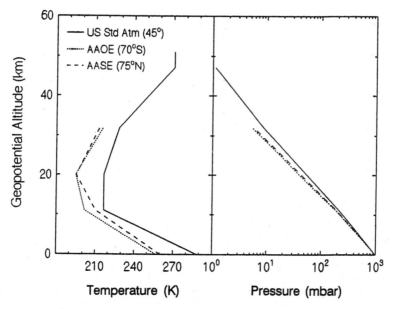

Figure 2. The temperature and pressure distribution of the stratosphere. The solid line is from reference 7, and the dashed lines are from measurements made by the Meteorological Measurement System (MMS) instrument on the NASA ER–2 high-altitude aircraft during the AAOE mission in 1987 (8) and the AASE mission in 1989 (9). The Arctic was colder in 1989 than usual.

coordinates throughout the stratosphere, but potential temperature becomes the most meaningful coordinate in the lower stratosphere, where heating and cooling rates are small. Potential temperature is the temperature that an air parcel would have if it were adiabatically compressed to 1 atm (101 kPa). Air parcels tend to follow isentropic trajectories (constant potential temperature), so the goal of measurements is often to follow constant-potential trajectory surfaces to map out the meridional and zonal components of a trace gas constituent field. This approach implies that both pressure and temperature will vary along the path. Such trajectories are meaningful for about a week, after which diabatic effects become important and mixing of air parcels from different trajectories creates a new air parcel with a new average trajectory.

The vertical distributions of some trace gases for some regions of the stratosphere would be expected to be smooth and slowly varying. For other trace gases in other regions, extreme vertical stratification and high temporal variability would be more likely. An example of this latter case is the Antarctic

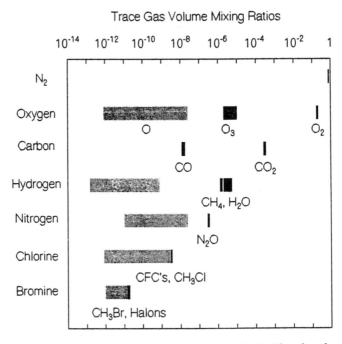

Figure 3. Abundances of gases in the stratosphere (12). The abundances of source gases marked by heavy vertical bars are the abundances at the base of the stratosphere, with the exception of H$_2$O. The thick vertical bar indicates the range of H$_2$O abundances throughout the stratosphere. Shaded bars represent ranges of observed or calculated reactive or reservoir trace gases for midday midlatitude conditions.

ozone hole, where extreme gradients in trace gas abundances are established. As the polar air mixes with midlatitude air, highly variable three-dimensional structures in the trace gas distributions evolve (*10, 11*).

The trace gases in this highly variable physical environment can be grouped into chemical families. The four most prominent in photochemistry are those of oxygen, nitrogen (other than N_2), hydrogen, and chlorine and bromine (Figure 3). Gases emitted in the troposphere that migrate to the stratosphere are the primary sources for these chemical families. The main source gases for nitrogen are N_2O and to a lesser extent NO and NO_2; for hydrogen H_2O or CH_4; and for the halogens CH_3Cl, chlorofluorocarbons, halons, and CH_3Br. These stable gases are broken down in the stratosphere by either sunlight or chemical products of sunlight and become the reactive species that interact with each other but also destroy ozone. Figure 3 shows the long-lived source gases as solid bars, and the range of the resulting reactive gases—including everything from free radicals to acids—is given by the shaded bars.

How trace gases are distributed with altitude can be illustrated for midlatitude conditions. The abundances of trace gases in the stratosphere as a function of altitude are given in Figure 4 in terms of volume mixing ratio and in Figure 5 in terms of concentration. The concept of mixing ratio is important in the consideration of the transport of the trace gases because

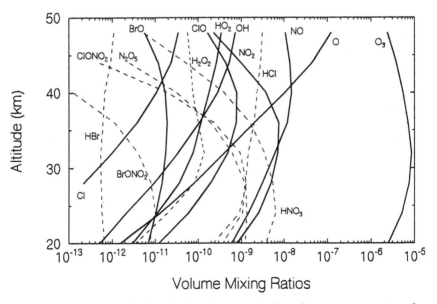

Volume Mixing Ratios

Figure 4. Calculated altitude distributions of the volume mixing ratios of several trace gases (12). The reactive trace gases that directly affect ozone are given by dark lines. Conditions are for the equinox at 30°N.

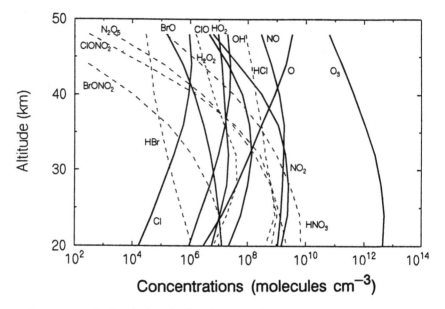

Figure 5. Calculated altitude distributions of the concentrations of several trace gases (12). The reactive trace gases that directly affect ozone are given by dark lines. Conditions are for the equinox at 30°N.

mixing ratios are preserved as the air parcels descend and contract or ascend and expand. On the other hand, many measurement techniques use absorption or fluorescence, which are dependent upon the concentration of the trace constituent. The mixing ratios of many free radicals increase substantially with height, but their concentrations are somewhat more constant with altitude. Near the equator, the source region for most trace gases, the .abundances of the tropospheric source gases are larger that those at midlatitude for comparable vertical coordinates. Near the polar regions, just the opposite is true. Although this is a one-dimensional view of trace gas distributions, the three-dimensional view is actually required for careful comparison of model results and observations.

Ozone Photochemistry at Midlatitudes

The thrust of much of stratospheric research has been to understand the production and loss of stratospheric ozone. The chemical trail in this process has been marked in Figure 6 as the top four arrows. The production of ozone is almost exclusively by this mechanism:

$$O_2 + \text{ultraviolet sunlight} \longrightarrow 2O \tag{1}$$

$$O + O_2 + M \longrightarrow O_3 + M \tag{2}$$

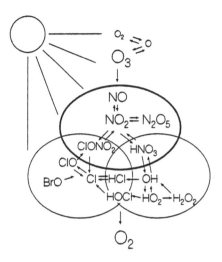

Figure 6. Schematic of the major gas-phase cycles in the stratosphere. The Chapman mechanism (oxygen reactions) is indicated by the top four arrows. The ovals represent the odd-nitrogen, odd-hydrogen, and inorganic halogen chemical families. Arrows indicate conversion of species by reaction or photolysis, but reaction partners are not shown. The overlap area in the center represents heterogeneous and unknown photochemistry.

The destruction of ozone occurs by a number of mechanisms that result in

$$O_3 + sunlight \longrightarrow O + O_2 \tag{3}$$

$$O + O_3 \longrightarrow 2O_2 \tag{4}$$

These four reactions constitute the Chapman mechanism for establishing the abundance of ozone in the stratosphere. However, the abundance of ozone is dictated also by other loss mechanisms that mimic reaction 4.

The involvement of reactive nitrogen, reactive hydrogen, and reactive chlorine in catalytic cycles that destroy ozone has been known for about 20 years. These cycles have the form

$$O_3 + X \longrightarrow XO + O_2 \tag{5}$$

$$O + XO \longrightarrow X + O_2 \tag{6}$$

where X = NO, OH, Cl, or Br. In each couplet of reactions, one is much slower than the other and dictates the speed of the net reaction. For all of these species, the rate-limiting step is reaction 6. Catalytic cycles of reactive hydrogen are also possible; in the lower stratosphere both OH and HO_2 react with O_3, and in the upper stratosphere both OH and HO_2 react with O. The importance of NO, OH, and Cl catalytic cycles for destroying ozone is given in Figure 7, along with the pure oxygen production and losses. The

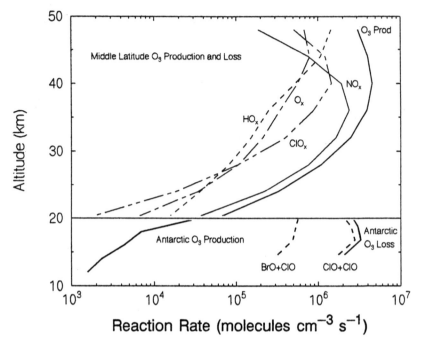

*Figure 7. Calculated ozone production and loss rates for two different con-
ditions from the AER two-dimensional model. Production and loss rates above
20 km are diurnally averaged loss rates for the spring equinox at 30°N. Midday
loss rates are approximately two times larger. Production and loss rates for
midday below 20 km are calculated for the chemically perturbed region over
Antarctica on September 16, 1987. The catalytic cycles responsible for the loss
are explained in the text. Although ozone loss occurs at higher altitudes over
Antarctica, in situ observations extend only to ~19 km.*

reactive nitrogen cycle dominates ozone destruction throughout most of the
stratosphere, although chlorine and hydrogen are equally important higher
in the stratosphere near 40 km, and hydrogen is more important lower in
the stratosphere near the tropopause.

The change in the ozone abundance at any place or time can be written
as

$$\partial O_3/\partial t = \text{Production} - \text{Loss} - \nabla \cdot \phi_{O_3} \tag{7}$$

or

$$\partial O_3/\partial t = 2J_{O_2}[O_2] - 2k_1[O][O_3] - 2k_2[NO_2][O] \tag{8}$$

$$- 2k_3[HO_2][O] - 2k_4[ClO][O]$$

$$- (\text{other smaller terms}) - \nabla \cdot \phi_{O_3}$$

where J is the photolysis rate of O_2, the brackets indicate concentration, and ϕ_{O_3} is the flux of ozone through the volume and represents the transport of ozone (12). For a given location in the stratosphere, the abundance of ozone can be said to be either under photochemical control or dynamical control depending on whether the chemical terms or the flux term in equation 8 is larger. In this view, much of the stratosphere above 30 km is under photochemical control, and much below 30 km or in the wintertime polar region is under dynamical control (6). However, this view is not entirely valid for the real stratosphere, and it predates the understanding of the rapid chemical loss in the springtime Antarctic stratosphere.

In addition to the reactive nitrogen species, NO_2 and NO, the odd-hydrogen species, HO_2 and OH, and the inorganic chlorine species, ClO and Cl, there are the other family members identified in Figure 6. For the sake of clarity and brevity the photochemical scheme given in Figure 6 illustrates the subset of reactions that to a large extent determines the effects of these compounds on ozone. The identity of the reactive partners (or sunlight) involved in each reaction and reactions of lesser importance are given in numerous other publications (references 6 or 12, for example).

The additional species shown within each ellipse are, in each case, reservoir species, so-called because they effectively store the active radical species in molecular forms that do not catalytically destroy ozone. The presence of reservoir species dramatically affects the distribution of active radical species in a given family and thus the effectiveness with which they can destroy ozone. The molecules within the overlapped boundary areas, $ClONO_2$, HNO_3, HOCl, and HCl, identify crucial reservoir species that interlink the families. In this picture, the interlinking reservoir species serve notice that, although the division of stratospheric photochemistry into the NO_x, Cl_x, and HO_x species is a very useful construct for understanding stratospheric photochemistry, in reality the photochemistry must be understood as a complete system. Another example of important interfamily reactions not explicit in Figure 6 is the reactions of ClO and HO_2 with NO to form Cl and OH, respectively.

An example of the complexity implied by the interlinking of the chemical families in Figure 6 is the effect of increasing Cl_x in the lower stratosphere, below 30 km (13, 14). Presently in this region the NO_x, Cl_x, and HO_x species account for ~70, 20, and 10% of ozone loss, respectively, as seen in Figure 7. If Cl_x were to increase significantly, then the natural buffering effect of NO_x on Cl_x by $ClONO_2$ formation would be depleted as NO_x is more or less completely titrated to $ClONO_2$. Then the excess Cl_x could destroy significantly more ozone than the NO_x system it replaces. In addition, the loss of HO_x through the formation of HNO_3 would decrease with the NO_2 decrease, thus raising the OH concentration and liberating more Cl_x from the HCl reservoir because of the reaction of HCl with OH that forms Cl. This type of nonlinear atmospheric response to increases in one of the photochemical families makes it imperative that the current understanding of stratospheric photochemistry be tested.

The High-Latitude Lower Stratosphere in Winter and Spring

Although many measurements of trace species have been made in the middle altitudes over the last 20 years, relatively few measurements were made in the polar regions until about 1986, particularly in the coldest times of the year in winter and spring. As a result, although researchers thought they had a good understanding of the photochemistry at middle latitudes (15, 16), little was known about the polar regions. It is really no surprise that no one recognized the possible consequences of the formation of polar stratospheric clouds, which had been observed by satellites since 1980 (17), and that the first report that rapid ozone loss was occurring came in 1985 from Farman et al. at the British Antarctic Survey (18). This loss, apparent in the early 1980s, began in September and became greatest in October. In 1986, 35% of the total column of ozone was chemically destroyed over an area the size of Antarctica. In 1987, the loss for October was 50% of the 1979 column amount (19). Although the ozone loss in 1988 was substantially less, the losses in 1989, 1990, and 1991 have been equal to that in 1987 (20, 21). Essentially all of the ozone between the altitudes of 13 and 23 km is removed in an air mass that remains located roughly over the Antarctic continent (22) in a volume of air inside the circumpolar vortex that is called the "chemically perturbed region".

When the ozone hole was announced in 1985, several theories sprang up immediately to explain the loss. Only the theory that chlorine from chlorofluorocarbons—alone (23) and in combination with bromine (24) and to a lesser extent HO_2 (25)—was responsible has survived the process of scientific investigation. What is required for ozone between the altitudes of 13 and 22 km and south of ~65°S latitude to be destroyed at the observed rate of 2% per day appears in retrospect to be a chemical conspiracy. First, the polar stratospheric clouds (PSCs) form at temperatures below 195 K, which is 4 to 7 K above the frost point of water vapor (26, 27). Reactive nitrogen and water cocondense on background sulfuric acid aerosols, presumably as nitric acid trihydrate ($HNO_3 \cdot 3H_2O$), although other forms also appear likely. Reactive nitrogen, called NO_y (NO_y = NO + NO_2 + NO_3 + N_2O_5 + HONO + HNO_3 + HO_2NO_2 + $ClONO_2$ + other reactive nitrogens) is thus removed from the gas phase and converted into HNO_3.

This reactive nitrogen may then be actually removed from the air parcel when the particles containing it grow large enough, at least several micrometers in diameter, to gravitationally settle out in less than a day (26). The mechanisms for this particle growth are not completely understood, but one possibility is that water vapor condenses on the nitric acid–water core as the temperature decreases below the frost point. These particles then become large and heavy enough that they can fall to lower altitudes, perhaps out of the stratosphere. From measurements of NO_y, NO, and column abundances of HNO_3 and NO_2, it is known that much of the total NO_y is actually removed from the stratosphere by the sedimentation of large particles containing the reactive nitrogen. Over Antarctica, the stratospheric air is both denitrified and dehydrated, and these observations support the mechanism of sedimentation presented (for examples of these measurements, see many papers in reference 28). However, a number of other possible mechanisms can accomplish the same effect (29). Although most of the reactive nitrogen is removed, some remains, presumably in the form of HNO_3, which is photolyzed only slowly back into NO_2 in the weak ultraviolet sunlight of the springtime polar region.

At the same time, the PSCs are excellent sites for the conversion of chlorine compounds from the relatively inactive reservoir forms of HCl and $ClONO_2$, which make up 99% of the chlorine budget in the lower stratosphere, to photolytically labile species such as Cl_2, HOCl, and ClONO (30–32):

$$ClONO_2 + HCl \longrightarrow Cl_2 \text{ (gas)} + HNO_3 \text{ (solid)} \tag{9}$$

$$ClONO_2 + H_2O \longrightarrow HOCl \text{ (gas)} + HNO_3 \text{ (solid)} \tag{10}$$

$$N_2O_5 + HCl \longrightarrow ClONO \text{ (gas)} + HNO_3 \text{ (solid)} \tag{11}$$

In the weak sunlight of polar spring, these gas-phase chlorine species release their chlorine atoms, which attack ozone almost exclusively. The catalytic cycle that requires a reaction between ClO and O is not very effective, because few oxygen atoms exist in these cold, relatively dark regions. Instead, ClO reacts with another ClO molecule, forming Cl_2O_2, which can then be easily photolyzed by the weak visible sunlight that penetrates the atmosphere.

$$ClO + ClO + M \longrightarrow Cl_2O_2 + M \tag{12}$$

$$Cl_2O_2 + \text{sunlight} \longrightarrow Cl + ClOO \tag{13}$$

$$ClOO + M \longrightarrow Cl + O_2 + M \tag{14}$$

$$Cl + O_3 \longrightarrow ClO + O_2 \tag{15}$$

A second catalytic cycle involves the ClO and BrO radicals, which react; some form Cl and Br atoms, which can then react with ozone to form ClO and BrO again.

$$ClO + BrO \longrightarrow Br + ClOO \tag{16a}$$

$$ClO + BrO \longrightarrow Br + OClO \tag{16b}$$

$$ClO + BrO \longrightarrow BrCl + O_2 \tag{16c}$$

$$ClOO + M \longrightarrow Cl + O_2 + M \qquad (17)$$

$$BrCl + sunlight \longrightarrow Br + Cl \qquad (18)$$

$$Cl + O_3 \longrightarrow ClO + O_2 \qquad (19)$$

$$Br + O_3 \longrightarrow BrO + O_2 \qquad (20)$$

Approximately one-half of the total reaction of ClO and BrO results in the destruction of ozone. Other mechanisms exist, such as a catalytic cycles that are rate-limited by the reaction between ClO and O and between ClO and HO_2 (25), but the contribution from these reactions is small in the polar regions.

A large number of observations, both remote and in situ, confirm this qualitative picture of the loss of ozone over Antarctica. The in situ data have come from instruments carried on small balloons and the NASA ER–2 high-altitude aircraft. Small-balloon measurements are of particle distributions and sizes, ozone, and water vapor (23, 33). ER–2 measurements, listed in Table I, are of particle size and composition; atmospheric parameters such as temperature, pressure, lapse rate, and winds; and trace gas abundances of O_3, N_2O, NO_y or NO, ClO and BrO, and stable gases, including CH_4, chlorofluorocarbons, halons, and others (34–45).

Data from the ER–2 were collected during the AAOE that was based in Punta Arenas, Chile, in August and September 1987 (46). Twelve flights were made from Punta Arenas (54°S) to the base of the Palmer Peninsula, Antarctica (72°S), and back during the six-week-long mission. Flight paths were restricted to a narrow range of longitudes, and altitudes were chosen so that the potential temperature remained constant during each leg of the flight. Potential temperature surfaces between 420 and 470 K were flown. Measurements over a large range of potential temperatures were taken when the pilot executed a dive to 340 K before recovering to a higher potential temperature surface and flying north. Such a program allowed sampling of air parcels from 470 to 340 K inside the chemically perturbed region and also permitted the observation of strong gradients in trace gas abundances on a given potential temperature surface. Another goal of the mission was to sample as deeply into the chemically perturbed region as possible, and measurements as deep as 15° latitude inside the vortex were achieved on some flights. These measurements give a picture of the extent and evolution of the ozone loss over Antarctica.

The presence of large amounts of reactive chlorine, 500 times the concentration at midlatitudes, and the absence of reactive nitrogen, 5 times less than that at midlatitudes, is but one result of these measurements. Another result is the evolving anticorrelation between ClO and O_3, of which one snapshot is given in Figure 1 (5). To establish quantitatively that the observed abundances of ClO and BrO can result in the observed decline in ozone is somewhat more difficult. With the assumption that the abundances of ClO

Table I. In Situ Measurements on the ER–2 Aircraft

Measuring Device	Method	Ref.
MMS (pressure, temperature, and wind vector)	temperature and pressure sensors; inertial navigation system	34
Microwave temperature profiler	passive microwave radiometry of O_2 thermal emission	35
Aerosol and cloud spectrometer: 0.1- to 3.0-μm particles 0.3- to 20.0-μm particles	laser scattering: inside a cavity in the free air stream	36
Condensation nuclei counter	growth in alcohol-saturated chamber; optical particle counting	37
Particle chemistry impactor	particle impaction; postflight X-ray analysis	38
Multifilter sampler (total nitrate, sulfate, and acidic chloride and fluoride)	filter collection; postflight aqueous extraction and ion chromatography	39
Whole-air sampler (CO_2, CH_4, N_2O, CO, CFCs, and halons)	pressurized canisters; postflight analysis with gas chromatography	40
Airborne tunable laser absorption spectrometer (ATLAS) (N_2O)	infrared absorption by tunable diode laser spectroscopy	41
Lyα hygrometer (H_2O vapor)	photodissociation by hydrogen (Lyα) emission at 121.6 nm; detection of OH ($A^2\Sigma^+ \rightarrow X^2\Pi$) emission	42
Dual-beam UV absorption ozone photometer	UV absorption at 254 nm; comparison of signals from scrubbed and unscrubbed airstreams	43
NO and NO_y detector	NO: chemiluminescence reaction of NO + O_3 and NO_2^* detection NO_y: catalytic conversion to NO and chemiluminescence NO detection	44
ClO–BrO detector	chemical conversion with reagent NO to Cl or Br; resonance fluorescence detection of atoms	45

and BrO are zonally uniform and that air parcels are neither heated nor cooled very rapidly during the six weeks of observations, then calculations of ozone loss, from the observed ClO and BrO amounts and the mechanisms given, can be compared to the observed ozone loss. The calculated ozone loss rates are shown in Figure 7 in comparison with the midlatitude ozone loss rates, and the time evolution of ozone loss during August and September for three potential temperature surfaces is given in Figure 8 (47). Within the combined uncertainties of the measurements and calculations, the calculated loss matches the observed loss. Improving this comparison will require a complete understanding of the air parcel trajectories and the variability of the trace gas abundances along those trajectories (48), but the large involvement of halogen photochemistry has been verified.

The studies over Antarctica have shown that the motions of the stratospheric air parcels must be well known and that tracers of some sort must

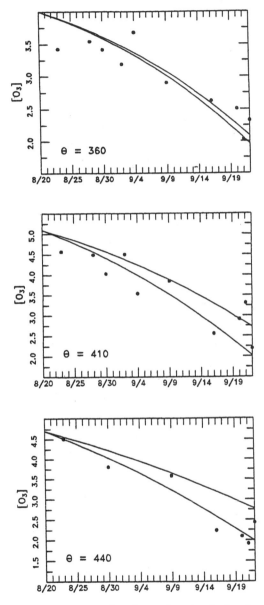

Figure 8. Comparison between the observed and calculated ozone loss over Antarctica during the four-week period of the AAOE mission in 1987 (5). The units on the y-axis are 10[12] molecules cm[-3]. Calculations were based on the observed abundances of ClO and BrO and reactions 12 through 15 alone (ClO + ClO) in the upper of each set of two curves and on the sum of this reaction sequence and reactions 16 through 20 (ClO + BrO) in the lower of each set of two curves. Dots are observations of ozone abundances.

be discovered to mark that motion. One tracer that proved to be very useful was N_2O, which has its source in the troposphere and is chemically destroyed in the stratosphere with a time constant longer than the time constants for dynamical transport. The interpretation of the values of this tracer is not well established, but in the polar, lower stratosphere, the tracer's gradients mimic those of potential vorticity, a dynamical tracer that provides a convenient marker of motion of stratospheric air (49, 50). The abundance of N_2O also provides a surrogate for NO_y, as empirically determined by regressing one data set against another for a large number of flights (51). This surrogate is called "NO_y*". The simultaneous measurement of N_2O and NO_y provides the strongest evidence that reactive nitrogen is actually removed from air parcels inside the Antarctic and Arctic polar vortices, because the measured abundances of NO_y fall significantly below that expected from the near-linear relationship with N_2O that has been frequently observed (52). The analysis of such relationships is crucial for the understanding of the coupling between trace gas transport and photochemistry throughout the stratosphere.

If ozone is being lost in the Antarctic stratosphere in the austral spring, then why hasn't rapid ozone loss also been detected in the Arctic stratosphere in the Northern Hemisphere spring? The differences apparently are in the meteorology (reference 53 and references therein). The Arctic stratosphere in winter less frequently gets cold enough for the formation of polar stratospheric clouds, and it is cold enough over a smaller area than the Antarctic stratosphere. Further, the Arctic polar vortex that somewhat isolates the polar air from midlatitude air rich in NO_x usually breaks apart in February, a month before the spring equinox, unlike the polar vortex over Antarctica that breaks apart in late October, a month after spring equinox. The less extensive PSC formation and the shorter lifetime of the Arctic polar vortex have thus far prevented the large-scale loss of ozone over the Arctic that is observed over Antarctica.

Nonetheless, the Arctic stratosphere was found to have as much ClO and Cl_2O_2 as was found in the Antarctic stratosphere, and the reactive nitrogen compounds were found to be converted from NO_x forms to HNO_3, and some of that was removed (54). These measurements were made during the Airborne Arctic Stratospheric Experiment (AASE), based in Stavanger, Norway, in January and February 1989 (55). The ER–2 aircraft was carrying the same instruments that it carried for the AAOE mission, as listed in Table I. A comparison of the average data for the two polar regions that are directly relevant to ozone loss are presented in Figure 9. The ClO mixing ratios are not distributed over quite the altitude range in the Arctic as over Antarctica, although the ClO profile for the Arctic is only from one flight, that of February 10, because the profiles from all other flights occurred when the solar zenith angle was greater than 90° and reactive chlorine was mostly in the form of Cl_2O_2. The abundances of BrO for both of the chemically perturbed regions were the same.

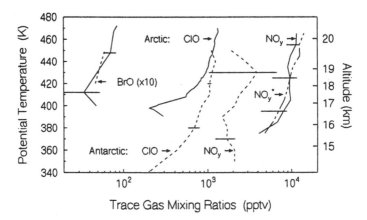

Trace Gas Mixing Ratios (pptv)

Figure 9. Comparison of Antarctic and Arctic in situ data, taken during the AAOE and AASE expeditions, respectively (54). Arctic data are represented by solid lines, Antarctic data by dashed lines. The dot–dash line represents the NO_y* mixing ratio for the Arctic and is only slightly different ($\sim$1 part per billion by volume (ppbv) less NO_y* at a given potential temperature) for the Antarctic. All data have been averaged over the flights except for ClO over the Arctic, which are data from only February 10, 1989. Error bars are the variability of the results for all flights at the 1σ confidence level. (Reproduced with permission from reference 54. Copyright 1991 American Association for the Advancement of Science.)

The NO_y values were significantly different, however, and much more NO_y was removed in the Southern Hemisphere compared to the Northern Hemisphere. The curve of NO_y marks the profile of NO_y predicted by measurements of the N_2O. However, in patches and later in the mission, as much as 35% of the NO_y was observed to be removed even in the Arctic. The NO abundance, not shown in Figure 9, was measured to be below the detection limit of the instrument in both chemically perturbed regions.

The possibility for ozone depletion in the Arctic polar stratosphere depends only on the abundances of reactive chlorine and bromine, the temperature (Cl_2O_2 rapidly decomposes back into 2ClO at higher temperatures), and the presence of sunlight. However, ClO, Cl_2O_2, and BrO must remain a large fraction of the total available chlorine and bromine for several weeks in the spring in order for the loss to be readily detectable by satellite-borne instruments. This requirement means that NO_x must remain small, because any NO_2 will rapidly react with ClO to form $ClONO_2$. Thus, although the sequestering of NO_x into HNO_3 or its possible removal by sedimentation does not dictate the rate of ozone loss, it does establish the length of time that ozone will be destroyed by halogen reactions. The Arctic polar vortex has on occasion been stable and cold enough for PSC formation well into March. If such a year were to occur with the current abundances of chlorine and bromine in the stratosphere, substantial ozone depletion in the Arctic polar vortex would be possible.

Heterogeneous chemistry occurs not only on polar stratospheric clouds but also on the global sulfate aerosols. The primary heterogeneous reaction is $N_2O_5 + H_2O \rightarrow 2HNO_3$ (56, 57), although the reaction $ClONO_2 + H_2O \rightarrow HOCl + HNO_3$ may also play a role. Model calculations that include these heterogeneous reactions better simulate the observations of ClO (58) and many of the nitrogen compounds, including NO (59), NO_2, and HNO_3 (56), than models containing only gas-phase chemistry. However, the agreement of the latitudinal and seasonal variation of these reactive trace gases is not always well simulated by models. Thus, although the evidence is strong that N_2O_5 is converted to HNO_3 by heterogeneous reactions on sulfate aerosols, these discrepancies indicate that more needs to be learned about the photochemical processes of the lower stratosphere.

Measurement Strategies for Stratospheric Trace Gases

Although researchers now have a working knowledge of the stratosphere, they do not have all the answers. The challenge is to go beyond a working knowledge to a predictive knowledge that can track the changing states of the stratosphere and accurately portray what will happen when any other changes in either the stratospheric environment or trace gas abundances occur. A predictive knowledge requires not only the ability to match computer model output to average values of observations of a number of species, but also to correctly match the observations when the environmental conditions of the stratosphere change.

There is already one excellent example of our failure to make such a predictive leap—the Antarctic ozone hole. The reason for the failure to anticipate the rapid loss of ozone in the lower stratosphere was a failure to appreciate the potential role of the subtle photochemistry, in particular, the heterogeneous chemistry. Nor did researchers have a full appreciation for the consequences of the air parcels inside the polar vortices being relatively isolated from midlatitude air. Some of these same issues are important in the Arctic region in wintertime, but researchers lack the predictive capability to determine how ozone will ultimately be affected.

The capability to predict what will happen to ozone in the future rests firmly on the ability to model recent ozone changes. These ozone changes have been carefully studied in a number of recent reports, although perhaps the most significant is the Ozone Trends Panel Report (15, 53, 56). This study of ground-based Dobson ozone-monitoring stations for the past two decades revealed an ~1% per decade summertime decrease in ozone in the Northern Hemisphere between 30 and 55°N that is roughly consistent with the predictions of computer model simulations. Further, losses of ozone in the upper stratosphere above 30 km due to increases in stratospheric chlorine are also roughly consistent with model results. However, computer models fail to explain the wintertime losses of ozone in the Northern Hemisphere by a factor of 3 and fail to explain the decrease in ozone at lower altitudes, as detected by the SAGE (Stratospheric Aerosol and Gas Experiment) I and II satellite instruments. Of course, assessment models without heterogeneous chemistry also fail to predict the massive losses in the Southern Hemisphere associated with the Antarctic ozone hole. All of these ozone losses must be explained.

Throughout the global stratosphere, many of the photochemical mechanisms remain untested. Although certain reactions are clearly occurring, they may not be the only reactions. A simple example is a test for the balance between the production and the destruction of ozone, as represented by equation 8. No experiment has yet been performed during which the abundances of all the rate-limiting components for ozone loss and ozone production, NO_2, HO_2, ClO, BrO, and O, have been measured.

However, even if such measurements were possible, would the uncertainty of the result be small enough to establish that production does indeed balance observed loss of ozone? The calculation of ozone loss in the Antarctic ozone hole was shown to have an uncertainty of 35 to 50%. The uncertainty for analyzing whether production balances loss in the midlatitude stratosphere is similarly 35 to 50%. About half of the uncertainty is in the measurements of stratospheric abundances, which are typically 5 to 35%, and half is in the kinetic rate constants, which are typically 10 to 20% for the rate constants near room temperature but are even larger for rate constants with temperature dependencies that must be extrapolated for stratospheric conditions below the range of laboratory measurements. In addition to uncertainties in the photochemical rate constants, there are those associated with possible missing chemistry, such as excited-state chemistry, and the effects of transport processes that operate on the same time scales as the photochemistry. Thus, simultaneous measurements, even with relatively large uncertainties, can be useful tests of our basic understanding but perhaps not of the details of photochemical processes.

An alternative strategy is to make a large number of in situ measurements and examine how the balance between production and loss changes for different conditions. Particularly important are variations with altitude, because the relative importance of the various chemical families to ozone loss changes between 20 and 45 km. Seasonal and latitudinal effects are important for testing both the production and the transport of ozone. At present, the main source of data for these variations comes from satellite instruments, although in situ instruments have been used for several altitudinal studies, primarily by balloon-borne instruments, and some limited seasonal and latitudinal studies.

The in situ measurement of several related trace gas species in the stratosphere has for a long time proven to be difficult. Part of the problem stems from the fact that it is difficult to get a larger number of instruments to work well together on the same platform at the same time. This problem has been largely conquered in the 1980s, and a number of multiple instrument programs using either aircraft or balloons as instrument platforms have been successfully completed. The remaining problem is that even more simultaneous measurements need to be made to resolve many issues. Measurements that were combined with the simplest photochemical relationships have had to be used to establish the validity of other photochemical relationships that are more complex.

An example of the promise and problems of combining observations and photochemical mechanisms to derive other trace gas abundances is provided by simultaneous in situ measurements of NO, ClO, and O_3 in the lower stratosphere over California in October 1988 during sunrise (*60*). These measurements were made by instruments mounted on the NASA ER–2 aircraft, which was flown in a tight 2°-latitude square at a constant altitude near 20 km. From these measurements and a knowledge of the solar zenith angle, the air density, and the air parcel trajectories, the abundances of a number of related species can be derived. From the measurements of NO ($\pm 25\%$), ClO ($\pm 30\%$), and O_3 ($\pm 5\%$) during sunrise, the abundances of NO_2 ($\pm 60\%$), $ClONO_2$ ($\pm 70\%$), N_2O_5 ($\pm 75\%$), NO_y ($\pm 50\%$), and OH ($\pm 150\%$) can be derived, where the numbers in parentheses are the uncertainties calculated by propagation of errors for the methods of determination shown in Table II. Much of the uncertainty in these calculations again stems from the uncertainties in the laboratory photochemical constants. A comparison of these derived abundances with other measurements in a sense validates the photochemistry assumed, and all the derived abundances are similar to direct measurements primarily from the ATMOS instrument. However, the uncertainties are large. Direct measurements are clearly better when possible—they have lower uncertainty and they test the photochemical mechanisms.

An example of using simultaneous measurements to check midlatitude stratospheric photochemical mechanisms is the use of the balloon-borne measurements of OH, HO_2, O_3, and H_2O to examine HO_x photochemistry (*61, 62*). This data set is a daytime, snapshot set of high vertical resolution (~ 100 m) measurements extending from 23 to 36 km. Over this altitude interval the data have been used to check production and loss balance of HO_x (OH and HO_2), to check the ratio HO_2 to OH, and to infer the abundance of H_2O_2. These results have provided the best check to date of HO_x photochemical theory with stratospheric measurements. For example, the comparison of measurement and theory for the ratio of HO_2 to OH is shown in Figure 10. There is excellent agreement above ~ 30 km but a marked divergence below this altitude by a factor of 2 in the 25-km region. However, the shaded region enclosing the calculated curve and indicating primarily uncertainty due to error in laboratory rate constant measurements overlaps the error in the measured ratio. An interpretation of this result is that present HO_x photochemical theory adequately explains the stratosphere, and a more discriminating check of HO_x chemistry will require additional low-temper-

Table II. Calculating Other Trace Gas Concentrations from Aircraft Measurements of NO, ClO, and O_3

Trace Gas	Derivation	Uncertainty[a]
NO	measured	0.15
O_3	measured	0.05
ClO	measured	0.30
NO_2	$\dfrac{[NO_2]}{[NO]} = \dfrac{k_{NO+O_3}[O_3] + k_{NO+ClO}[ClO]}{J_{NO_2}}$	0.60
$ClONO_2$	$\dfrac{d[ClO]}{dt} = J_{ClONO_2}[ClONO_2] - k_{ClO+NO_2}[M][ClO][NO_2]$	0.70
N_2O_5	$\dfrac{d([NO] + [NO_2])}{dt} = \dfrac{d[ClO]}{dt} + 2J_{N_2O_5}[N_2O_5]$	0.75
NO_y	measured on flights on October 6 and 15, 1988	0.30
HNO_3	$[HNO_3] = [NO_y] - ([NO] + [NO_2] + 2[N_2O_5] + [ClONO_2])$	0.50
OH	$[OH] = \dfrac{J_{HNO_3}[HNO_3]}{k_{OH+NO_2}[M][NO_2] - k_{OH+HNO_3}[HNO_3]}$	1.50

[a]Uncertainty is reported as the fraction of the observed value at the 1σ confidence level.
SOURCE: Reference 60.

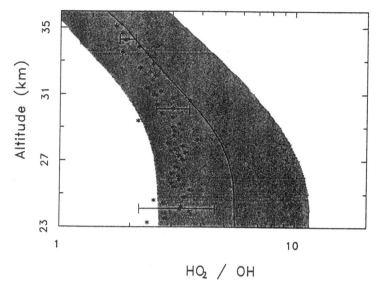

Figure 10. Calculated (solid) and measured (*) ratio of [HO₂] to [OH] (62). The shaded area represents the 1σ estimate of the uncertainty in the calculated ratio. Error bars for the measured ratio are 1σ. (Reproduced with permission from reference 62. Copyright 1990.)

ature, laboratory kinetics studies of some notable reactions controlling HO_x chemistry.

These examples illustrate that an increasing number of trace gases must be measured simultaneously if even limited subsets of stratospheric photochemistry and transport are to be understood. The combined uncertainties will also become less of a constraint as simultaneous measurements of trace gas abundances can be compared to values derived from other observed abundances and simple photochemical relationships. As important is the improved measurement of photochemical parameters from laboratory studies as well as the search and study of other mechanisms that may be occurring in the stratosphere. Concerted effort in all of these categories is required to avert future failure in predicting shifts in stratospheric photochemistry, like the Antarctic ozone hole.

Platforms and Instruments for In Situ Measurements of Reactive Trace Gases

This chapter has so far concentrated on the conditions and photochemistry of the stratosphere and what strategies are used to understand them. In this section, some challenges of actually making the in situ measurements are discussed. In particular, the measurements of some of the reactive trace gases that have a direct affect on ozone—NO_2, NO, OH, HO_2, ClO, Cl, BrO, and O—are discussed. No measurements of Br have been made. Two research platforms that have been used for in situ stratospheric measurements are the helium-filled research balloon and the NASA ER–2 high-altitude aircraft. Each has strengths and weaknesses that are discussed. Abundances of NO_2, NO, ClO, and BrO have been measured from both balloons and aircraft; OH, HO_2, Cl, and O have been measured only from balloons. Some of the techniques and some of the resultant measurements are discussed in detail.

Balloon-Borne Measurements.   To illustrate the versatility possible with balloon-borne platforms, the in situ techniques that have recently made important contributions to our understanding of stratospheric reactive trace gases are highlighted. Each technique is based on a fundamentally different physical principle, providing measurements with unique and characteristic spatial and temporal scales. But first the advantages and disadvantages offered (and suffered) in balloon-borne experimentation are reviewed. Some unique facets of balloon behavior that are relevant to a specific experiment are discussed with that experiment.

Most stratospheric research ballooning in the United States is supported directly by NASA through the National Scientific Balloon Facility (NSBF), which has its headquarters and main launching site in Palestine, Texas (32°N). Another launch site at Ft. Sumner, New Mexico, (34°N) is being used with increasing frequency, and NSBF has supported launches out of other remote locations such as Antarctica and Greenland.

The foremost advantage of balloon-borne deployment is that it provides a relatively inexpensive vehicle that accesses an exceedingly large altitude interval of the atmosphere. That interval extends from ground level up to 40 km, but most experimenters use balloons to operate exclusively in the stratosphere. There is considerable flexibility in controlling the amount of time spent at a given flight level, or rates of ascent and descent, by controlled release of ballast, typically iron or glass shot or helium. The duration of a flight may extend from a few hours to several days. Balloons provide very gentle, vibration-free flights. The balloons are expendable because they are made of very fragile materials and are normally rendered unusable by flight termination procedures. The lift capacity of a large balloon, 30 to 40 million cubic feet (800,000 to 1.1 million m^3) in volume, can exceed 2000 kg; thus balloons can fly relatively large experimental gondolas supporting instrumentation designed to detect several different species. The scientific objective of simultaneous measurements of photochemically linked species may be met, albeit at the cost of some gondola complexity.

An experiment operating remotely (modern-era balloon-borne experiments never carry passengers) in the temperature and pressure environment of the stratosphere presents several engineering challenges that must be met. Remote operation means that the balloon instrument is by design more autonomous than a typical DC–8-borne instrument, for example. All experimental power sources and heating or cooling subsystems must be on board. Most experiments require real-time control from the ground, that is, telemetry links to and from the experiment.

Ballooning is of course not without disadvantages that are sometimes considerable. There is relatively little control on the trajectory a balloon will follow after it is launched. What little control there is comes from controlling altitude or ascent rates to take advantage of different wind directions and speeds at different altitudes. For this reason balloon launches are always preceded by several weather balloon sondes to forecast flight paths. These forecasts are important scientifically when they involve issues such as the trade-off between telemetry range and flight time, for example. For NSBF personnel, who continuously track altitude and position of each balloon flight and control all balloon command procedures from launch to landing, accurate forecasting is critically important. There are some safety issues involved that, depending on the payload weight and mission complexity, impose restrictions on areas over which a flight is allowed to take place. Thus flight opportunities can be constrained by overhead meteorology and the presence of densely populated areas under flight trajectories. In general, balloon overflight restrictions have now limited the opportunity to fly large gondolas out of Palestine to the summer months when the winds at altitude are blowing west, leading to the increasing use of the Ft. Sumner location. In addition, the local meteorological conditions at the launch site and at the projected landing site can cause unavoidable delays in carrying out a flight. Large balloon launches require relatively calm winds during the inflation and launch procedure.

Ballooning has always presented risk with respect to balloon performance. In the mid-1980s there was an abnormal number of balloon failures due to subtle changes in the manufacturing process used in polyethylene film production. That problem has since been corrected, but there still remains the risk of mechanical or related electrical failure, which, while low, still occurs with some frequency. In the worst case a free fall can occur that results in total destruction of an instrument. A normal recovery takes places by instrument deployment on a parachute followed by tracking and recovery by NSBF personnel. The consequences of landing can be benign or something considerably less so, depending on the presence of obstacles and sometimes the wind at the landing site. Most experiments are sufficiently complex and sensitive that it is not a simple matter to go through a launch and recovery operation without incurring significant refurbishment time before the next launch attempt. This situation is clearly a disadvantage of ballooning: considerable effort goes into preparing and flying a balloon-borne experiment in which the balloon is expendable, and the amount of data recovered can seem small in comparison to the effort.

Measurements of NO and NO₂ by Chemiluminescence. Most balloon-borne in situ measurements of NO and NO_2 published to date have been made with the chemiluminescence technique (6, 63, 64). Reagent ozone is added to the ambient flow in a reaction chamber, and the chemiluminescence observed from the reaction $NO + O_3 \rightarrow NO_2^* + O_2$ is detected by a photomultiplier tube that views the volume. The amount of light observed is proportional to the amount of NO in the ambient air. NO_2 is detected by exposing the ambient air going into the detection cell to a strong light source; this exposure photolyzes NO_2 to NO, which can then be detected.

This technique has both advantages and disadvantages for balloon-borne measurements. It is sensitive, able to detect less than 50 parts per trillion by volume (pptrv) of either NO or NO_2 (6), and its characteristics are well understood. Further, one type of instrument with different inlets can be used to detect both NO and NO_2. On the other hand, any other trace species that chemiluminesce in the reaction chamber are detected as well as NO and lead to artifact signals that must be measured and subtracted from the total signal. The problem of an artifact signal is exacerbated for the NO_2 detection because of the photochemistry of nitrogen-containing species that occurs in the photolysis cell. In the troposphere, where the instrument can be kept at a relatively constant temperature and pressure, the artifact signals can be tracked over time and can be well characterized. For a balloon flight that lasts a matter of hours and spans wide ranges in temperature and pressure, stabilizing the instrument and its artifacts is a considerable experimental challenge.

An example of the use of this technique is the measurement of NO and NO_2 made near 50°N latitude by Ridley and co-workers. Summertime measurements of NO, NO_2, O_3, temperature, and the photolysis rate of NO_2 showed that NO and NO_2 were in photochemical steady state (65). However, the abundances of NO_x (NO + NO_2) were observed to be ten times smaller in winter than in summer at altitudes between 20 and 28 km (63). At altitudes above 28 km, the abundances of NO_x were similar in both winter and summer. Considering the trajectories of air at different altitudes, they were able to determine that N_2O_5 must be the wintertime reservoir species, as was predicted in a number of previous studies.

Ridley and co-workers dismissed the possibility of heterogeneous conversion of N_2O_5 to HNO_3 proposed by Evans et al. (66) because the estimated time constant, about 18 days, was too slow. Current estimates of the aerosol surface areas in the lower stratosphere (56) give a conversion time constant of about two days, which is comparable to the time constant for the conversion of NO_2 to N_2O_5. Their observations of NO may have been affected by heterogeneous chemistry, and simultaneous measurements of HNO_3 and aerosol properties would have allowed them to discover the effects. Thus the importance of simultaneous measurements of the relevant trace species for developing an understanding of stratospheric chemistry is clear.

The Balloon-Borne Laser In Situ Sensor. An instrument was developed by Webster and co-workers at the Jet Propulsion Laboratory that uses single-mode, tunable diode lasers (TDL) to detect IR-active molecules by absorption (67–69). Typically four different lasers are used, each spanning a small range within the 1000- to 3000-cm^{-1} interval. The following species have been detected: NO, NO$_2$, N$_2$O, O$_3$, CH$_4$, CO$_2$, H$_2$O, HNO$_3$, and HCl. Upper limits, that is, no detectable signals, have been reported for HOCl and H$_2$O$_2$. This method benefits from being able to detect a large number of species besides NO and NO$_2$ with a single instrument.

In a balloon-borne configuration this technique might be labeled a semi-remote one because a long absorption path length is attained by lowering a tethered retroreflector below the gondola by up to 500 m, which yields a 1-km path length. In practice a retroreflector distance of 200 to 300 m is used to optimize the observed signal, trading off less fractional absorption for more returned laser power. A somewhat sophisticated aiming and tracking system using a He–Ne laser is used to find and track the retroreflector.

The basic simplicity of absorption is an advantage for this system. After an absorption feature is identified, the absorber density can be calculated with knowledge of the integrated absorption line strength, line broadening parameters, and the path length. The spectroscopic information is usually known or can be further investigated in laboratory studies when required. The atmosphere is optically thin for all of the detected species. Temperature and pressure are measured directly. Only one TDL may be used at a time, but the time required to change lasers is about a minute, so essentially simultaneous measurements of all of the above species can be made. The laser optical path and absorption detection hardware are identical for each laser.

The single-mode output from a TDL, 1×10^{-4} to 5×10^{-4} cm^{-1}, is considerably less than the Doppler- or pressure-broadened widths of the absorbers, yielding high-resolution absorption spectra. The TDL is tuned with a 2-kHz dither, allowing phase-sensitive detection techniques to measure small changes in the returning laser power. Absorbances as low as 10^{-5} are detectable. Because the detected species are mostly stable molecules, in-flight line identification can be verified by passing a beam-split fraction of the laser beam through on-board samples of the detected gases. In some cases, a surrogate molecule serves as a wavelength standard, such as NH$_3$ for O$_3$.

Balloons offer several different scenarios for mission planning. Slow-descent flights are possible for generating altitude-dependent measurements. However, for observing rapidly varying species such as NO$_2$ around sunrise and sunset it is obviously important to remain at one altitude, because the kinetics of this process is dependent on density and temperature. Thus the typical balloon flight remains for extended periods of time at a single altitude to record short- and long-term changes in NO$_x$ species. Some flights have provided data from two different altitudes.

Perhaps the greatest challenge with this instrument is the design and fabrication of a system that can lower a retroreflector and reflect a 0.5-mW laser beam off of it and back to the detection optics. A balloon-borne platform is a particularly benign vehicle from which to operate. Of all of the detected species, only the H_2O measurement can significantly be affected by the presence of the gondola. In fact, water contamination was noted on one flight when a heater was turned on to warm up a translation stage on the gondola. The presence of the perturbation was obvious because of the large absorption signal, equivalent to 16 ppm, a factor of 3 greater than that expected. The correlation of the signal with heater activation provides a sound demonstration of an instrumental influence. Another diagnostic, if required, could have included checking the absorbance dependence on path length by varying the retroreflector position, because water outgassing from the gondola must be strongly localized in the vicinity of the gondola. It is highly unlikely that any of the other detected species could be affected by the gondola, tether, or retroreflector.

Balloon lift capacity has allowed flying a gondola that can support the BLISS instrument with other instrumentation for detecting J_{NO_2} and O_3 density. Perhaps the most important contribution of this instrument thus far has been its comprehensive check of NO_x chemistry. In particular, rapid conversion of NO to NO_2 at sunset and slow conversion of NO_2 to N_2O_5 at night is detailed by the diurnal measurements of NO_2 illustrated in Figure 11 (70). The NO_2 diurnal dependence, the ratio of NO_2 to NO, and the

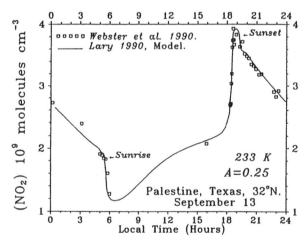

Figure 11. Comparison of [NO₂] against time from BLISS September 1988 flight measurements at 30 km with model predictions of the NO₂ diurnal behavior; an albedo of 0.5 was used (70). The peak at sunset is due to a slight descent of the balloon at that time and is accurately captured by the model calculations that mimic the motion. (Reproduced with permission from reference 70. Copyright 1981.)

mixing ratios of NO, NO_2, and HNO_3 confirm that the NO_x photochemistry indicated in Figure 8 is essentially correct. New measurements of some precursor species, CH_4, H_2O, and N_2O, have served as valuable independent checks of previous measurements by other methods.

OH Detection by Laser-Induced Fluorescence. The OH (and HO_2) instrument is deployed on a large gondola that has flown with instrumentation for O_3 and H_2O detection (61, 62, 71). The gondola has been configured to include NO and ClO, but a flight including that complement has not been carried out. High-resolution, altitude-dependent measurements are made as the balloon descends by controlled release of helium through a large valve mounted at the apex of the balloon. Descent is initiated by cutting open ducts at the side of the balloon to release a large amount of He rapidly.

OH is detected with the laser-induced fluorescence technique. OH is optically excited in the A-X(1,0) band via the $Q_1 1$ line at 282 nm with a pulsed Cu vapor laser-pumped dye laser with a 17-kHz repetition rate. The primary signal is composed of fluorescence of OH in the A-X(0,0) band at 309 nm, observed with a filtered photomultiplier tube. Detection takes place in the core of a flowing sample of ambient air as it passes through a cylindrical detection chamber that has a 12.5-cm diameter at it narrowest point. The OH fluorescence is distinguished from the background signal by successively step tuning the laser on and off resonance with the OH absorption line. Sources of background signal are solar scattering off the atmosphere below the instrument; Rayleigh, Raman, and chamber-scattered laser light; and non-OH fluorescence. HO_2 is detected by conversion to OH by the addition of NO to the sample in a sequence that is synchronized with the laser tuning cycle.

The reactive nature of the OH radical presents a significant experimental challenge. Foremost, the detection method must be nonintrusive, demonstrating by auxiliary diagnostic measurements that the atmospheric sample is not perturbed by the presence of the instrument. OH is detected within 20 ms after it enters the shrouded detection chamber, a time scale much shorter than its stratospheric lifetime of ~20 to 10 s over the 36- to 25-km interval. It is assumed that OH will not survive collisions with surfaces or with contaminants that outgas from the detection volume walls, the gondola, or the balloon. Avoiding contamination from the gondola and balloon requires that measurements be taken when the gondola is descending at velocities of 3 to 5 m/s. More rapid descent velocities are detrimental only in the sense that they would decrease the altitude resolution of the measurements. A typical altitude profile extends from 37 to 23 km. In practice, 37 km represents an upper limit that a balloon of 30 million cubic feet can reach with a 1600-kg payload. The lower limit is decided by operational factors on the day of the flight, such as an appropriate landing site that avoids possibly hazardous locations.

The balloon descent period takes place in a relatively short amount of time, ~50 min to descend 15 km. In order to achieve a result with good altitude resolution the instrument must have high absolute OH sensitivity in concert with low total background signal. In the August 25, 1989, flight, observed signal-to-noise ratios achieved in 20-s integration periods at 35 km were 135 and 80, for OH + HO$_2$ detection (with NO addition) and OH detection (no NO addition), respectively, and at 23 km the ratios were 11 and 5. The absolute OH detection sensitivity of the flight instrument is calibrated prior to flight by inserting the core of the detection module into a laboratory flow tube apparatus. Known OH densities are produced by titrating NO$_2$ with excess H atoms with standard laboratory chemical kinetics investigation techniques.

The most serious possible problem with this measurement technique is the possible loss of OH prior to detection. The primary diagnostic indicator for the presence of contamination is the temperature of the sample air flow after passage through the detection volume. If boundary-layer air has intruded into the detection volume, the sample air temperature will rise above ambient because the walls of the detection module are always warmer than the ambient air temperature. For example, at float altitude the temperature difference is ~15 °C.

While at float the instrument is brought to a fully operational state before the critical part of the flight, the descent, is initiated. The float period provides an opportunity to observe instrument behavior under decidedly nonideal conditions because it is highly likely that the sampled air will be perturbed either by the presence of the gondola in the local environment or by poorly developed flow in the pod itself. Absence of a concerted gondola descent velocity leads to more easily perturbed flow in the detection module by influences such as relative wind shear at the pod entry point. Also, the fan used as a flow impeller at the exit of the detection module operates less efficiently at 6 mbar (37 km) than it does at greater pressures. The orientation of the gondola with respect to the sun may also be important in the absence of vertical motion if the sample has a long residence time in the shadow of the gondola. Moreover, at float the balloon undergoes a slight oscillatory motion of ~100-m amplitude in an ~3-min period. This motion can affect the purity of the sampled air if the gondola has just passed through the column of air it is sampling.

In any case, at float dramatic instances of contaminated sampling on a short time scale associated with poor flow in the pod are observed, and variability on perhaps a longer scale due to varying influence of the gondola on the air sample is observed. An example is shown in Figure 12, a plot of unaveraged detector count-rate data and the air flow temperature versus time. The signal shows a high correlation with the observed temperature trace. The intrusion of warm air causes variability in the ratio of HO$_2$ to OH, variability in total HO$_x$, and increases the off-resonant background signal.

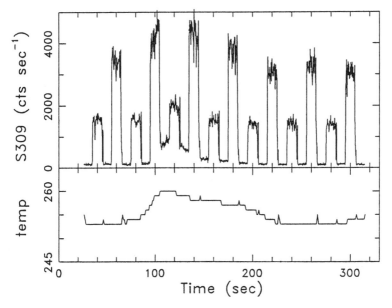

Figure 12. Plot of unaveraged detector count-rate data (top) and airflow temperature (bottom) versus time for the August 25, 1989, flight. Data are observed during the balloon float period at an altitude of 37 km. The 20-s period modulation in the count-rate data is due to tuning the laser on- and off-resonance with the OH absorption. The alternating high and low OH signal is due to measurements with and without NO injection.

Evolving boundary-layer separations or excessive growth that intrudes into the core of the flow are observed on the axis at the deepest distance into the detection module first as flow disturbances grow into the flow stream lines. The background increase is thought to be due to organic compounds in the paint used to blacken the interior of the detection pod. The background signal is an excellent marker of boundary-layer air intrusion into the detection volume.

In contrast, data recorded during the descent are shown in Figure 13. The pod flow temperature matches the ambient air temperature, and the background fluorescence is stable. There are no instances of poorly developed flow affecting either the detection volume in the temperature or the background fluorescence data. The gondola is descending at a sufficient velocity that neither the gondola nor the balloon can affect the sample. The derived HO_2 mixing ratio from this flight is shown in Figure 14.

Measurements of O, Cl, ClO, and BrO by Resonance Fluorescence. The resonance fluorescence technique that is used for the detection of OH has also been used for the detection of O (72), Cl (73), ClO (74–76), and BrO (76). Detection of the free radicals was made on the descent of the

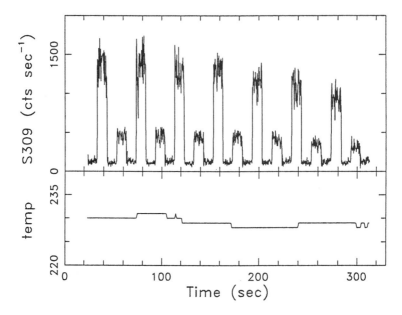

Figure 13. Same plot as in Figure 12. Data were observed during the balloon
descent through the 30.5- to 28.8-km interval.

balloon to minimize the effects of contamination and wall loss, just as for
OH. But the light sources were small, low-pressure helium lamps containing
trace amounts of the atom of interest. To separate the signal due to the
radical from the background scattering, a reagent chemical is added to the
flow to either convert ClO and BrO to their respective atoms or to remove
Cl by chemical reaction.

Aircraft measurements of ClO and BrO are discussed in some detail in
the next section. Because the aircraft and balloon-borne techniques are
essentially the same, further discussion about these two reactive trace gases
is deferred to this later section. A discussion of some of the challenges of
measuring atomic trace gases in the stratosphere follows.

Determining the abundance of Cl atoms is difficult because of the small
abundance of this species (Figure 5), which is predicted to be between 10^4
and 10^6 atoms/cm^3 between 20 and 40 km, respectively. With the sensitivity
achieved for the detection of chlorine on balloon-borne instruments, about
10^5 atoms/cm^3 could be detected in a minute above an altitude of 35 km.
However, the background signal increases rapidly below 35 km because of
the increase in Rayleigh scattering by air, and the signal-to-noise ratio is no
longer sufficient for the measurement. On the other hand, most recent
balloon descents have been initiated either by opening a valve in the top of
the balloon or by cutting ducts. These procedures usually require a few
kilometers to achieve the desired descent rates. Thus, even for a balloon
starting at an altitude of 40 km, a good flow would not be established in the

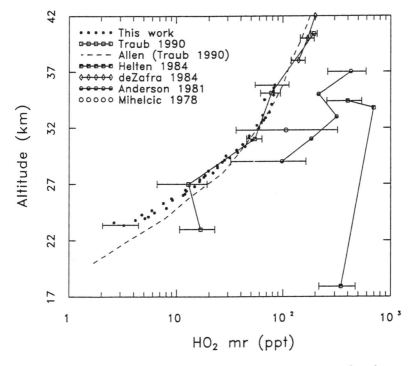

Figure 14. Summary of HO$_2$ mixing ratio measurements versus altitude at a 54° solar zenith angle (61). References to other measurements can be found in reference 71. The dashed line is a model calculation. (Reproduced with permission from reference 61. Copyright 1990.)

measurement flow tube until below an altitude of ~37 km, where Cl is difficult to detect. As a result, the only successful atomic chlorine measurements have been from balloon payloads that descended through the stratosphere on a parachute.

The abundances of oxygen atoms are orders of magnitude larger than the abundances of chlorine atoms in the stratosphere (Figure 5). Because the detection sensitivity and the background signals are comparable for O and Cl, the abundances of O can be detected to below 30 km, where the combination of smaller signals from smaller O abundances and larger background signals from the increasing Rayleigh scattering again reduces the signal-to-noise ratio. The main difficulty with measurements of O is the rapid conversion into ozone in the absence of solar UV light. Even at 35 km, the time required for conversion is substantially less than a second. Thus, any shadowing of the sampled air mass by the instrument, the gondola, or the balloon, even for a second, will distort the measurement. This constraint has forced the development of O detection systems that were open to the stratosphere, although the background signal in these systems is larger than that for closed system. As a result, just as for Cl, abundances of O atoms have only been detected in situ by balloon-borne instruments that were dropped on parachutes through the stratosphere.

The measurements of Cl and O atoms agree with model calculations to within about a factor of 2 (6, 77). This agreement is reasonable in light of the uncertainty of the measurements and the lack of simultaneous measurements of other trace gases and meteorological parameters. Although these measurements suggest that researchers have a general understanding of the photochemical balance that affects these trace reactive species, a more critical evaluation of this fast photochemistry awaits new, more sensitive measurements.

In conclusion, these balloon-borne techniques use very different detection strategies to produce measurements that, when taken together, span a wide range in spatial and temporal resolution for reactive trace gases. Progress in scientific instrument design is pushing forward scientific goals from simply checking model agreement with observed average abundances to actually constraining photochemical models by making simultaneous or diurnal measurements of several different species. However, future progress with expanded simultaneous measurement data sets and perhaps also instrument intercomparisons will be limited by the logistical difficulty in mounting large balloon campaigns.

It is difficult to envision that balloon-borne techniques will be able to satisfy the demand for more complex and frequent measurements. Therefore it is critical for maximizing the scientific return of balloon-borne flights to include simultaneous measurements of the "right" mix of species and photolysis rates. In the absence of frequent balloon-borne measurements it is nonetheless very gratifying to have achieved the coalescence of results by independent techniques as a zero-order substitute for intercomparisons. A possible direction for future stratospheric research with a new platform, remotely piloted aircraft, that alleviates some disadvantages of balloon-borne platforms is discussed in the last section of this chapter.

Measurements from the NASA ER–2 Aircraft. Measuring trace gas abundances from high-altitude aircraft is fundamentally different from measuring trace gases from balloons. The vertical range of measurements is restricted to below 18 to 21 km, the ceiling for the aircraft, but the horizontal range is ~4000 km. Thus, while the high-altitude aircraft has limited use for studying the evolution of the global stratosphere, it is invaluable for studies of the lower stratosphere and issues such as stratosphere–troposphere exchange, the Antarctic ozone hole, and the potential for ozone loss in the Arctic. Another strength of the high-altitude aircraft is that it can carry and power a large number of trace gas sensors. The NASA ER–2 aircraft (78) has been the most widely used high-altitude aircraft for measurements of trace gases in the last six years. Its characteristics are representative of other subsonic, stratospheric aircraft, and so a discussion of ER–2 measurements provides a good basis for stratospheric aircraft measurements in general.

One of the strengths of the ER–2 is that the flight trajectory is controlled by the pilot on board. The pilot is able to perform tasks related to instrument operation, such as the timing of certain measurements with respect to some position or event in the stratosphere. The pilot is also able to restart instruments that have suffered momentary failure. However, the disadvantages of having a pilot on board are as significant as the advantages. First, for pilot safety, the flight path cannot be further than 200 miles from land without an escorting rescue aircraft, nor can the flight venture into a region where engine performance would become questionable. Second, pilot fatigue limits the duration and thus horizontal extent of the flight to ~8 hours or less. Third, at unfamiliar airfields both takeoff and landing must occur during daylight. And fourth, all instruments need to be engineered with the safety of the pilot as highest priority. These restrictions, while subject to negotiations should the situation demand it, are really a matter of common sense.

The reliability of the high-altitude aircraft is very high, with few potential flight days lost to aircraft malfunction, and unlike balloon-borne instruments, aircraft-borne instruments may be used repeatedly without needing to be extensively refurbished after each landing. Surface winds on both takeoff and landing are often an issue, however, because the surface winds must not exceed 15 knots in crosswind or 30 knots along the runway during either takeoff or landing. For instance, during the Airborne Arctic Stratospheric Experiment in Stavanger, Norway, 13 potential flights out of 34 planned flights were canceled shortly before takeoff by strong surface winds at the scheduled time for takeoff or predicted strong surface winds at the scheduled time for landing. Only three were canceled by aircraft problems. For missions to both the Arctic and the Antarctic, strong surface winds were invariably coupled to close proximity of the polar vortices; these conditions prevented flights that could have penetrated deep into the chemically perturbed regions from taking off.

The speed of the ER–2 aircraft is approximately constant at 0.7 Mach, and this speed presents some difficulties for measurements. First, fast sensors are required to measure subkilometer-scale variations of trace gas abundances. A speed of 0.7 Mach translates into 200 m/s, so data rates exceeding 1 Hz are required. This requirement is difficult to meet in the detection of some trace radical species, such as ClO, BrO, and even NO under some conditions. Another consequence of this aircraft speed is that sampling of a particular air parcel for several hours is difficult, because the aircraft does not remain in one place very long. As a result, diurnal studies of trace gases must be attempted by either having the pilot fly in a tight box pattern or in the direction of the prevailing wind to minimize spatial gradients in measured abundances.

Second, if the air sample captured by the instrument inlet is slowed appreciably from the 200-m/s air speed, then the temperature of the gas will rise ~20 °C and the pressure will be increased as much as 50% by ram force. For measurements of volatile particles, such a temperature rise may evaporate some unknown fraction of the particles and cause an undercounting. Another difficulty occurs when particles contain the trace gas of interest. An example is the detection of NO_y in the presence of polar stratospheric clouds. If the inlet is not isokinetic, then as most of the airstream passes around the inlet, the heavier particles will have too much momentum to deviate their course and will enter the inlet. If then the particles evaporate, as the polar stratospheric clouds do, then the measured abundance of NO_y will be increased enormously. These large spikes mask the gas-phase component of reactive nitrogen, but they also provide information about the solid-phase component (44). Careful modeling and calibration of the inlet is required to determine both the size range of the particles accepted and the mixing ratio of the NO_y in the PSCs.

A final challenge that must be solved is the instrument operation and data collection. Although the pilot can perform some operations, the primary responsibility is flying the aircraft. Thus, instruments that operate autonomously, with the pilot only turning them on and off, have the highest probability of successful data collection. Most instruments are computer-operated; the sequencing of operations and calibrations are performed automatically. One problem with this approach, in which the program for sequencing is dictated before flight, is that calibrations and diagnostic tests can mask interesting events in the data. Another problem is that the data are collected on board with either tapes or hard disk memories and are not known until the aircraft returns. Recent efforts to change this mode of operation are being implemented so that commands can be sent to the instrument and data can be received from it by a scientist on the ground (S. Wegener, private communication). This change will make some aspects of choosing a flight plan more responsive to the data being collected.

The measurements of trace species—both gases and particles—and of atmospheric parameters that were measured during the Airborne Antarctic Ozone Expedition and the Airborne Arctic Stratospheric Experiment are given in Table I. These techniques have quite different requirements that are dictated by the detection technique and the way an air sample is handled.

Of all the trace gases, particularly the reactive trace gases, some of the most difficult to measure are the trace free radicals. At present, NO, ClO, and BrO have been measured from the NASA ER–2 high-altitude aircraft. The challenges of measuring NO (44, 79) from the ER–2 are similar to those of measuring from balloons, as discussed earlier in this chapter and in Chapter 9. Those discussions are not repeated here, but some examples of NO measurements are given. Instead, the measurement of ClO and BrO from the aircraft platform is discussed.

Measurements of ClO and BrO by Resonance Fluorescence.

The ClO–BrO instrument was designed to test the theory that chlorine and bromine were the cause of the rapid loss of ozone over Antarctica. Its pedigree includes several versions of balloon-borne halogen radical measuring instruments that have been discussed in a number of publications (74–76). However, measurements from an instrument that is mounted in a pod under the wing of an aircraft traveling at 200 m/s are substantially different from measurements from balloons that are descending at a much lower speed. A brief description of instrument operation is given here. More complete descriptions are presented elsewhere (45, 80–82).

ClO and BrO abundances are detected simultaneously and continuously as the airstream passes through the instrument. They are not detected directly but are chemically converted to Cl and Br atoms by reaction with reagent nitric oxide gas that is added to the airstream inside the instrument. The Cl and Br atoms are then detected directly with resonance fluorescence in the $^2D_{5/2} \rightarrow {}^2P_{3/2}$ transitions in the vacuum ultraviolet region of the spectrum. In resonance fluorescence, the emissions from the light sources are resonantly scattered off of the Cl and Br atoms in the airstream and are detected by a photomultiplier tube set at right angles to both the light source and the flow tube. The chemical conversion reactions

$$ClO + NO \longrightarrow Cl + NO_2 \qquad (21)$$

$$BrO + NO \longrightarrow Br + NO_2 \qquad (22)$$

are fast, with bimolecular rate constants of $\sim 3 \times 10^{-11}$ cm^3 molecule^{-1} s^{-1} (76), so NO concentrations of 10^{13} molecules/cm^3 will convert all the halogen oxides to halogen atoms in less than 10 ms. The signals from resonance fluorescence are distinguished from background signals, consisting of Rayleigh scattering and scattering from the instrument optics and surfaces, by periodically adding NO to the flow. The difference in signal between when NO is added and when it is not present is proportional to the ClO or BrO abundance.

The resonance fluorescence technique is the same as that used to detect OH by the balloon-borne in situ instrument. The only difference is that the resonant radiation is obtained from a low-pressure helium light source, excited by 25 W of radio frequency power, that contains a trace amount of chlorine or bromine. The chlorine light source, when filtered through molecular oxygen, has one emission line for chlorine at 118.9 nm and an impurity line from hydrogen at 121.6 nm. The hydrogen line is $\sim 5\%$ of the chlorine line for the light sources used. The bromine light source has several lines with wavelengths between 115.0 and 320.0 nm that illuminate the detection volume.

The chemical conversion is not without some difficulty because the reagent NO also reacts with the product Cl to form ClNO. In the lower stratosphere, this termolecular reaction is about 10% as fast as the forward reaction. In addition to this removal of Cl is the reaction

$$Cl + O_3 \longrightarrow ClO + O_2 \qquad (23)$$

which shifts some of the converted chlorine back into ClO. As a result, three different flow rates of NO are added to the airstream in the 32-s NO-on–NO-off cycle to find the maximum conversion. An additional constraint for detecting free radicals is to prevent exposure to wall surfaces, on which they are destroyed with nearly every collision.

The minimum detectable ClO and BrO abundances are dependent on the detection sensitivity, the background signals, and the chemical conversion efficiency. All of these factors are affected by the air density and thus the flight altitude. The sensitivity is affected by the small but measurable absorption of the resonant emissions by molecular oxygen, which has a higher density at lower altitudes, and by increased quenching of the $^2D_{5/2}$ excited state by collisions with nitrogen molecules at the higher densities of the lower altitudes. The Rayleigh scattering, which contributes more than half of the total background signal, also increases with increasing density. This Rayleigh scattering is used to help track the instrument sensitivity in flight.

The instrument sensitivity is calibrated in the laboratory by making a known amount of Cl or Br atoms and recording the signal and all relevant parameters. Rayleigh scattering from known amounts of air, in the laboratory and in flight, is used to maintain the calibration of the instrument to Cl atoms. This technique is discussed in some detail in references 80 and 82. The portion of the background signal due to Rayleigh scattering is determined when the pilot performs a vertical dive with the aircraft. Because the change in calibration during each seven-hour flight is usually less than 10%, one dive during the flight provides the in situ calibration of the instrument.

The third factor that is important in determining the detection limit is the conversion efficiency of the kinetics. A conversion efficiency of ~1.0 requires that the airstream have a velocity substantially less than 200 m/s because uniform mixing of NO is very difficult. At the same time, collisions of the sample airstream with wall surfaces in slower inlet systems may cause a chemical loss of ClO and BrO, because they are both reactive with wall surfaces. The solution to this problem was suggested by Soderman (83). Soderman's novel design consists of two nested ducts in which the air speed is decreased from 200 m/s to 60 m/s in a 14-cm-diameter outer duct that protrudes 60 cm in front of the left wing pod and is reduced to ~20 m/s inside a smaller 5-cm-square duct in which the measurements are made. The entrance to the smaller measurement duct is 60 cm downstream of the entrance to the outer duct, and the NO injector tubes, the two ClO detection axes, and the one BrO axis are 25 cm, 37.5 cm, 55 cm, and 72.5 cm downstream of the entrance of the measurement duct. Ninety percent of the air that enters the outer duct bypasses the measurement duct through additional duct work, and only the center 10% of the airstream is captured and sampled by the measurement duct. These two flows are recombined downstream of the instrument and are vented out the side of the wing pod that houses the instrument.

The measurement duct samples a laminar airstream traveling at 20 to
30 m/s that has not come into contact with any wall surfaces. The temperature
of the airstream is increased ~20 °C by adiabatic compression, and the
pressure is higher because of dynamic pressure, but the trace chemical
content of reactive chlorine is intact. As a test of the system, and to prevent
it from getting too cold, the temperatures of the wall surfaces were maintained near 0 °C. Arrays of eight 0.010-inch (0.025-cm) diameter bead thermistors in both the outer and the measurement ducts (10 cm downstream
of the BrO detection axis) are then used to distinguish between warmer air
that has been in the boundary layer of the walls and air that is ambient. At
the same time, the velocity of the air in the measurement duct can be varied
from 0 to 40 m/s for additional flow and chemical kinetic diagnostics. Temperature, pressure, and potential temperature for a typical flight are shown
in Figure 15. The potential temperature in the core of the flow does not
change as the air stream passes through the instrument. Air from the boundary layer of the outer duct has never been seen to enter the measurement

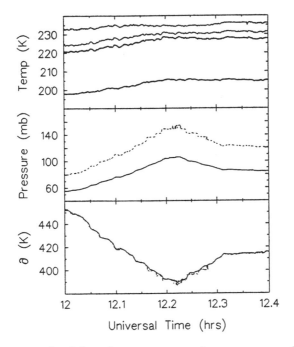

*Figure 15. Example of flow characteristics in the measurement duct of the
ClO–BrO instrument during a flight over Norway on January 6, 1989. The
lowest solid line in each panel is the ambient temperature, pressure, and
potential temperature as measured by the MMS instrument (9). In the top
panel, temperatures 1, 2, and 3 (bottom to top) are taken inside the 5-cm-
square measurement duct 0.0, 1.0, and 1.5 cm from the center of the duct.
The velocity of the flow was roughly 22 m/s in the measurement duct.*

duct. However, at a velocity of 20 m/s, the boundary layer of the measurement duct is beginning to converge in the detection volume of the BrO axis, as evident by the slightly higher temperatures of thermistors 2 and 3, located away from the center of the flow (thermistor 1) by 1.0 and 1.5 cm, respectively. Thus the measured air flow is well behaved despite the differences between wind and aircraft direction.

This solution to a difficult sampling problem is not perfect, unfortunately. When the instrument is sampling air in the lower stratosphere, near the tropopause, the mixing of reagent NO into the flow is not complete, and intermingled regimes of underconversion and removal are established. This effect has been carefully documented in the laboratory, during one flight over Antarctica in 1987, and more recently in test flights. As a result, the conversion efficiency drops from values near 0.9 at altitudes of 20 km to 0.4 at altitudes near 13 km, although this drop in conversion efficiency can be corrected by injecting the reagent NO into the flow tube at higher velocities.

The change in the typical calibration with a change in density (and thus altitude) is given in Figure 16 for both ClO and BrO. The dominant effect for ClO is the chemical conversion efficiency, which can be greatly improved. The signal-to-noise ratios for ClO and BrO can be determined from the 1-s averaged data shown in Figure 17a, which were taken from the flight of

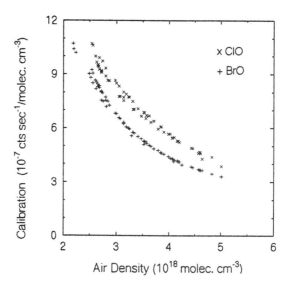

Figure 16. Variation of the calibration for ClO and BrO with air density from a flight during the AASE mission on February 10, 1989. The calibration number when multiplied by the ClO or BrO concentration gives the observed count rate and accounts for changes in chemical conversion efficiency and detection sensitivity of the Cl or Br atoms.

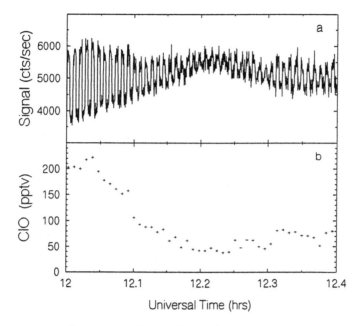

Figure 17. The detector signal and calculated ClO mixing ratio for a portion of a flight on January 6, 1989, during the AASE mission. The rise in the background is caused by the increase in Rayleigh scattering as the aircraft descended from the 460- to the 390-K potential temperature surface. This change in Rayleigh scattering provides the in-flight calibration.

January 6, 1989, in the Arctic polar vortex. At 20 km, ClO levels of 4 pptrv can be detected with a signal-to-noise ratio of 2 in a 32-s NO on–off cycle. BrO abundances of 2 pptrv can be detected with a signal-to-noise ratio of 2 in 40 min. For both ClO and BrO the added NO reagent absorbs a small fraction of the resonant emissions and thus reduces the Rayleigh and chamber-scattered signals slightly. The effect is that an offset of 0.5 ±0.3 pptrv and 1.5 ± 0.8 pptrv must be added to the calculated ClO and BrO abundances. The uncertainty in the sensitivity for ClO, when this effect is included, is ±25% at the 2σ confidence level and for BrO is ±35% at the 2σ confidence level. The conversion of the detector signal to ClO mixing ratios is shown in Figure 17b.

Examples of the quality of the data provided by the ER–2 ClO–BrO instrument are provided in Figures 1 and 9, but numerous other examples also exist, for example, data from the flight outside the Arctic polar vortex from January 25, 1989. Shown in Figure 18 are the solar zenith angle and measurements of potential temperature (9), NO (79), O_3 (84), ClO, and BrO. As the solar zenith angles decrease, more ClO and NO are photolytically released from their nighttime reservoirs, presumably $ClONO_2$ and NO_2,

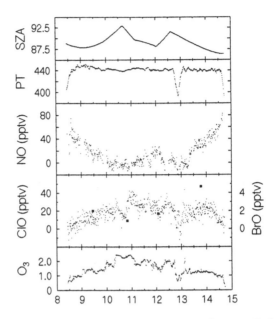

Figure 18. Data taken by instruments on the ER–2 during a flight on January 25, 1989, outside the Arctic polar vortex. All results are averaged over 32 s, which is the measurement rate of the ClO–BrO instrument. PT is the potential temperature, and SZA is the solar zenith angle. (Unpublished data from K. R. Chan, D. W. Fahey, and M. H. Proffitt.)

respectively. The abundance of ozone is measured with both high precision and accuracy, as can be seen in Figure 18. The statistical variability of the NO measurement is small, but an offset of ~10 pptrv exists in the data. The statistical variability of the ClO data is larger than that of the NO data, but the offset is much smaller, roughly 0.5 pptrv. In contrast to these three measurements that provide some spatial and temporal resolution are the measurements of BrO. These are sufficient to determine roughly what the abundance of BrO is, but they are not useful for testing photochemical mechanisms. Thus, fast-response detection of radicals is a requirement for understanding photochemical mechanisms in the stratosphere.

A last example of how these ClO data can be used to derive information about the abundances of other trace gases comes from data taken during the AASE mission. ClO abundances were measured, but the total amount of chlorine that is converted from HCl and ClONO$_2$ resides as ClO and Cl$_2$O$_2$ inside the Arctic polar vortex. The total amount of chlorine that is in the form of ClO and Cl$_2$O$_2$ can be estimated from the observed ClO abundance and temperature under certain circumstances (54). First, the air parcel being measured must have been in darkness for the preceding 12 h so that thermal equilibrium has time to be established. Second, the temperature of the air parcel must not vary significantly in those previous 12 h. If these conditions are met, then

$$[Cl_2O_2] = [ClO]^2 \, A \, \exp(B/T) \qquad (24)$$

where the measured values of A and B are 3.0×10^{-27} cm^{-3} molecule^{-1} and 8450 ± 850 K, respectively (57). The total amount of reactive chlorine is equal to $[ClO] + 2[Cl_2O_2]$. When this calculation is done for flights that had dives in darkness, the results from four of those flights—January 6, 20, and 30 and February 8—emerge, as shown in Figure 19. The total amount of inorganic chlorine is shown as a solid line (L. E. Heidt, unpublished data). Chlorine in the form of ClO and Cl_2O_2 is thus calculated to be a large fraction of the amount of available chlorine for all flights that occurred after mid-January. The uncertainty in the calculations is a factor of 2; the greatest contribution to the uncertainty is the uncertainty in the temperature dependence of the equilibrium constant. This result shows that the heterogeneous processes are efficient at converting chlorine from relatively dormant reservoir forms to reactive forms even in the Arctic, where the PSC activity is much less prevalent than in the Antarctic. This example also illustrates that measurements of one species and of the stratospheric con-

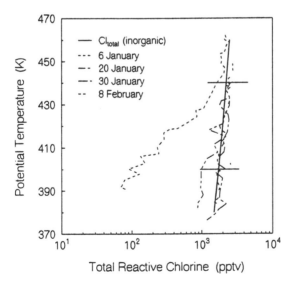

Figure 19. Calculated total reactive chlorine ($ClO + 2Cl_2O_2$) plotted against potential temperature for four Arctic flights (54). Cl_{total} is the total inorganic chlorine estimated from the whole-air sampler measurements of Heidt (unpublished data). Uncertainty in the calculations is a factor of 2 with 1σ confidence. (Reproduced with permission from reference 54. Copyright 1991 American Association for the Advancement of Science.)

ditions can be used with laboratory data to derive a result that is startling despite the large uncertainty.

Future Measurement Strategies. The photochemistry of the stratosphere is evolving as the atmosphere changes, and this evolution will continue as long as the source strengths for trace gases continue to evolve. Climate changes due to greenhouse warming will affect the stratosphere. According to current predictions, as the troposphere warms, the stratosphere will cool. Thus, the most destructive steps for ozone catalysis will slow slightly because of the positive temperature dependence of the rate constants. But on the other hand, the incidence of polar stratospheric clouds may increase because of larger volumes of air being exposed to lower temperatures. The net effect on stratospheric ozone can only be discerned if the effects on ozone transport are also taken into account.

Details in the abundances of trace gases need some work in several photochemical families. In the reactive nitrogen family, the connection between the low values of NO_x in the wintertime high latitudes and heterogeneous chemistry of the sulfate aerosol layer needs to be quantitatively explored. Also, data from satellites indicate that nitric acid abundances are larger than predicted in the wintertime near the polar regions (85). This alteration of NO_2 abundances has been explained as a result of formation of N_2O_5 (86) but may in part be the result of heterogeneous processes occurring on the background sulfate aerosol layer. A further question is the effect of increased aerosol surface area, such as occurred after the eruption of Mt. Pinatubo in the Philippines in June 1991. The effects throughout the stratosphere need to be known, not just near 20 km where the ER–2 can fly.

The understanding of the odd-hydrogen family has improved dramatically with the simultaneous in situ OH, HO_2, O_3, and H_2O measurements. However, because these measurements have been made infrequently only at middle latitudes, they cannot be construed as providing anything more than a first-order check of HO_x photochemistry. Additional simultaneous in situ measurements are needed to check their reproducibility and thus demonstrate whether they are representative of summer midlatitude conditions. Continued incremental improvements, to include simultaneous measurements of other species, for example NO, O, and HNO_3, will be required to check more thoroughly HO_x photochemistry. An in situ H_2O_2 measurement would be helpful to verify the photochemistry of this important HO_x reservoir species, and this measurement requires new instrument development. The need to carry out simultaneous HO_x-related measurements during other seasons and at other latitudes will follow to test that the theory is sufficiently robust to correctly model the response of HO_x in a variety of different, local photochemical environments.

For the halogens, perhaps the most pressing concern is how the substitution of hydrohalocarbons for the fully halogenated chlorofluorocarbons

(CFCs) and halons over the next decade, as mandated by the Montreal Protocol for the Protection of the Ozone Layer, will affect the chlorine burden of the stratosphere. Hydrochlorofluorocarbons (HCFCs) can be used as substitutes for the CFCs for a few decades without having a substantial impact on the chlorine burden of the stratosphere because they are primarily destroyed in the troposphere by reactions with OH before they are able to deliver the chlorine to the stratosphere. The elimination of CFCs and the temporary use of HCFCs into the early part of the next century must be carefully orchestrated to minimize the peak chlorine loading and promote the most rapid reduction of the chlorine burden of the stratosphere (56, 87). Another issue is the effects that perturbations to the reactive nitrogen abundances will have on the abundances of reactive chlorine. A better understanding and clarification of the direct heterogeneous conversions of chlorine species on both PSCs and sulfate aerosols are also needed.

However, using OH in the troposphere as a filter for HCFCs has its dangers because calculations of the OH distribution have not been directly tested and knowledge of stratospheric tropospheric exchange is incomplete. Clearly a careful monitoring program for these trace gases is essential, and in situ measurements can be a large part of this program.

A number of issues remain to be resolved for a full understanding of the polar, lower stratosphere as well. Much of the information about these regions has been derived from instruments flown either on balloons or on the NASA ER–2 high-altitude aircraft. Yet even with the 13 instruments on the aircraft, a complete picture of the photochemical mechanisms that were occurring could not be derived, nor could it be rigorously proved that heterogeneous reactions were responsible for the observed conversions of both the reactive chlorine and reactive nitrogen species. The measurements required for a better understanding of both the photochemical mechanisms and the conditions they require to occur include those of $ClONO_2$, NO_2, HCl, HNO_3, OH, HO_2, aerosol chemical composition, and BrO with better signal-to-noise ratios. Instruments are being developed for all of these measurements.

The understanding of four aspects of the lower stratosphere must be improved. The first is the processes of heterogeneous conversion. Does it occur in three hours or in three days? Which type of aerosols or particles are most favorable for conversion and what are the conditions required for their formation? How do the particles form, and what are their compositions and structures? What are the most important mechanisms for denitrification? These issues can be settled only by a combination of laboratory and stratospheric measurements.

A second issue is the transport of trace gases in the stratosphere. How do the polar vortices form? How readily are air parcels exchanged between the vortex and midlatitudes and between altitudes at different latitudes and seasons? What are the diabatic heating and cooling rates at different times of the year? When the polar vortex breaks up, how fast is the polar air mixed with the midlatitude air? Is mixing inhibited between the tropics and the middle latitudes? The transport of air between the troposphere and stratosphere needs to be understood. What are the important mechanisms in which locations and seasons? Which air from which altitudes and latitudes is transported?

A third issue is the estimates of ozone loss associated with the polar vortex. Will the Antarctic ozone hole expand and will the Arctic ozone hole begin? Can the loss of ozone be better quantified, and what are the zonal asymmetries in the ClO and BrO fields? How much additional ozone is lost when the polar vortex breaks up?

The fourth issue is the observed, unexplained decadal wintertime decline in ozone as reported by the Ozone Trends Panel and the loss of ozone in the lower stratosphere. Does the loss result from the perturbed photochemistry initiated by PSCs inside the polar vortex? Or is it a result of photochemistry associated with the ubiquitous sulfate aerosol layer? Are observed seasonal and latitudinal trends in reactive trace species consistent with the observed seasonal and latitudinal declines in ozone?

Another potential impact is that of high-speed civil transport aircraft that will be designed to fly supersonically in the stratosphere (88). One of the important exhaust products, but by no means the only one, is NO_x. For present engines, the amount of NO_x produced per kilogram of fuel consumed is roughly 40–60 g. Even if the levels of NO_x from these aircraft engines can be reduced a factor of 8 to 10, the globally averaged ozone depletion would be 2% to 3% for a 500-aircraft fleet flying at 20 km. As pointed out by Johnston and co-workers (88), this amount of ozone depletion is comparable to that predicted for the indefinite use of CFCs at 1985 production levels. Direct increases in ozone loss due to increases in NO_x is one issue. Of course, the high levels of NO_x emitted can react with the nonmethane hydrocarbon cycles in the stratosphere to actually produce ozone through photochemical smog mechanisms. This production is greater at lower altitudes, so that a turnover point, where the production of ozone just matches the loss of ozone, will occur, probably near 15 km.

The heterogeneous reactions of N_2O_5 on sulfate aerosols to form HNO_3 reduce the effects of NO_x from the jet exhaust on the ozone abundances by shifting reactive nitrogen from NO_x to HNO_3. It is tempting to suggest without further consideration that aircraft flying at these altitudes will cause little damage to the stratospheric ozone layer. However, water vapor and the total reactive nitrogen, NO_y, from the exhaust may increase the frequency and distribution of polar stratospheric clouds and thus lead to rapid ozone loss.

These issues that have direct bearing on stratospheric ozone, and thus on life on this planet, will not go away. The chlorine burden of the stratosphere will remain sufficient to produce the Antarctic ozone hole for the next 75 years and will increase by 20% over the next two decades no matter what action is taken to ban chlorofluorocarbons. If the system is further perturbed, by climate changes or other pollutants, more unknown effects may appear. Clearly the current information with the current measurement capability is not enough.

The introduction of a number of new measurement techniques and new instruments is required. Instruments on satellites such as the Upper Atmosphere Research Satellite (UARS) and the Earth Observing System (EOS) will provide excellent global coverage and fair altitude resolution of much of the stratosphere, although not perhaps below 20 to 25 km. Ground-based measurements, and in particular the Network for the Detection of Stratospheric Change (89), with five or six heavily instrumented sites, will give good temporal coverage and trends with good to fair altitude resolution for a number of key species such as O_3, ClO, N_2O, and HO_2 above 30 km. Stratospheric column abundances for O_3, NO_2, HCl, CH_4, HNO_3, and $ClONO_2$ will also be measured. The globally dispersed sites will permit quasi-global coverage that can be linked with satellite and tropospheric aircraft remote measurements. However, in situ measurements of reactive trace gases are still an important measurement component for gaining an understanding of the photochemical processes and the small-scale variability and interactions.

Large helium-filled research balloons are too difficult to launch and too inflexible in their capabilities. The current fleet of high-altitude aircraft has too low an altitude ceiling and too many necessary constraints for ensuring the safety of the pilots to attack many of the current problems. A technique is required that merges the favorable qualities of each of these two proven techniques. One possibility is the increased use of small balloons carrying small payloads. Measurements of ozone, water vapor, and particles have been successfully made on such platforms for a number of years (*see*, for example, reference 90). However, the payloads that are deployed may be considered to be expendable, depending on the launch location. In general, instruments using novel technologies for detection of species other than those already measured will be too expensive to be used only once.

A possibility for complementing scientific balloons is the use of remotely piloted vehicles (RPVs), that can carry a number of instruments for simultaneous measurements (*91*). These aircraft almost by design must carry smaller and lighter instruments than those currently being deployed. Among them are the Boeing Condor, the Ames Research Center High-Altitude Aircraft Research Program (HAARP), and the suite of aircraft including Perseus and Theseus from Aurora Flight Sciences. The concept is to use light-weight materials, large wing spans, and flight at low Reynolds numbers to achieve high-altitude performance up to 30 km. Such aircraft are powered by internal combustion engines that turn large propellers to achieve speeds that approach 0.4 Mach (100 m/s). A major constraint of such aircraft is that the payload capability is lower, in some cases substantially, than the current high-altitude aircraft, the ER–2 (1170 kg). However, the need to reduce the weight and size of instruments is consistent with the miniaturization in electronics and light sources and will pay rich dividends in the future.

The result will be platforms that can cover large areas, perhaps stay aloft for more than a day, and bring instruments back to earth without the need for extensive refurbishment before reflight. Different RPVs have different capabilities. For instance, the Boeing Condor is capable of transglobal flight, but only at altitudes up to 21.4 km. The HAARP, Perseus, and Theseus aircraft are being designed for shorter horizontal distances, 1,000 to 10,000 km, but with altitude ceilings approaching 30 km. Some or all of these aircraft will eventually play a role in the definition of stratospheric photochemistry in the near future.

Summary

The last five years have been a particularly exciting time in the history of stratospheric measurements and research. Measurement techniques were optimized to perform truly constraining simultaneous measurements of the photochemistry. Enough measurements have been made to validate the basic (gas-phase) tenets of stratospheric photochemistry, to solidly link chlorofluorocarbons to the destruction of ozone in the Antarctic ozone hole, and to signal a warning for the Arctic stratosphere as well. The Montreal Protocol now defines a pathway for the reduction of the use and production of chlorofluorocarbons and halons that will cause the chlorine burden of the stratosphere to begin dropping within the next 20 years. Despite political successes in getting international agreement on the pathway to reduction of chlorine, now is not the time to slow the response to the challenges of stratospheric photochemistry. Enough remains unknown that it cannot be predicted how changes in climate and pollutants will effect stratospheric ozone and how stratospheric ozone will affect climate and life. Researchers must continue to rise to the challenges, both scientific and technological, to gain a fuller, predictive understanding of the stratosphere.

Acknowledgments

We thank J. M. Rodriguez and Atmospheric and Environmental Research, Inc., for providing the altitude distributions of trace gases and ozone reaction rates and K. R. Chan, D. W. Fahey, M. H. Proffitt, and L. E. Heidt for providing data prior to publication. We are also thankful for helpful discussions and encouragement from C. R. Webster and for help with publication from P. S. Stevens and D. W. Toohey.

References

1. Gille, J. C.; Russell, J. M., III *J. Geophys. Res.* **1989**, *89*, 5125.
2. Farmer, C. B. *Mikrochim. Acta* **1987**, *3*, 189.
3. Albritton, D. L.; Fehsenfeld, F. C.; Tuck, A. F. *Science (Washington, D.C.)* **1990**, *250*, 75.
4. deZafra, R. L.; Jaramillo, M.; Barrett, J.; Emmons, L. K.; Solomon, P. M.; Parrish, A. *J. Geophys. Res.* **1989**, *94*, 11,423.
5. Anderson, J. G.; Brune, W. H.; Proffitt, M. H. *J. Geophys. Res.* **1989**, *94*, 11,465.
6. World Meteorological Organization. *Atmospheric Ozone: Assessment of Our Understanding of the Processes Controlling Its Present Distribution and Change*; World Meteorological Organization: Washington, DC, 1985; Report 16; Global Research and Monitoring Project, Geneva.
7. *U.S. Standard Atmosphere, 1976*; U.S. Government Printing Office: Washington, DC, 1976.
8. Chan, K. R.; Scott, S. G.; Bui, T. P.; Bowen, S. W.; Day, J. *J. Geophys. Res.* **1989**, *94*, 11,573.
9. Chan, K. R.; Bowen, S. W.; Bui, T. P.; Scott, S. G.; Dean-Day, J. *Geophys. Res. Lett.* **1990**, *17*, 341.
10. Juckes, M. N.; McIntyre, M. E. *Nature (London)* **1987**, *328*, 590.
11. Prather, M. J.; Jaffe, A. H. *J. Geophys. Res.* **1990**, *93*, 3437.
12. Brasseur, G.; Solomon, S. *Aeronomy of the Middle Atmosphere*; D. Reidel: Boston, 1986.
13. Prather, M. J.; McElroy, M. B.; Wofsy, S. C. *Nature (London)* **1984**, *312*, 227.
14. Cicerone, R. J.; Walters, S.; Liu, S. C. *J. Geophys. Res.* **1983**, *88*, 3647.
15. Watson, R. T.; Ozone Trends Panel; Prather, M. J.; Ad Hoc Theory Panel; Kurylo, M. J.; NASA Panel for Data Evaluation. *Present State of the Knowledge of the Upper Atmosphere 1988: An Assessment Report*; National Aeronautics and Space Administration: Washington, DC, 1988; NASA Reference Publication 1208.
16. McElroy, M. B.; Salawitch, R. J. *Science (Washington, D.C.)* **1989**, *243*, 763.
17. McCormick, M. P.; Steel, H. M.; Hamill, P.; Chu, W. B.; Swissler, F. S. *J. Atmos. Sci.* **1982**, *39*, 1387.
18. Farman, J. C.; Gardiner, B. G.; Shanklin, J. D. *Nature (London)* **1985**, *315*, 207.
19. Krueger, A. J.; Schoeberl, M. R.; Stolarski, R. S.; Sechrist, F. S. *Geophys. Res. Lett.* **1988**, *15*, 1365.
20. Stolarski, R. S.; Schoeberl, M. R.; Newman, P. A.; McPeters, R. D.; Krueger, A. J. *Geophys. Res. Lett.* **1990**, *17*, 1267.
21. Newman, P.; Stolarski, R.; Schoeberl, M.; McPeters, R.; Krueger, A. *Geophys. Res. Lett.* **1991**, *18*, 661.
22. Hofmann, D. J. *Nature (London)* **1989**, *337*, 447.
23. Molina, L. T.; Molina, M. J. *J. Phys. Chem.* **1987**, *91*, 433.
24. McElroy, M. B.; Salawitch, R. J.; Wofsy, S. C.; Logan, J. A. *Nature (London)* **1986**, *327*, 759.
25. Solomon, S.; Garcia, R. R.; Rowland, F. S.; Wuebbles, D. W. *Nature (London)* **1986**, *321*, 755.
26. Toon, O. B.; Hamill, P.; Turco, R. P.; Pinto, J. *Geophys. Res. Lett.* **1986**, *13*, 1284.
27. Poole, L. R.; McCormick, M. P. *Geophys. Res. Lett.* **1988**, *93*, 8423.
28. *J. Geophys Res.* **1989**, *D9* and *D14*, special issues on the Airborne Antarctic Ozone Experiment.

182   MEASUREMENT CHALLENGES IN ATMOSPHERIC CHEMISTRY

29. Toon, O. B.; Turco, R. P.; Hamill, P. *Geophys. Res. Lett.* **1990**, *17*, 445.
30. Molina, M. J.; Tso, T. L.; Molina, L. T.; Wang, F. C. Y. *Science (Washington, D.C.)* **1987**, *238*, 1253.
31. Tolbert, M. A.; Rossi, M. J.; Malhotra, R.; Golden, D. M. *Science (Washington, D.C.)* **1987**, *238*, 1258.
32. Leu, M.-T. *Geophys. Res. Lett.* **1988**, *15*, 851.
33. Hofmann, D. J.; Rosen, J. M.; Harder, J. W.; Hereford, J. V. *J. Geophys. Res.* **1989**, *94*, 11,253.
34. Scott, S. G.; Bui, T. P.; Chan, K. R.; Bowen, S. W. *J. Atmos. Oceanic Technol.* **1990**, *7*, 525.
35. Denning, R. F.; Guidero, S. L.; Parks, G. S.; Gary, B. L. *J. Geophys. Res.* **1989**, *94*, 16,757.
36. Ferry, G. V.; Neish, E.; Schultz, M.; Pueschel, R. F. *J. Geophys. Res.* **1989**, *94*, 16,459.
37. Pueschel, R. F.; Snetsinger, K. G.; Goodman, J. K.; Toon, O. B.; Ferry, G. V.; Oberbeck, V. R.; Livingston, J. M.; Verma, S.; Fong, W.; Starr, W. L.; Chan, K. R. *J. Geophys. Res.* **1989**, *94*, 11,271.
38. Wilson, J. C.; Loewenstein, M.; Fahey, D. W.; Gary, B.; Smith, S. D.; Kelly, K. K.; Ferry, G. V.; Chan, K. R. *J. Geophys. Res.* **1989**, *94*, 16,437.
39. Gandrud, B. W.; Sperry, P. D.; Sanford, L.; Kelly, K. K.; Ferry, G. V.; Chan, K. R. *J. Geophys. Res.* **1989**, *94*, 11,285.
40. Heidt, L. E.; Vedder, J. F.; Pollock, W. H.; Lueb, R. A.; Henry, B. E. *J. Geophys. Res.* **1989**, *94*, 11,599.
41. Loewenstein, M.; Podolske, J. R.; Chan, K. R.; Strahan, S. E. *J. Geophys. Res.* **1989**, *94*, 11,589.
42. Kelly, K. K.; Tuck, A. F.; Murphy, D. M.; Proffitt, M. H.; Fahey, D. W.; Jones, R. L.; McKenna, D. S.; Loewenstein, M.; Podolske, J. R.; Strahan, S. E.; Ferry, G. V.; Chan, K. R.; Vedder, J. F.; Gregory, G. L.; Hypes, W. D.; McCormick, M. P.; Browell, E. V.; Heidt, L. E. *J. Geophys. Res.* **1989**, *94*, 11,317.
43. Proffitt, M. H.; Steinkamp, M. J.; Powell, J. A.; McLaughlin, R. J.; Mills, O. A.; Schmeltekopf, A. L.; Thompson, T. L.; Tuck, A. F.; Tyler, T.; Winkler, R. H.; Chan, K. R. *J. Geophys. Res.* **1989**, *94*, 16,547.
44. Fahey, D. W.; Murphy, D. M.; Kelly, K. K.; Ko, M. K. W.; Proffitt, M. H.; Eubank, C. S.; Ferry, G. V.; Loewenstein, M.; Chan, K. R. *J. Geophys. Res.* **1989**, *94*, 16,665.
45. Brune, W. H.; Anderson, J. G.; Chan, K. R. *J. Geophys. Res.* **1989**, *94*, 16,649.
46. Tuck, A. F.; Watson, R. T.; Condon, E. P.; Margitan, J. J.; Toon, O. B. *J. Geophys. Res.* **1989**, *94*, 11,181.
47. Anderson, J. G.; Brune, W. H.; Toohey, D. W. *Science (Washington, D.C.)* **1991**, *251*, 1.
48. Solomon, S. *Nature (London)* **1990**, *347*, 347.
49. Loewenstein, M.; Podolske, J. R.; Chan, K. R. *Geophys. Res. Lett.* **1990**, *17*, 477.
50. Hartmann, D. L.; Chan, K. R.; Gary, B. L.; Schoeberl, M. R.; Newman, P. A.; Martin, R. L.; Loewenstein, M.; Podolske, J. R.; Strahan, S. E. *J. Geophys. Res.* **1989**, *94*, 11,625.
51. Fahey, D. W.; Solomon, S.; Kawa, S. R.; Loewenstein, M.; Podolske, J. R.; Strahan, S. E.; Chan, K. R. *Nature (London)* **1990**, *345*, 698.
52. Fahey, D. W.; Kelly, K. K.; Kawa, S. R.; Tuck, A. F.; Loewenstein, M.; Chan, K. R.; Heidt, L. E. *Nature (London)* **1990**, *344*, 321.
53. World Meteorological Organization. *Scientific Assessment of Stratospheric Ozone: 1989*; World Meteorological Organization: Washington, DC, 1990; Report 20; Global Ozone Research and Monitoring Project, Geneva.

54. Brune, W. H.; Anderson, J. G.; Toohey, D. W.; Fahey, D. W.; Kawa, S. R.; Jones, R. L.; McKenna, D. S.; Poole, L. R. *Science (Washington, D.C.)* 1991, 252, 1260.
55. *Geophys. Res. Lett.* 1990, 17(4), special issue on the Airborne Arctic Stratospheric Experiment.
56. World Meteorological Organization. *Scientific Assessment of Stratospheric Ozone, 1991*; World Meteorological Organization: Washington, DC, 1992; Report 25; Global Ozone Research and Monitoring Project, Geneva.
57. DeMore, W. B.; Sander, S. P.; Golden, D. M.; Hampson, R. F.; Kurylo, M. J.; Howard, C. J.; Ravishankara, A. R.; Kolb, C. E.; Molina, M. J. *Chemical Kinetics and Photochemical Data for Use in Stratospheric Modeling, Evaluation 10*; Jet Propulsion Laboratory: Pasadena, CA, 1992; JPL Publication 92–20.
58. King, J. C.; Brune, W. H.; Toohey, D. W.; Rodriguez, J. M.; Starr, W. L.; Vedder, J. F. *Geophys. Res. Lett.* 1991, 18, 2273.
59. Kawa, S. R.; Fahey, D. W.; Heidt, L. E.; Solomon, S.; Anderson, D. E.; Loewenstein, M.; Proffitt, M. H.; Margitan, J. J.; Chan, K. R. *J. Geophys. Res.* 1991, 97, 7905.
60. Kawa, S. R.; Fahey, D. W.; Solomon, S.; Brune, W. H.; Proffitt, M. H.; Toohey, D. W.; Anderson, D. E.; Anderson, L. C.; Chan, K. R. *J. Geophys. Res.* 1990, 95, 18,597.
61. Stimpfle, R. M.; Wennberg, P. O.; Lapson, L. B.; Anderson, J. G. *Geophys. Res. Lett.* 1990, 17, 1905.
62. Wennberg, P. O.; Stimpfle, R. M.; Weinstock, E. M.; Dessler, A. E.; Lloyd, S. A.; Lapson, L. B.; Schwab, J. J.; Anderson, J. G. *Geophys. Res. Lett.* 1990, 17, 1909.
63. Ridley, B. A.; McFarland, M.; Schmeltekopf, A. L.; Proffitt, M. H.; Albritton, D. L.; Winkler, R. H.; Thompson, T. L. *J. Geophys. Res.* 1987, 92, 11,919.
64. Kondo, Y.; Aimedieu, P.; Pirre, M.; Matthews, W. A.; Ramaroson, R.; Sheldon, W. R.; Benbrook, J. R.; Iawa, A. *J. Geophys. Res.* 1990, 95, 22,513.
65. McFarland, M.; Ridley, B. A.; Proffitt, M. H.; Albritton, D. L.; Thompson, T. L.; Harrop, W. J.; Winkler, R. H.; Schmeltekopf, A. L. *J. Geophys. Res.* 1986, 91, 5421.
66. Evans, W. F. J.; McElroy, C. T.; Galbally, I. E. *Geophys. Res. Lett.* 1985, 12, 825.
67. May, R. D.; Webster, C. R. *J. Geophys. Res.* 1989, 94, 16,343.
68. Webster, C. R.; May, R. D. *J. Geophys. Res.* 1987, 92, 11,931.
69. Webster, C. R.; May, R. D.; Toumi, R.; Pyle, J. A. *J. Geophys. Res.* 1990, 95, 13,851.
70. Lary, D. J.; Pyle, J. A.; Webster, C. R.; May, R. D. *Geophys. Res. Lett.* 1991, 18, 2261.
71. Stimpfle, R. M.; Lapson, L. B.; Wennberg, P. O.; Anderson, J. G. *Geophys. Res. Lett.* 1989, 16, 1433.
72. Anderson, J. G. *Geophys. Res. Lett.* 1975, 2, 231.
73. Anderson, J. G.; Margitan, J. J.; Stedman, D. H. *Science (Washington, D.C.)* 1977, 198, 501.
74. Anderson, J. G.; Grassl, H. J.; Shetter, R. E.; Margitan, J. J. *J. Geophys. Res.* 1980, 85, 2869.
75. Weinstock, E. M.; Phillips, M. J.; Anderson, J. G. *J. Geophys. Res.* 1981, 86, 7273.
76. Brune, W. H.; Anderson, J. G. *Geophys. Res. Lett.* 1986, 13, 1391.
77. World Meteorological Organization. *The Stratosphere, 1981: Theory and Measurements*; World Meteorological Organization: Washington, DC, 1981; Report 11; Global Ozone Research and Monitoring Project, Geneva.

78. *ER–2 Investigators Handbook*; High Altitude Missions Branch, Airborne Missions and Applications Divisions, NASA Ames Research Center: Moffett Field, CA, 1988.
79. Fahey, D. W.; Kawa, S. R.; Chan, K. R. *Geophys. Res. Lett.* **1990**, *17*, 489.
80. Brune, W. H.; Anderson, J. G.; Chan, K. R. *J. Geophys. Res.* **1989**, *94*, 16,649.
81. Brune, W. H.; Anderson, J. G.; Chan, K. R. *J. Geophys. Res.* **1989**, *94*, 16,639.
82. Toohey, D. W.; Anderson, J. G.; Brune, W. H. *J. Geophys. Res.* **1992**, in preparation.
83. Soderman, P. T.; Hazen, N. L.; Brune, W. H. "Aerodynamic Design of Gas and Aerosol Samplers for Aircraft" NASA Technical Memo 103854, 1991.
84. Proffitt, M. H.; Margitan, J. J.; Kelly, K. K.; Loewenstein, M.; Podolske, J. R.; Chan, K. R. *Nature (London)* **1990**, *347*, 31.
85. Gille, J. C.; Russell, J. M.; Barley, P. L.; Remsberg, E. E.; Gordley, L. L.; Evans, W. F. J.; Fischer, H.; Gandrud, B. W.; Girard, A.; Harries, J. A.; Beck, S. A. *J. Geophys. Res.* **1984**, *89*, 5179.
86. Solomon, S.; Garcia, R. R. *J. Geophys. Res.* **1983**, *88*, 5497.
87. Prather, M. J.; Watson, R. W. *Nature (London)* **1990**, *344*, 729.
88. Johnston, H. S.; Kinnison, D. E.; Wuebbles, D. J. *J. Geophys. Res.* **1989**, *94*, 16,351.
89. *Network for the Detection of Stratospheric Change*; NASA Upper Atmosphere Research Program and NOAA Aeronomy Laboratory: Washington, DC, 1990.
90. Hofmann, D. J.; Rosen, J. M.; Harder, J. W.; Rolf, S. R. *Geophys. Res. Lett.* **1986**, *13*, 1252.
91. Prather, M. J.; Wesoky, H. L.; Miake-Lye, R. C.; Douglass, A. R.; Turco, R. P.; Wuebbles, D. J.; Ko, M. K. W.; Schmeltekopf, A. L. *The Atmospheric Effects of Stratospheric Aircraft: A First Program Report*; National Aeronautics and Space Administration: Washington, DC, 1992; NASA Reference Publication 1242.

RECEIVED for review March 20, 1991. ACCEPTED revised manuscript September 15, 1992.

6

Probing the Chemical Dynamics of Aerosols

Richard C. Flagan

Department of Chemical Engineering, California Institute of Technology, Pasadena, CA 91125

Atmospheric aerosols are complex mixtures of particles emitted into the atmosphere and secondary particles formed as a result of gas-phase chemical reactions. Secondary aerosols are formed by condensation of the products of gas-phase reactions onto particle surfaces or by homogeneous nucleation. Particle formation and growth are often very rapid; this rapidity places severe demands on the instrumentation used to monitor the aerosol evolution. These demands are particularly evident in smog chamber studies that are designed to elucidate the fundamental processes that take place in the atmosphere. Many reacting systems produce several condensible species; this situation further complicates the analysis of aerosol formation and growth. This chapter reviews the present aerosol instrumentation available and gives results from smog chamber studies, showing some of the recent advances that are helping further the understanding of atmospheric aerosol chemical dynamics. Examples from smog chamber studies are used to illustrate needed improvements in instrumentation for following the evolution of the composition and size distributions of atmospheric aerosols.

ATMOSPHERIC AEROSOLS ARE COMPLEX MIXTURES of particles derived from diverse sources. Soot from diesel engines, fly ash from coal combustion, and sulfates, nitrates, and organic compounds produced by atmospheric reactions of gaseous pollutants all contribute to the aerosol. Particle size and composition depend upon the conditions of aerosol formation and growth and determine the effects of atmospheric aerosols on human health, ecosystems, materials degradation, and visibility. Much of the research on environmental aerosols has focused on fine particles ranging from a few micrometers in

0065–2393/93/0232–0185$07.50/0
© 1993 American Chemical Society

diameter to the submicrometer size range. Submicrometer particles can penetrate deep into the respiratory tract when inhaled and they deposit where clearance is slow and can be assimilated by the body. Observations made in the early 1970s that submicrometer particles are frequently enriched with toxic species have increased concerns about the consequences of such particles on human health. The optically absorbing component of the atmospheric aerosol, primarily black carbon soots formed in combustion, is also concentrated in the submicrometer size range, as are the products of atmospheric reactions of gaseous pollutants. Thus, both particle size and composition are important parameters insofar as the environmental consequences of the atmospheric aerosol are concerned.

The many studies of the atmospheric aerosol over the past two decades have advanced the ability to probe and predict many features of the atmospheric aerosol; particle size distributions, light scattering and absorption efficiencies, and, to a lesser extent, the distribution of chemical composition with respect to particle size are much better understood than they were 20 years ago. Ambient studies have revealed the multimodal nature of the atmospheric aerosol along with estimates of typical atmospheric lifetimes of the various modes. Mechanically generated particles contribute most of the mass of particles larger than 1 μm in diameter, particles with diameters smaller than 0.1 μm result from homogeneous nucleation, and the accumulation mode in the 0.1- to 1-μm size range arises from reactions of gaseous pollutants to form secondary aerosols by condensation on nucleation-mode particles.

Many of the particles in the nucleation mode are emitted to the atmosphere from high-temperature processes, primarily combustion. Such particles can also be produced by homogeneous nucleation of the products of atmospheric photochemical reactions, as was demonstrated by McMurry and Friedlander (1) in smog chamber experiments. They explored the competition between homogeneous nucleation and heterogeneous condensation theoretically and elucidated the demarcation between the two modes of gas-to-particle conversion. Rigorous experimental validation of their predictions in smog chambers or in the atmosphere has remained elusive because of (1) incomplete knowledge of the physical properties of the condensing species and their rates of formation and (2) inadequate time and size resolution of the available aerosol instrumentation. Smaller scale experiments have filled this gap insofar as homogeneous nucleation and binary nucleation of carefully prepared vapors at controlled supersaturation in the presence of foreign particles are concerned (2), and recent instrumental developments have advanced the ability to probe reacting systems.

Atmospheric aerosols are frequently assumed to be internal mixtures in which all particles of the same size are assumed to have the same composition. This interpretation is a natural outgrowth of the methods used to measure the distribution of chemical composition in aerosol systems and is consistent with the level of detail possible in the most advanced theoretical descriptions of aerosol evolution by solution of the multicomponent general dynamic equation. Such uniformity is approached after long times, because coagulation mixes particles derived from different sources, but this uniformity does not accurately represent the aerosol either initially or if new material is inserted into the aerosol system. Greater detail has been obtained through analysis of individual particles, primarily in the supermicrometer size range, as is discussed in this chapter.

The size and time resolution of current instruments are, at best, marginal for the measurement of environmental aerosols. Figure 1 illustrates the time scales on which important changes take place for different aerosols of environmental concern. New particle formation by homogeneous nucleation usually takes place in a brief burst, ranging from milliseconds or shorter times in combustion systems to minutes for the slower reactions and lower vapor concentrations of the atmosphere. Combustion aerosols form and are emitted within a matter of seconds. Smog evolves over hours, although in extreme episodes the aerosol can survive for days. Particles grow from the nanometer size range to near-micrometer sizes relatively rapidly, but the larger particles persist for longer times. On the other hand, deposition of particles much larger than 1 μm is rapid, so their atmospheric lifetimes

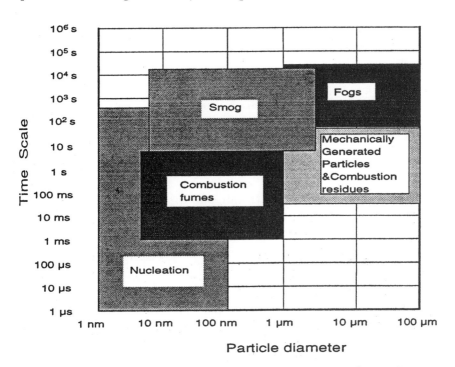

Figure 1. Size and time scales of interest for environmental aerosols.

diminish with increasing size. Thus, for an understanding of environmental aerosols a broad range of particle sizes must be measured on time scales that generally increase with particle size.

This chapter shows the needs for further developments in instrumentation for following the evolution of the composition and size distribution of atmospheric aerosols. Because no instrument currently exists that approximates this ideal, the needs are illustrated by experimental observations made with admittedly inadequate instruments rather than by specific instrumental approaches. The examples are chosen from the vast literature on atmospheric aerosols and are not meant to be comprehensive but rather to illustrate approaches that have yielded new insights and to point to possible directions for future work. The chemical dynamics that result from gas- and particle-phase reactions in the atmosphere are the special concern of this chapter. The smog chamber is the preferred platform for the study of secondary aerosol dynamics because it reproduces the essential features of the atmosphere with a captive parcel of air. It also represents a particular challenge because a well-defined onset of reaction leads to abrupt transitions, as is shown in the chapter. For these reasons, smog chamber experiments are used to illustrate the present state of the art and the challenges to be met in future work. Examples illustrating the need for improved particle size, time, and composition resolution are drawn from smog chamber experiments performed in my laboratory and elsewhere and from atmospheric observations. Many of the efforts to probe the composition inhomogeneity of the atmospheric aerosol have focused on elucidating the contributions of particular sources to visibility degradation, cloud condensation nuclei, and the like, so some examples from that literature are used as well. The methods currently used for aerosol measurements are briefly reviewed. For a more detailed survey of the available instruments, the reader is referred to the excellent review of aerosol instrumentation by Pui and Liu (3).

Current Aerosol Instruments

Particle Number Concentration and Size Distribution. The development of aerosol science to its present state has been directly tied to the available instrumentation. The introduction of the Aitken condensation nuclei counter in the late 1800s marks the beginning of aerosol science by the ability to measure number concentrations (4). Theoretical descriptions of the change in the number concentration by coagulation quickly followed. Particle size distribution measurements became possible when the cascade impactor was developed, and its development allowed the validation of predictions that could not previously be tested. The cascade impactor was originally introduced by May (5, 6), and a wide variety of impactors have since been used. Operated at atmospheric pressure and with jets fabricated by conventional machining, most impactors can only classify particles larger than a few tenths of a micrometer in diameter. The introduction of the low-pressure cascade impactor extended the accessible size range to about 50 nm (7), but later examination revealed that condensation and evaporation within the high-velocity and low-pressure regions of the impactor could cause significant biases (8). An alternative approach to the classification of sub-micrometer-sized particles was developed by Marple et al., namely, the use of very small jets in an impactor (9). With this micro-orifice impactor, the problems associated with high-velocity jets and low-pressure operation are avoided and particle size resolution is extended to about 50 nm. Impactor measurements are usually made by analysis of the material collected on the impactor stages some time after collection. Several methods have, however, been applied to on-line measurement of aerosol mass distributions through the use of inertial classification. Vibrating mass sensors measure the mass of deposited particles by the changes in the acoustic response of a transducer. In the quartz crystal microbalance, the natural vibration frequency of the crystal shifts with deposited mass (3). With a sensitivity of about 5 ng, the quartz crystal microbalance is useful for measuring particles at concentrations of 50 $\mu g/m^3$ to 5 mg/m^3.

A related technique that is suitable for measurement of aerosols at lower mass loadings is the aerodynamic particle sizer (3, 10). In this instrument the aerosol is rapidly accelerated through a small nozzle. Because of their inertia, particles of different aerodynamic sizes are accelerated to different velocities, and the smallest particles reach the highest speeds. The particle velocity is measured at the outlet of the nozzle. From the measurements of velocities of individual particles, particle size distributions can be determined. The instrument provides excellent size resolution for particles larger than about 0.8 μm in diameter, although sampling difficulties limit its usefulness above 10 μm.

The single-particle optical counter, introduced during the 1950s, allowed real-time size distribution measurements of particles larger than several hundredths of a micrometer in diameter. In this instrument, light scattered from individual particles as they are conveyed through a small view volume is used to infer the particle size. Light scattered from each particle is focused onto a photodetector, generating a voltage pulse with an amplitude related to particle size. Some of the many different configurations used in these instruments are described in detail by Pui and Liu (3). The sizing range of optical counters and sizers has been extended below 0.1 μm through the use of laser light sources; the minimum detection limit is achieved by passing the particles through the resonant cavity of the laser to take advantage of the high intensities there. The amount of light scattered from a particle is a strong function of particle size because of resonant coupling of the electromagnetic radiation with particles in the so-called Mie scattering size range. For small particles the scattered intensity is a single-valued function of particle size, but for particles in the Mie regime, particles of different sizes can produce the same scattered intensity. This situation limits the resolution of optical particle sizing above about 1.8 μm; the precise size range depends on the optical design of the particular instrument.

The electrical mobility analyzer developed by Whitby and co-workers at the University of Minnesota in the 1960s (11) was a practical implementation of concepts that had seen limited use since they were first introduced in the laboratory in the 1920s, and it made possible for the first time rapid measurements of the particle size distribution in the 10 nm to 1 μm size range. Data generated with that instrument revealed new structure in the atmospheric aerosol size distribution and catalyzed numerous theoretical and experimental investigations into the origins of that structure. A variant of the mobility analyzer, the differential mobility classifier (12, 13), and the Berglund–Liu vibrating orifice aerosol generator (14), both developed at Minnesota, provided precision calibration aerosols that greatly improved the accuracy of aerosol size distribution measurements. These sources of calibration aerosols do not satisfy the entire need, however, because the differential mobility analyzer (DMA) is well suited to the generation of monodisperse particles at sizes below 0.1-μm diameter and the vibrating orifice aerosol generator is useful in routine operation only at particle sizes greater than 1 to 2 μm.

Higher resolution in particle size distribution measurements is possible with the DMA. However, this instrument has seen little use for environmental analysis because of the long times required to acquire a complete particle size distribution, on the order of tens of minutes to 1 h or more depending on the size resolution sought. Like the earlier electrical aerosol analyzer (EAA), the DMA classifies charged particles by migration in an electric field between concentric cylinders. In the EAA, the current carried by all particles that are not deposited on the cylinders is measured. In contrast to this cumulative measurement of all particles with mobilities below a threshold value, a small flow extracted through holes in the central cylinder of the DMA carries particles with mobilities in a narrow range out of the instrument for analysis, either with an electrometer as in the EAA or with a condensation nuclei counter. Thus, by making measurements at a sequence of field strengths, a differential mobility distribution can be determined directly. Under typical operating conditions, the range of mobilities extracted at any time amounts to about 10% of the mean value.

Determination of the particle size distribution from mobility distribution measurements requires knowledge of the charging and transmission efficiencies of the instrument as a function of particle size. The commercially available EAA utilized a unipolar diffusion charger that generated singly charged particles below about 50 nm in diameter. Larger particles could have one, two, or more charges; hence, a single mobility interval contained particles of a number of sizes. In conventional use the DMA incorporates a bipolar diffusion charger that shifts the range of multiple charging to larger sizes, greater than about 100 nm. Still, multiple charging creates data analysis problems for accumulation-mode aerosols. To determine the particle size distribution, the data must be inverted by solving a set of Fredholm integral equations for a best estimate of the actual particle size distribution. Various techniques have been developed, notably methods based on the Twomey algorithm (*15*) and on regularization methods (*16*).

A recent development has transformed the differential mobility analyzer into a powerful instrument for the analysis of atmospheric aerosols (*17*). The long times required to measure a particle size distribution are primarily due to the long flow times in the analyzer column. Between voltage steps, all particles in the column must be flushed out of it before a measurement can be made. The critical feature of the mobility analysis of the particle size distribution is that each measurement corresponds to particles of a given mobility. The particles need not be classified at constant field strength, but each measurement must sense only a limited range of mobilities. If the field strength is changed continuously but monotonically, the particles migrate in the analyzer column along characteristic trajectories such that the particles arriving at the detector at a particular time correspond to a narrow range of mobilities. Wang and Flagan (*17*) developed a modified mobility analyzer called the scanning electrical mobility spectrometer (SEMS) that operates in this way. In the initial implementation, this type of migration was accomplished by computer control of the commercially available TSI Model 3070 Differential Mobility Analyzer with a TSI Model 3760 Clean Room Condensation Nuclei Counter as a detector. Through the use of an exponential ramp on the field strength, 100-point particle size distributions have been acquired in as little as 30 s.

Examining the time and size resolution of the current suite of particle measurement techniques reveals that most of the techniques for analysis of extracted aerosol samples are relatively slow (Figure 2). The EAA can complete a size distribution measurement in 5 to 10 min, but the DMA can require 1 h or more, depending on the counting time and the number of channels sampled. The difference in size resolution that can be achieved is indicated by the shading of the plot. The EAA is designed to determine concentrations in four size intervals per decade of particle diameter. The DMA can measure 10 or more intervals per decade, although each size measurement increases the sampling time. At a size resolution comparable to that of the EAA, the DMA can achieve comparable time resolution.

Although the SEMS represents a marked advance in the state of the art for measurement of aerosol size distribution, an important gap remains in current measurement technology, namely, the ability to make rapid, high-resolution measurements of the accumulation-mode aerosols on-line. The limitation of the DMA or SEMS for measurement of particles larger than 0.2 μm in diameter is the multiple charging that allows particles of two or more different sizes to contribute a given mobility fraction. Regardless of

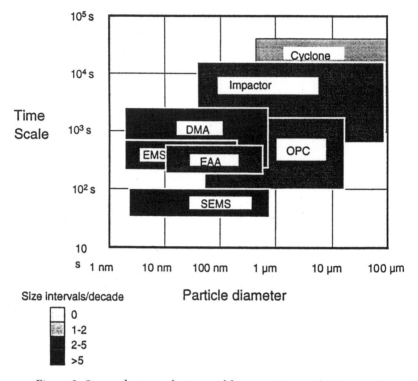

Figure 2. Size and time scales covered by current aerosol instruments.

the upper bound on mobility that is probed, multiply charged particles of sizes larger than the maximum nominal (singly charged) size will contribute to the measurement. Because these particles cannot be detected directly with resolution comparable to that of the DMA or SEMS, they introduce uncertainty into the inferred size distributions that is particularly worrisome if the mass or volume concentrations at the upper end of the measurement range are of interest.

Recent work at the University of Minnesota has provided two options for extending the upper bound of DMA operation: (1) operation of an impactor at the inlet of the DMA to prevent the entrance of particles larger than the nominal size being measured (*18*) and (2) use of a low-efficiency unipolar diffusion charger to extend the range of singly charged particles to a larger size than can be achieved with the bipolar charger (*19*). Ideally, in the first approach the size cut of the impactor would be tuned for each mobility measurement to eliminate the entrance of particles that would be multiply charged and that would thus contribute to the output aerosol from entering the DMA. Thus, the impactor is well suited to conventional operation of the DMA but is incompatible with the SEMS.

In the use of the low-efficiency unipolar charger, the time available for charging (or more precisely the product of that time and the gas ion concentration) is limited, so the probability of particles of the target size acquiring multiple charges is low. This measurement is predicated upon beginning the charging with an aerosol consisting only of neutral particles. All charged particles in the aerosol to be analyzed must, therefore, first be removed with an electrostatic precipitator. Assuming that the initial charge distribution is the Boltzmann equilibrium distribution, all but about 13% of particles of 1-μm size would be removed upstream of the charger. The fraction of the initial aerosol available for measurement will decrease further with increasing particle size. Thus, the counting efficiency is low for large particles that are present at relatively low number concentrations, limiting the method to the submicrometer size range. These two modifications to the use of the DMA do fill one important need in aerosol measurements— the generation of monodisperse aerosols in the accumulation size range— but they do not fully meet the needs for measurement of the accumulation-mode aerosol.

Chemical Composition.   Aerosol composition measurements have most frequently been made with little or no size resolution, most often by analysis of filter samples of the aggregate aerosol. Sample fractionation into coarse and fine fractions is achieved with a variety of dichotomous samplers. These instruments spread the collected sample over a relatively large area on a filter that can be analyzed directly or after extraction. Time resolution is determined by the sample flow rate and the detection limits of the analytical techniques, but sampling times less than 1 h are rarely used even when the analytical techniques would permit them. These longer times are the result of experiment design rather than feasibility. Measurements of the distribution of chemical composition with respect to particle size have, until recently, been limited to particles larger than a few tenths of a micrometer in diameter and relatively low time resolution. One of the primary tools for composition–size distribution measurements is the cascade impactor.

The fact that fine atmospheric particles are enriched in a number of toxic trace species has been known since the early 1970s. Natusch and Wallace (*20, 21*) observed that the fine particles emitted by a variety of high-temperature combustion sources follow similar trends of enrichment with decreasing particle size as observed in the atmosphere, and they hypothesized that volatilization and condensation of the trace species was responsible for much of the enrichment. Subsequent studies of a number of high-temperature sources and fundamental studies of fine-particle formation in high-temperature systems have substantiated their conclusions. The principal instruments used in those studies were cascade impactors, which fractionate aerosol samples according to the aerodynamic size of the particles. A variety of analytical methods were used to analyze impactor samples, including PIXE (particle-induced X-ray emission), X-ray fluorescence, gas chromatography–mass spectrometry, thin layer chromatography, and many others. Sample times for these measurements have been determined from the detection limits of the analytical methods; the averaging intervals for ambient measurement ranged from 1 to 24 h or more. Size resolution has been limited to two to four size fractions per decade of particle diameter.

Less-specific chemical information has been derived from measurements that couple a probe for one aspect of aerosol composition with more traditional aerosol instruments. Roberts and Friedlander (22) electrically heated stainless steel strips on which aerosol samples had been collected and used a cascade impactor as a rapid analysis method for total elemental sulfur. Only 15 min was required to collect enough material for measurement of the distribution of sulfur with respect to particle size in the atmospheric aerosol. The optically absorbing portion of the aerosol has been probed by using the integrating plate method to analyze impactor samples. Further developments along this line can be expected as microanalytical techniques are coupled to cascade impactors. An example of this is the use of a Fourier transform infrared (FTIR) microscope system to analyze individual impactor spots obtained with a single-jet low-pressure cascade impactor (23). By concentrating the sample onto a small region of a zinc selenide disk, the time required to obtain sufficient material for FTIR analysis in atmospheric and smog chamber experiments was reduced to 5–10 min.

On-line measurements of the sulfur content of atmospheric aerosols have been made by removing gaseous sulfur species from the aerosol and then analyzing the particles for sulfur with a flame photometric detector (24) or by using an electrostatic precipitator to chop the aerosol particles from the gas so that the sulfur content could be measured by the difference in flame photometric detector response with and without particles present. These and similar methods could be extended to the analysis of size-classified samples to provide on-line size-resolved aerosol composition data, although the analytical methods would have to be extremely sensitive to achieve the size resolution possible in size distribution analysis.

Aerosol Heterogeneity.   The variation of the chemical composition from particle to particle within an aerosol size class has been probed in a number of ways. Single-particle chemical analysis has been achieved by using the laser Raman microprobe (25) and analytical scanning electron microscopy (26). With the electron microscope techniques, the particle can be sized as well as analyzed chemically, so the need for classification prior to sample collection is reduced. Analyzing hundreds to thousands of particles provides the information necessary to track the particles back to their different sources but is extremely time consuming.

On-line size-resolved chemical analysis has been an elusive goal of many research efforts. At usual sampling rates, the quantity of material in any size fraction is so small as to represent a serious challenge to analytical chemistry unless coarse size resolution and long sampling times are accepted. The greatest successes to date have been in partial characterization of the aerosol. In 1978, Liu et al. (27) demonstrated that the differential mobility analyzer could be used to measure the variation in the hygroscopic nature of particles in a narrow size interval. They used two DMAs to produce what they called an aerosol mobility chromatograph. The first DMA provided a size-classified aerosol that was then humidified. The second DMA was used to measure the size distribution of the modified aerosol, providing sufficient resolution to discriminate between potassium sulfate and sulfuric acid aerosols. This approach was taken further by Rader and McMurry (28) in their development of the tandem differential mobility analyzer (TDMA). In the TDMA the aerosol is passed through a conditioner that has been carefully designed to maximize resolution. The TDMA was used in the measurement of the sensitivity of atmospheric aerosols to relative humidity (29), and a number of laboratory studies of fundamental aspects of aerosol thermochemistry, including rates of the reaction between ammonia gas and sulfuric acid aerosols, water activity over salt solutions, and vapor pressures of low-volatility organic compounds, were conducted.

The size-classified aerosol streams produced with the DMA have been used to probe other aspects of aerosol heterogeneity. Covert et al. (30) have used an optical particle counter (OPC) to determine the optical size of particles that had been size-classified with a DMA. For particles smaller than the wavelength of light, the scattered intensity is greater for nonabsorbing particles than it is for optically absorbing soot particles. Thus, the DMA–OPC system allows one to measure the relative contribution of soot to the atmospheric aerosol. Time resolution was limited in their measurements, and 3-h averages were reported. There is, however, no fundamental reason why the measurements could not be accelerated by using the OPC as a detector for the SEMS.

More ambitious attempts at measuring the heterogeneity of the atmospheric aerosol have been undertaken as well. Single-particle analysis by mass spectrometry was demonstrated by Sinha and co-workers (31, 32). In this technique, an aerosol sample is introduced into a vacuum chamber in the form of a particle beam. The particles are injected into a Knudsen cell oven, where they undergo many collisions with the cell wall and are ultimately vaporized and ionized. The ions are then mass-analyzed with a quadrupole or sector mass spectrometer. So that individual particles can be analyzed, the flux of particles into the Knudsen cell is limited so that coincidence errors are minimized. Ion pulses from individual particles allow the determination of the amount of the species being analyzed in the particular particle. The sensitivity of the technique is limited. For sodium, the detection limit is about 10^{-15} g, corresponding to a pure sodium chloride particle about 0.1 μm in diameter. The instrument was used to measure sodium in the Pasadena, California, aerosol, which typically has concentrations too low to be detected if the material were uniformly distributed throughout the atmospheric aerosol. A small fraction of the particles was found to contain a detectable amount of sodium that was consistent with the total atmospheric concentration; thus, the aerosol is very heterogeneous.

Other approaches have been taken for on-line analysis of individual aerosol particles as well. Laser spark spectroscopy (33) vaporizes individual particles in the breakdown plasma created by a pulsed laser. Atomic emission spectra can then be used to deduce the elemental composition of the particle that was vaporized. The timing of the laser pulse is critical because the particle must be caught in the focal volume of the pulsed laser, so a second laser is used to detect the particle and trigger the pulsed laser. To date the technique has been applied to large particles, that is, coal particles on the order of 60 to 70 μm in diameter in combustion studies. The use of inductively coupled plasma would eliminate the complex triggering and might allow on-line analysis of smaller particles spectroscopically.

Aerosol Instrument Classification.   Friedlander (34) classified the range of aerosol instrumentation in terms of resolution of particle size, time, and chemical composition. This classification scheme is illustrated in Figure 3. The ideal instrument would be a single-particle counter–sizer–analyzer. Operating perfectly, this mythical instrument would fully characterize the aerosol, with no lumping of size or composition classes, and would make such measurements sufficiently rapidly to follow any transients occurring in the aerosol system.

Actual instruments represent compromises on this ideal. Many sizing instruments provide no way of obtaining composition data. Size and time resolution are frequently limited, and in some cases no resolution is possible at all. Filter samplers and impactors collect time-integrated samples, and unless the time variations of interest occur on scales of hours or days, these techniques do not allow one to follow transient phenomena. Condensation nuclei counters and filter samplers similarly lump all particle sizes into a single measurement, so size resolution is precluded unless the particles are preceded by an aerosol size classifier. Even then, relatively few size cuts are generally achieved, although the scanning electrical mobility spectrometer or differential mobility spectrometer can acquire high-resolution particle size distributions. The aerodynamic particle sizer is capable of good size resolution over a size range that is complementary to the mobility-based methods. The single-particle analysis methods described are far from being practical instruments for routine aerosol measurements. Measurement techniques have improved dramatically since Friedlander's initial assessment, but no technique yet approaches this ideal instrument. The data obtained

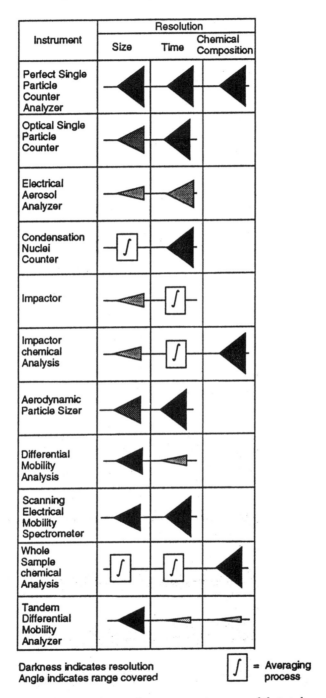

Figure 3. *Classification of aerosol instruments in terms of their inherent time, size, and composition resolution. (Adapted from reference 34.)*

with present instrumentation, however, clearly demonstrate the need for the additional degrees of resolution.

Atmospheric Aerosol Dynamics

Analysis of the evolution of particle size distribution functions and chemical composition of atmospheric aerosols has, over the years, revealed many of the processes that determine the nature of the atmospheric aerosol. Ion mobility analysis and impactors revealed the Junge distribution in the atmospheric aerosol. Later studies with the electrical aerosol analyzer and its predecessors revealed greater structure in the fine-particle distribution (35). From the nature of the observed particle size distributions and a fundamental understanding of aerosol dynamics, the processes that generated those structures were identified. Growth rates inferred from ambient measurements were larger than could be explained on the basis of coagulation alone (36), so it was suggested that gas-to-particle conversion accounted for the rapid increase in the accumulation-mode aerosol mass. Only rarely is the chemical nature of the atmospheric aerosol understood, so particle size distribution measurements only provide estimates of the quantity of aerosol-phase material. In ambient studies, the dynamics of the atmospheric aerosol is complicated by the introduction of reactive species along the entire trajectory of the air parcel under study and by the vast assortment of reactive species present in the atmosphere at any time.

Efforts to unravel the physics and chemistry of the atmospheric transformations of both gaseous and particulate pollutants have, for this reason, turned to controlled laboratory studies to gain a better understanding of the microscopic processes that take place in the atmosphere. Smog chamber studies make it possible to probe the transformations that occur in a captive parcel of air. Observations of particle formation have been reported since the earliest studies of atmospheric photochemical reactions. Haagen-Smit (37) observed aerosol formation in conjunction with studies of crop fumigation, noting that ring opening in the photooxidation of cyclic olefins produces species with very low vapor pressures. Sulfur dioxide was found to result in aerosol formation in reactions of all types of olefins (38, 39). Multiple additions of hydrocarbons to chamber photochemical experiments led to multiple bursts of new particle formation in other early studies (40). Many of the early measurements were limited to condensation nuclei counter (CNC) estimates of the total number concentration and photometer measurements of total light scattering by the product aerosol. Still, the data reveal important dynamics of the atmospheric aerosol. CNC data frequently show very rapid rises in the total number concentration, and increases of orders of magnitude occur within 5 to 10 min (40–42).

The introduction of the electrical aerosol analyzer allowed direct observations of the evolution of the size distribution of the fine aerosol particles (41). The growing aerosol was found to develop what appeared to be an equilibrium surface area. Subsequent theoretical analysis of these and other data by McMurry and Friedlander (*1*) revealed that the asymptotic distribution is actually a slowly increasing distribution that results from particle growth by a combination of coagulation and condensation. Other studies used the optical particle counter to generate similar data for particles larger than 0.3 μm (*43*, *44*). Observations of particle growth rates supported observations made from earlier ambient studies that gas-to-particle conversion was a major contributor to particle growth in the atmosphere.

A comprehensive study of the formation of aerosols in the photochemical oxidation of SO_2 in clean air systems, hydrocarbon–NO_x systems, and a combination of the two revealed complex aerosol dynamics that is still not quantitatively understood (*45*). When SO_2 was photochemically oxidized in clean air, particle formation occurred abruptly after an incubation time that varied from less than 1 min to 30 min, depending on the SO_2 concentration. The number concentration increased rapidly, peaking within about 10 min of the onset of particle formation, and then decayed slowly. Late in the experiment, the lights were turned off to stop the photochemical reactions. The number concentration decayed more rapidly thereafter; this observation indicates that new particle formation had continued long after the number concentration had peaked. The photochemical reactions of a variety of hydrocarbon–NO_x systems produced particles after a much longer delay than was observed in the SO_2–clean air system, and these reactions had a more rapid increase in the volume concentration of particles than in the former system but slower increases in the number concentration. Photochemical reactions of SO_2–hydrocarbon–NO_x systems produced number concentration profiles that appeared to be a linear combination of the two systems. Particles were rapidly formed early in the process, and a second, slower burst of new particle formation took place after a long delay. Although the number combination profile appeared to be a linear combination of the two separate systems, the volume of aerosol was greater than the sum of the two systems. The influence of the hydrocarbon–NO_x chemistry on the photooxidation of SO_2 was not understood at the time, but the consequences of that coupling on the aerosol dynamics were clearly outlined in reference 45.

Other studies of the aerosol dynamics in smog chamber experiments that used the EAA, OPC, and CNC have reproduced the general observations of the earlier studies, with some improvements in data analysis. Stern et al. (*46*) examined aerosol formation and growth in the photooxidation of aromatic hydrocarbons in an outdoor smog chamber. One modification from the previous experimental design was the intentional introduction of particles into the initial reactant mixture. Growth of those seed particles by condensation provided a direct measure of the excess vapor pressure of condensible reaction products as soon as they began to form. In some experiments the number concentration of seed particles was sufficient to suppress homogeneous nucleation entirely, whereas nucleation occurred in spite of the seed aerosol in other experiments. The seeded experiments thus provide a way to probe the competition between homogeneous nucleation and condensational growth. Theoretical predictions of nucleation required estimates of the physical properties of the condensible reaction products. Because the aerosol composition and physical properties were not known, the physical property data were estimated from the condensational growth data obtained in the seeded growth experiments.

In later experiments, Izumi et al. (47, 48) examined aerosol formation during photooxidation of a variety of hydrocarbons in an evacuable smog chamber. No seed particles were used in these experiments, but good estimates of the yield of aerosol from photochemical oxidation of the hydrocarbon precursors were obtained by using EAA data. In some cases, the volumetric yield was found to decrease with decreasing precursor concentration (Figure 4), so the finite vapor pressure of the reaction products limited nucleation, particle growth, or both.

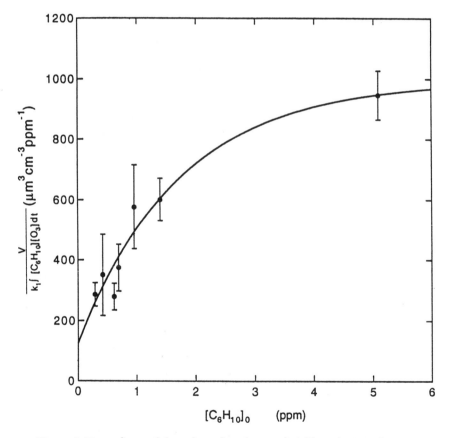

Figure 4. Dependence of the volume-based aerosol yield on the initial concentration of cyclohexene. (Reproduced from reference 47. Copyright 1988 American Chemical Society.)

The attempts to derive fundamental data on secondary aerosol formation by using EAA data were hampered by the limited time and size resolution of the instrument and the inability to measure the physical and chemical properties of the aerosol species. Conventional differential mobility analysis eliminates many of the size uncertainties of the EAA measurements. Time resolution limited the application of differential mobility analysis in smog chamber experiments until very recently. Three approaches have allowed these higher resolution instruments to follow the rapid transients: (1) By following the time evolution of the concentrations of particles at fixed size, the resolvable time is reduced to the counting time required to obtain good signal-to-noise ratio. This operating mode was used by Flagan et al. (*49*) with both the DMA with a CNC detector and a University of Vienna electrical mobility spectrometer, which couples an extremely sensitive electrometer to a differential mobility analyzer column that has been designed to cope with particles as small as 3.5 nm in diameter. Because the size-classified particles could be measured in as few as 1 to 20 s, abrupt changes in the concentrations of fine particles were observed. These changes are illustrated by the extremely rapid increase in the number concentrations of particles of different sizes observed in the photooxidation of dimethyl disulfide (Figure 5). This figure shows a wave in particle size space as the small particles formed within the first 3 min of the smog chamber experiment grow. The time lag before the peaks in concentration increases with particle size.

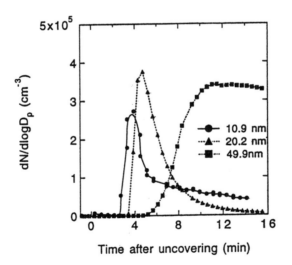

Figure 5. Transients observed in the concentrations of ultrafine particles in smog chamber studies of the photooxidation of dimethyl disulfide. Particles were measured with the electrical mobility spectrometer operating at fixed analyzer column voltages for the 11- and 20-nm sizes and with the differential mobility analyzer similarly operated for the 50-nm particles. (Reproduced from reference 49. Copyright 1991 American Chemical Society.)

(2) Operation of the EMS or DMA with widely spaced steps in the analyzer column voltage reduces the number of data obtained in a single-size distribution scan, thereby reducing the measurement time to several minutes. (3) More dramatic improvements are obtained with the scanning electrical mobility spectrometer, which achieves better time resolution without sacrificing size resolution.

The potential of rapid, high-resolution particle size distribution measurements with the SEMS to elucidate the dynamics of atmospheric aerosols is demonstrated in studies conducted with the outdoor smog chamber facility at Caltech. Figure 6A shows the particle size distribution in the photochemical oxidation of methylcyclohexane in the presence of NO_x and foreign seed particles. After 20 min of reaction, the ammonium sulfate seed aerosol has grown at approximately constant number concentration; this observation indicates the presence of a condensible species that deposits on the seeds. This growth at approximately constant number concentration provides the data necessary to estimate the excess vapor pressure of condensing species in the smog chamber. One minute after the second size distribution was

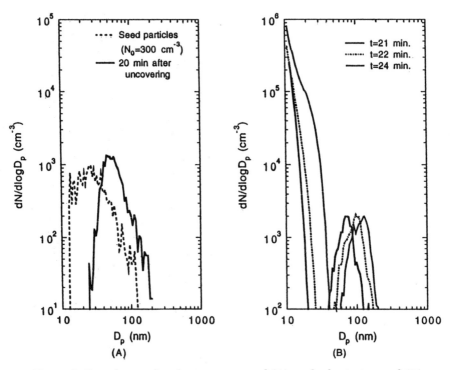

Figure 6. Particle size distributions measured (A) at the beginning and (B) after initial particle growth in a seeded methylcyclohexane–NO_x photooxidation experiment conducted in the Caltech outdoor smog chamber. (Reproduced from reference 49. Copyright 1991 American Chemical Society.)

measured, a large number of particles are present at very small sizes, as shown in Figure 6B, the result of homogeneous nucleation. Over the next several minutes, these particles grow rapidly, ultimately establishing the well-known self-preserving particle size distribution. The data presented have only been corrected to account for the measured counting efficiency of the SEMS and have not been smoothed or corrected for multiple charging at larger sizes in any way. The new particles are formed so rapidly that it would not be possible to follow these transients with stepping-mode differential mobility analyzers without extreme reductions in the numbers of sizes measured. Thus, the SEMS offers resolution in both particle size and time that is far superior to any measurement methods previously available.

A second example is the photochemical oxidation of 1-octene, both alone and with small amounts of SO_2. Seed particles grow with no apparent new particle formation when no SO_2 is present, leading to the seed particle growth shown in Figure 7A. The aerosol behavior in the presence of SO_2 differs dramatically, as is illustrated by the results of an experiment conducted simultaneously with the first experiment. After only 8 min, particles are formed by homogeneous nucleation. The number of particles produced in this nucleation burst is relatively small, so coagulation does not contribute significantly to particle growth. The particles grow rapidly nonetheless. The particle size distribution narrows, however, an indication that the growth is primarily the result of condensation. After several hours, a second burst of nucleation occurs, as is shown in Figure 7B. The extremely sharp particle size distributions that result from condensational growth represent a serious challenge for aerosol measurements. Attempts to accelerate measurements by sampling a limited number of size intervals risk missing the peak in the distribution entirely. On the other hand, all particle mobilities are probed in the SEMS because particles are continuously counted throughout the voltage scan.

One can speculate on the nature of the material that contributed to each burst of nucleation and the growth of the initial nuclei. The early nucleation did not occur under the same conditions without SO_2, so it is probable that it results from the primary oxidation product of that species, namely, H_2SO_4. The second nucleation burst is probably the same material that condensed without the initial SO_2, that is, the condensible hydrocarbons that result from the 1-octene photooxidation. Because the initial SO_2 concentration was much smaller than that of the hydrocarbon, much of the growth of the early nuclei is likely due to hydrocarbon condensation, that is, condensation of species that did not nucleate until much later in the first experiment. An examination of the quantity of aerosol produced in the two experiments supports this interpretation. As shown in Figure 8, particle formation occurs before significant hydrocarbon reaction in the SO_2-containing experiment. Once the hydrocarbon reaction begins in earnest, the aerosol yield increases by an amount that is comparable to that in the SO_2-free experiment. Two

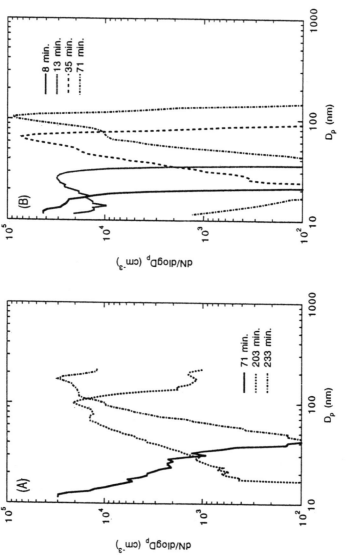

Figure 7. Particle size distributions measured with the SEMS showing the formation and growth of nuclei in the 1-octene photooxidation. The distributions are shown (A) without SO_2 and (B) with added SO_2. (Reproduced from reference 49. Copyright 1991 American Chemical Society.)

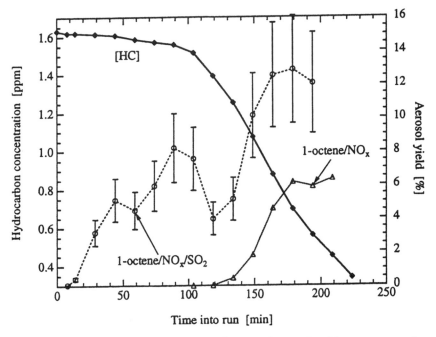

Figure 8. Hydrocarbon concentration decay and aerosol yield as a percent of reacted hydrocarbon during 1-octene photooxidation with and without added SO₂. (Reproduced with permission from reference 50. Copyright 1992 Pergamon.)

different materials are condensing, as evidenced by the variation of the aerosol yield with the amount of hydrocarbon reacted (Figure 9). With no SO₂ present, the yield drops dramatically if less than 0.1 ppm of hydrocarbon has reacted. When SO₂ was present, the yield remained significant to substantially lower amounts of reacted hydrocarbon, a situation consistent with the production of nonvolatile H_2SO_4. The scatter in the data for different SO₂-free experiments can be at least partially attributed to different operating temperatures in the outdoor smog chamber experiments.

The dynamics of the aerosols produced by photochemical reactions under simulated atmospheric conditions are complex functions of the chemical nature of the aerosols generated. Particle size distributions yield insufficient data for the complete characterization of the aerosol. Seeded smog chamber experiments have proven useful in estimating some of the physical properties of the secondary aerosols produced in smog chamber studies, but the uncertainties in estimated vapor pressures remain large. A more complete understanding of the dynamics of aerosol formation in such multiple-reactant systems could be obtained if the variation of particle composition with time and particle size could be measured. Some progress toward the measurement of size-dependent particle properties has been attained with the TDMA. Conventional TDMA operation requires as long as 1 h to acquire the vapor

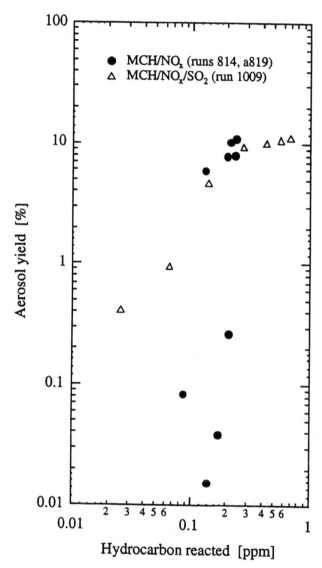

Figure 9. Aerosol yield as a function of the amount of hydrocarbon reacted during methylcyclohexane photooxidation with and without added SO₂. (Reproduced with permission from reference 50. Copyright 1992 Pergamon.)

pressure data. Recent experiments performed at Caltech have completed such measurements in a matter of minutes with the SEMS as the final analyzer. The TDMA–SEMS system is currently being used in this mode at Caltech to determine vapor pressure distributions for the aerosols produced by photochemical reactions at the end of our smog chamber studies.

Discussion

The arsenal of aerosol measurement methods has expanded dramatically over recent years, but a number of needs for fundamental research into the nature and origins of the atmospheric aerosol cannot be met by the current instrumentation. Instrumentation that has proven most valuable in the chemical characterization of the atmospheric aerosol is woefully inadequate either for following the chemical dynamics of aerosols because of the rapid changes that occur in homogeneous reacting systems or for unraveling the complexity of the aerosol products.

Techniques for measuring particle size distributions have improved dramatically, at least in the submicrometer size range. Differential mobility analyzers and the electrical mobility spectrometer are dramatic improvements over earlier electrical mobility analysis techniques, much as the EAA was over its predecessors. The further improvements achieved with the scanning electrical mobility spectrometer make it possible, for the first time, to follow the rapid transients of particle nucleation and growth in smog chamber experiments. Mobility analysis only covers a limited size range, however. From a few nanometers to a few tenths of a micrometer, mobility analysis provides excellent size resolution. Resolution for larger particles is limited because of multiple charging.

Alternative ways to obtain high-resolution particle size distribution data for particles larger than 0.2 μm are needed. However, the changes in the concentrations and sizes of particles near 1 μm in size are generally slower than those observed for particles in the nanometer size range, so time resolution is not so severe a constraint on the measurement technologies used for the larger particles. Optical particle counters can provide excellent data from 0.1 μm to 1 μm, although better data processing electronics are generally required than are typically incorporated in commercial instruments. Dilution systems are frequently required with optical particle counters to avoid coincidence errors at typical ambient concentrations, introducing additional uncertainties and increasing count noise for larger particles. The resolution of optical particle counters decreases for larger sizes because of the uncertainties caused by the Mie resonances in the light scattering signatures. On the other hand, the aerodynamic particle sizer can generate high-resolution particle size distribution data from particles perhaps 0.8 μm to several micrometers in size.

Size-resolved chemical information is much more difficult to obtain. The many applications of the differential mobility analyzer in measuring properties of size-classified particles are important tools for the characterization of aerosol systems, but the approaches demonstrated to date yield limited data. Vapor pressures, surface tension, and optical absorption have been measured on mobility-classified aerosols. Direct measurements of the distribution of chemical composition with particle size are needed. Elemental analysis by inductively coupled plasma emission spectroscopy or other analytical methods could replace the particle detectors used in size distribution analysis with the mobility classifier. The scanning mode of operation would optimize the use of the analytical technique and yield high-resolution composition–size distributions rapidly.

The mobility techniques only cover a fraction of the size range of concern to atmospheric studies. For particles above about 0.2 μm in diameter, multiple charging degrades the size resolution of the mobility analyzers. The low-ion-density charger of Gupta and McMurry (19) can extend the size range of singly charged particles up to about 1 μm without sacrificing the scanning capability, but for this method any charged particles present in the original aerosol must be removed with an electrostatic precipitator before charging can begin. At 1 μm, this requirement leads to a reduction in the number concentration to 13% of the original level, making demands on analytical sensitivity even more severe. Alternative approaches are needed for the classification of the larger particles. Aerodynamic methods can readily be applied to such particles, as evidenced by cascade impactors and the aerodynamic particle sizer. The ideal instrument would separate particles within a narrow range of aerodynamic diameters for analysis. Furthermore, it would facilitate measurements over a substantial range of particle sizes on a continuous basis. Thus, there are a number of challenges remaining for aerosol measurement methods that can follow rapidly changing populations and compositions of atmospheric aerosols.

Acknowledgments

This work was supported by National Science Foundation Grant ATM–9003186.

References

1. McMurry, P. H.; Friedlander, S. K. Atmos. Environ. 1979, 13, 1635–1651.
2. Warren, D. R.; Okuyama, K.; Kousaka, Y.; Seinfeld, J. H.; Flagan, R. C. J. Colloid Interface Sci. 1987, 116, 563–581.
3. Pui, D. Y. H.; Liu, B. Y. H. Phys. Scr. 1988, 37, 252–269.
4. Podzimek, J. Am. Meteorological Soc. 1989, 70, 1538–1545.
5. May, K. R. J. Aerosol Sci. 1982, 13, 37–47.
6. May, K. R. J. Sci. Instrum. 1945, 22, 187.
7. Hering, S. V.; Flagan, R. C.; Friedlander, S. K. Environ. Sci. Technol. 1978, 12, 667–673.
8. Biswas, P.; Jones, C. L.; Flagan, R. C. Aerosol Sci. Technol. 1987, 7, 231–246.
9. Marple, V. A.; Liu, B. Y. H.; Kuhlmay, G. A. J. Aerosol Sci. 1981, 12, 333.
10. Agarwal, J. K.; Remiarz, R. J.; Quant, R. J.; Sem, G. J. J. Aerosol Sci. 1982, 13, 222.
11. Whitby, K. T.; Clark, W. E. Tellus 1966, 18, 573–586.

12. Liu, B. Y. H.; Pui, D. Y. P. *J. Colloid Interface Sci.* **1974**, *47*, 155.
13. Knutson, E. O.; Whitby, K. T. *J. Aerosol Sci.* **1975**, *6*, 443.
14. Berglund, R. N.; Liu, B. Y. H. *Environ. Sci. Technol.* **1973**, *7*, 147.
15. Twomey, S. *J. Comput. Phys.* **1975**, *18*, 188–200.
16. Wolfenbarger, J. K.; Seinfeld, J. H. *J. Aerosol Sci.* **1990**, *21*, 227–247.
17. Wang, S. C.; Flagan, R. C. *Aerosol Sci. Technol.* **1990**, *13*, 230.
18. Romaynovas, F. J.; Pui, D. Y. H. *Aerosol Sci. Technol.* **1988**, *9*, 123–131.
19. Gupta, A.; McMurry, P. H. *Aerosol Sci. Technol.* **1989**, *10*, 457–462.
20. Natusch, D. F. S.; Wallace, J. R. *Science (Washington, D.C.)* **1974**, *186*, 695–699.
21. Davison, R. L.; Natusch, D. F. S.; Wallace, J. R.; Evans, C. A., Jr. *Environ. Sci. Technol.* **1974**, *8*, 1107–1113.
22. Roberts, P. T.; Friedlander, S. K. *Environ. Sci. Technol.* **1976**, *10*, 573–580.
23. Palen, E. J.; Allen, D. T.; Pandis, S. N. *Atmos. Environ. A* **1992**, *26*, 1239–1251.
24. Huntzicker, J. J.; Cary, R. A.; Ling, C. S. *Environ. Sci. Technol.* **1980**, *14*, 819–824.
25. Denoyer, E.; Natusch, D. F. S.; Surkyn, P.; Adams, F. C. *Environ. Sci. Technol.* **1983**, *17*, 457.
26. Post, J. E.; Buseck, P. R. *Environ. Sci. Technol.* **1984**, *18*, 35–42.
27. Liu, B. Y. H.; Pui, D. Y. P.; Whitby, K. T.; Kittelson, D. B.; Kousaka, Y.; McKenzie, R. L. *Atmos. Environ.* **1978**, *12*, 99–104.
28. Rader, D. J.; McMurry, P. H. *J. Aerosol Sci.* **1986**, *17*, 771–787.
29. McMurry, P. H.; Stolzenburg, M. R. *Atmos. Environ.* **1989**, *23*, 497–507.
30. Covert, D. S.; Heintzenberg, J.; Hansson, H.-C. *Aerosol Sci. Technol.* **1990**, *12*, 446–456.
31. Sinha, M. P.; Giffin, C.; Norris, D. D.; Estes, T. J.; Vilker, V. L.; Friedlander, S. K. *J. Colloid Interface Sci.* **1982**, *87*, 140–153.
32. Giggy, C. L.; Friedlander, S. K.; Sinha, M. P. *Atmos. Environ.* **1989**, *23*, 2223–2229.
33. Ottesen, D. K.; Wang, J. C. F.; Radziemski, L. J. *Appl. Spectrosc.* **1989**, *43*, 967–976.
34. Friedlander, S. K. *Aerosol Sci.* **1971**, *2*, 331–340.
35. Whitby, K. T.; Husar, R. B.; Liu, B. Y. H. *J. Colloid Interface Sci.* **1972**, *39*, 177–204.
36. Husar, R. B.; Whitby, K. T.; Liu, B. Y. H. *J. Colloid Interface Sci.* **1972**, *39*, 211–224.
37. Haagen-Smit, A. J. *Ind. Eng. Chem.* **1952**, *44*, 1342–1346.
38. Renzetti, N. H.; Doyle, G. J. *J. Air Pollut. Control Assoc.* **1960**, *2*, 327–345.
39. Praeger, J. J.; Stephens, E. R.; Scott, W. E. *Ind. Eng. Chem.* **1960**, *52*, 521.
40. Stevenson, H. J. R.; Sanderson, D. E.; Altshuller, A. *Int. J. Air Water Pollut.* **1965**, *9*, 367–375.
41. Clark, W. E.; Whitby, K. T. *J. Colloid Interface Sci.* **1975**, *51*, 477–490.
42. Grosjean, D.; Friedlander, S. K. *Environ. Sci. Technol.* **1980**, *9*, 435–473.
43. Shen, C. S.; Springer, G. S. *Atmos. Environ.* **1977**, *11*, 683–688.
44. Heisler, S. L.; Friedlander, S. K. *Atmos. Environ.* **1977**, *11*, 157–168.
45. Kockmond, W. C.; Whitby, K. T.; Kittelson, D. B.; Kemerjian, K. L. *Adv. Environ. Sci. Technol.* **1977**, *8*, 101–135.
46. Stern, J. E.; Flagan, R. C.; Seinfeld, J. H. *Aerosol Sci. Technol.* **1989**, *10*, 515–534.
47. Izumi, K.; Murano, K.; Mizuochi, M.; Fukuyama, T. *Environ. Sci. Technol.* **1988**, *22*, 1207–1215.

48. Izumi, K.; Fukuyama, T. *Atmos. Environ.* **1990,** *24A,* 1433–1441.
49. Flagan, R. C.; Wang, S. C.; Yin, F.; Seinfeld, J. H.; Reischl, G.; Winklmay, W.; Karsch, R. *Environ. Sci. Technol.* **1991,** *25,* 883–890.
50. Wang, S. C.; Paulson, S. E.; Grosjean, D.; Flagan, R. C.; Seinfeld, J. H. *Atmos. Environ.* **1992,** *26A,* 403–420.

RECEIVED for review May 30, 1991. ACCEPTED revised manuscript August 10, 1992.

Compositional Analysis
of Size-Segregated Aerosol Samples

Thomas A. Cahill and Paul Wakabayashi

Air Quality Group, Crocker Nuclear Laboratory, University of California, Davis, CA 95616–8569

Knowing both the size and composition of fine particles in the air is vital for understanding the sources, transport, transformation, effects, and sinks of atmospheric aerosols. However, compositional analysis of such size-segregated aerosol samples poses difficulties because of the small amount of mass available for analysis, the chemical complexity of the particles, and the nonuniform deposits characteristic of most impactors. Additional problems are posed by the need to measure both a wide range of elements and trace concentrations. Nevertheless, significant progress has been made in the past decade, especially in evaluation of the sources and nature of visibility degradation by fine particles. This chapter is a short summary of the difficulties of obtaining size-specific chemical information and the usefulness of such information once obtained.

THE AMBIENT ATMOSPHERIC AEROSOL consists of liquid and solid particles that can persist for significant periods of time in air. Generally, most of the mass of the atmospheric aerosol consists of particles between 0.01 and 100 μm in diameter distributed around two size modes: a "coarse" or "mechanical" mode centered around 10- to 20-μm particle diameter, and an "accumulation" mode centered around 0.2- to 0.8-μm particle diameter (1). The former is produced by mechanical processes, often natural in origin, and includes particles such as fine soils, sea spray, pollen, and other materials. Such particles are generated easily, but they also settle out rapidly because of deposition velocities of several centimeters per second. The accumulation mode is dominated by particles generated by combustion processes, industrial processes, and secondary particles created by gases converting to par-

0065–2393/93/0232–0211$06.00/0

ticles. The accumulation-mode particles are mostly anthropogenic. An example of a size–composition profile of ambient atmospheric aerosols is shown in Figure 1.

Figure 1 shows the fraction of each elemental component of particles that occurred in one of eight size ranges from 0.1 to about 15 μm in diameter. The element silicon, derived from the SiO_2 in soils, is present only in coarse size ranges. Thus, it is solely a coarse-mode particle derived from mechanical processes. Sulfur, present largely in the form of ammonium sulfate, occurs only in the accumulation mode around 0.3 μm. Potassium occurs in both modes: a coarse mode from soil and a fine mode derived from agricultural smoke. The chlorine is from sea salt, NaCl, which is a coarse-mode aerosol that lost its coarsest components during transport from oceanic sources about 100 km upwind of the site (Davis, California). This figure shows a typical bimodal distribution; such distributions may change. Thus, the challenge for analytical chemists is to generate such data accurately and inexpensively from collected aerosols that, as in the instance above, total no more than a few micrograms for each size range.

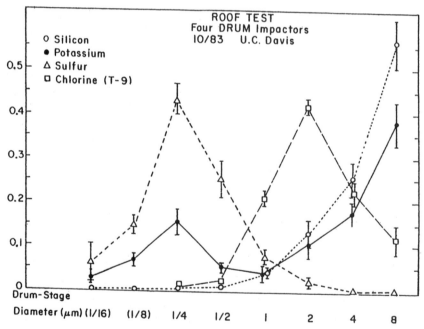

Figure 1. Example of compositionally resolved bimodal and monomodal distributions of aerosols. The ordinate gives the percent of the species found in the given size fraction of the impactor. The mode near 0.3 μm is the "accumulation mode", and that above 8 μm is the "coarse mode". The minimum of mass between 1 and 2 μm is typical; the chlorine distribution is anomalous. Chlorine is in fact a coarse-mode marine aerosol that has lost its larger particles during transport from the ocean to Davis, California, a distance of roughly 100 km. (Reproduced with permission from reference 15. Copyright 1988.)

Compositional Analysis of Atmospheric Aerosols

Because the atmospheric aerosol consists of a mixture of gases and particles and because the size and composition of the particles are usually of interest, the particles must be analyzed within the gases or the particles must be removed from the gases prior to analysis.

Compositional analysis of particles within a gas–particle system is highly desirable and physically possible, but very difficult. This analysis is highly desirable because it would provide a real-time size–composition analysis to complement real-time analyses of meteorology and gaseous pollutants as well as real-time analyses of certain prompt effects of aerosols, such as visibility degradation. This method is physically possible because the exciting source (optical, X-ray, nuclear, etc.) can be tuned to pick out particles without exciting the enormously abundant gases (N_2, O_2, Ar, CO_2, etc.). Lasers could vaporize particles of a certain size without exciting gases, giving rise to emission spectra. X-rays could excite all species but filter out soft X-rays from gases dominated by H, C, N, and O. However, for ambient sampling, potential analytical methods are very difficult and expensive, and they generally lack sensitivity. They are rarely, if ever, used for ambient sampling today, but the need is great and the challenges are clear. This, certainly, is one area that needs further work.

Almost all chemical analysis of atmospheric aerosols is based on removal of the particles from gases. Thus, the primary task for aerosol samplers is to separate the aerosol particles from the overwhelmingly larger mass of the gases in the atmosphere. Two procedures are commonly used. The first, and simplest, method is to draw air through filters, collecting the particles for future analysis. In the second, orifices accelerate the gas–particle stream to high velocity and then force it to make a sharp bend. Particles are removed from the air stream and impacted onto a surface. These samplers are called "impactors", and they are the instruments that pose the most difficult challenges to analytical chemists. Virtual impactors are a subset of these devices that avoid surface impaction by using filters.

Filters. For most situations, the most commonly used technique for collecting aerosols is pulling air through a filter that collects the particles but not the gases. Compositional analyses are then performed on the filter. In reality, a filter does not collect all particles in the aerosol because the inlet will miss large, wind-driven particles unless great care is taken to achieve isokinetic sampling. Modern samplers fix the maximum size of particles. The older high-volume sampler (Hi-Vol) has an effective upper cut point around 30 μm, but this value is strongly wind-dependent. The present standard is for 10-μm particles (PM–10), a size that is relatively constant with different wind velocities. Fine particles (diameters smaller than 2.5 μm) are often needed for health or visibility research, and many methods exist to perform such sampling (PM–2.5). If the filters are efficient, there are 0- to 30-μm, 0- to 10-μm, or 0- to 2.5-μm size regimes, but many others have been achieved by devices such as cyclones. With enough integrated samples, a full particle size spectrum can be derived from filter data (2). A price must be paid in high analytical costs and propagation of errors, but a good deal of mass can be collected, and standard filter analyses are possible.

If only particles larger than a certain size are of interest, this technique can be inverted by using diffusion to remove fine particles from the airstream and leaving the coarser particles to be collected on a filter. Again, techniques like this, of which the diffusion battery is the best known, also yield a standard filter and, thus, permit the use of standard analytical methods.

There are methods that collect particles from both the coarse and fine modes simultaneously. A clever way to achieve collection of both modes of particles onto filters is the virtual impactor (VI) (3). The gas–particle mixture is forced to make a sharp bend, and particles above 2.5 μm are ejected into a small portion (10%) of the gas stream onto a filter, so all the coarse particles (typically from the 10 μm limit set by the PM–10 inlet to 2.5 μm) and 10% of the fine particles are collected on a filter. The remainder is then filtered; this portion contains no coarse particles but 90% of the fine particles (2.5 to 0 μm). After analysis, the fine admixture in the coarse fraction is removed mathematically. Limitations in the process limit its usefulness below 1 μm, however.

A second way to achieve collection of coarse- and fine-mode particles onto filters is through tandem filtration through the "stacked filter unit" (SFU) (4, 5, 6). In these devices, the convenient filtration characteristics of filters (Nuclepore) allow a 2.5-μm cut point on the basis of pore size and the face velocity of the airstream. Such devices are very compact and inexpensive and have been heavily used in remote-area networks (7, 8, 9). Again, however, the limitations of the method limit the number and sharpness of the size cuts so that almost all units are operated at 2.5 μm and give coarse and fine fractions very much like those of the virtual impactor. Examples of the 2.5-μm cut points of VI, SFU, and impactors are shown in Figure 2. However, cyclones, virtual impactors, and stacked filter units cannot give the sharp, multiple cut points of impactors as shown in Figure 1.

Nevertheless, these methods all result in a filter that captures the atmospheric particles. The mass loading can be large, the deposit uniform, and the filter reasonably stable under transport to a central analytical laboratory. Numerous papers have treated analysis of such filters, so this information is not repeated. This chapter focuses on the problems of chemical analyses of impactor substrates, for which the problems are more serious and the solutions elusive.

Impactors.   Impactors work by forcing the gas–particle stream to make a sharp bend. This action causes the larger particles to impact onto a medium,

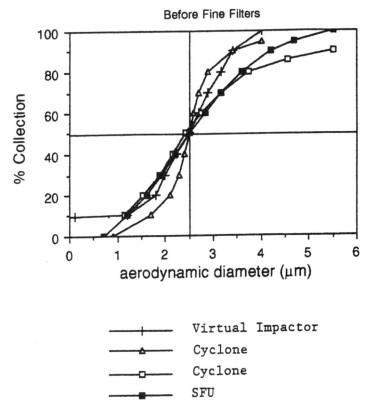

Figure 2. Comparisons of collection efficiencies of various types of aerosol samplers; although all have 50% capture efficiency points of roughly 2.5 μm, the shapes of the collection curves vary.

whereas the smaller particles and gases continue downstream. Impactors, by means of their construction, can sequentially segregate particulate matter to smaller and smaller sizes. By varying the orifice size, the number of orifices, the pressure, the velocity of the jet, and other specifications, the desired size-selected particles can be collected. These impactors have sharper size cutoffs than cyclones, virtual impactors, or stacked filter units.

Some examples of ambient air impactors include the low-pressure impactor (LPI) (*10*), the Battelle (*11, 12*), the Multi-Day (MD) (*13*), the Davis Rotating Unit for Monitoring (DRUM) (*14, 15*), the Berner Low-Pressure Impactor (BLPI) (*16*), and the Micro-Orifice Uniform Deposit Impactor (MOUDI) (*17*). Each has a different way of collecting particulate matter. Because of the small mass of aerosols, certain parameters of the design of the impactor are adjusted so that the aerosol can be analyzed with a maximum degree of sensitivity. Some of the parameters involve the flow rate at which the impactor operates: the material is concentrated to a smaller area or the period of sampling is extended. But increasing one parameter may decrease sensitivity. Two samplers, the BLPI and the MOUDI, rely on an increased flow rate to collect more material. Total collected mass per stage is the key parameter for these analytical methods, which remove the deposit from the collection substrate. Both of the samplers have more orifices to allow for the increased volume and to achieve the correct sizing of aerosols. Although increasing the number of orifices increases the flow rate, it also increases the spread of aerosols on a collection media. This spreading has the effect of decreasing the concentration of material and thereby decreasing the sensitivity of certain measurements. On the other hand, the DRUM, the LPI, and the Battelle utilize a single orifice to concentrate the material, an arrangement that increases sensitivity. High areal density of the deposit, in grams per square centimeter, is a key parameter for methods that analyze the deposit without removing it from the substrate. But problems arise from particle bounce-off and from layering of particulate matter, which cause problems in some analysis techniques.

Compositional Analysis of Particulate Samples

Once the particulate sample has been removed from the airstream and deposited on a filter or an impaction–diffusion surface, the analyst can either remove the deposit from the surface and analyze the resulting gas or liquid or leave the deposit on the surface and analyze the surface and deposit together.

In the first method, any and all analytical methods are available to the analyst; however, problems arise in two areas. First, the analyst must be sure that the particles are removed from the substrate and incorporated into the aliquot. Second, the mass of material is always limited, so extreme analytical sensitivity and very pure reagents are required. An example is the collection of fine particles with diameters less than 2.5 μm from a 10-μg/m³ fine aerosol for 4 h at a 20 L/min flow rate. The total particulate mass collected is 48 μg. This mass is removed from the filter or surface with 0.1 mL of solvent. The total dissolved particulate is 480 ppm in the solvent, and this concentration must be analyzed to about 1.1 ppm in accuracy. The analytical method needs to be sensitive at the 0.48-ppb level, and contaminants in the solvent must be held to such levels also.

For this reason, particulate samplers designed for particulate removal have to generate the maximum possible particulate mass. Modern examples include impactors based on the high-volume sampler (Hi-Vols), the MOUDI (17) of the University of Minnesota, and the BLPI (16). The Hi-Vols, in particular, collect 330 m³ of air in 4 h, giving 1100 μg of deposit for three size cuts below a particle diameter of 2.5 μm. Table I shows some key parameters for a few widely used ambient air impactors for multiple size cuts.

Table I. Mass-per-Stage Comparison of Several Impactors

Sampler Type	Flow (L/min)	Volume[a] (m³)	Stages <2.5 µg (n)	Average[b] Mass per Stage (µg)
DRUM	1.0	0.24	6	0.40
LPI	1.0	0.24	7	0.34
Battelle	1.0	0.24	6	0.40
MOUDI	30.0	7.20	6	12.00
BLPI	30.0	7.20	6	12.00
MD	30.0	7.20	2	36.00

[a]Volume per 4-h period.
[b]A density of 10 µg/m³ for particles smaller than 2.5 µm was assumed.

The second class of particulate samples is those that are analyzed without removal from the sampling substrate. For such samples, only limited classes of analytical methods can be used, and the substrate itself is critical. As an example, consider again a 20 L/min sample being collected from a 10 µg/m³ fine-particle ambient aerosol for 4 h. The 480 µg of mass are still collected, but now the area of the deposit is critical. If a 12-cm² Teflon filter of 480 µg/cm² thickness, such as stretched Teflon [poly(tetrafluoroethylene)], is used, an areal density of 40 µg/cm² of particulate deposit on a 480 µg/cm² substrate is produced. The total particulate filter sample is now 520 µg/cm², and a 1 ppm compositional analysis of the particulate deposit requires an analysis sensitivity to 80 ppb. In other words, the analytical method must be sensitive to 0.04 µg/cm². Clearly, the key parameter is areal density. If a filter of 6 cm² rather than 12 cm² were used, the deposit would have the same total mass but twice the areal density. For a given analytical method sensitive to area density, such as X-rays, laser absorption, or a beta gauge, halving the area gives roughly a factor of 2 gain in sensitivity. Table II shows a comparison of analytical sensitivity for a few widely used ambient air impactors for multiple size cuts.

The extreme examples of such samplers are the single-jet impactors such as the Battelle of Florida State University, the LPI, or the DRUM, all of

Table II. Analytical Sensitivity Comparisons for a 2-ng/cm² Detectable Limit

Sampler Type	Stages <2.5 µg (n)	Analysis Area (cm²)	Volume (m³)	Collection Sensitivity (m³/cm²)	Minimum Detectable Limit (ng/m³)
DRUM	6	0.084	0.24	2.90	0.7
LPI	7	0.071	0.24	3.40	0.6
Battelle	6	0.071	0.24	3.40	0.6
MOUDI	6	9.620	7.20	0.75	2.7
BLPI	6	13.040[a]	7.20	0.55	3.6
MD	2	18.600	7.20	0.39	5.1

NOTE: All parameters are identical to those in Table I.
[a]Area per stage. The number of orifices varies from 25 to 232 below 2.5 µm.

which operate at only 1.1 L/min. In 4 h in a 10 $\mu g/m^3$ aerosol, the collected mass for fine particles is only 2.4 μg, generally spread out over five separate stages. Yet, because the orifices are tiny, the deposit falls almost entirely within a 1.1-mm-diameter circle, so an areal density of 320 $\mu g/cm^2$ is produced. Good sensitivities despite the small mass can be obtained with a method, such as particle-induced X-ray emission (PIXE) (18), that uses a focused ion beam that only irradiates the deposit area plus 1.1 mm in all directions.

However, analytical methods that analyze the deposit and substrate together, such as PIXE and X-ray fluorescence (XRF), have a serious problem posed by a nonuniform deposit. For an ion beam, such as protons, incident upon an aerosol sample placed 45° to the beam, the ions pass through the sample and are collected in a Faraday cup to provide absolute concentrations. If the exiting radiation, whether it be ions, X-rays, electrons, or light, is uniform across the deposit or if the deposit itself is uniform, then the result is accurate. However, if both the beam and sample are nonuniform in either plane, a convolution integral is required to obtain the concentration on the substrate. In practice this integral is never done, so analytical accuracy is critically dependent on beam and sample uniformity, both of which are usually suspect.

The limits to the areal density of deposit for filters are set by clogging of the filter that sets in at typically 100 $\mu g/cm^2$. The limit of areal density for impactors is set by the problem of particle bounce. This is a serious problem for coarse, dry aerosols but less so for fine, wet, secondary aerosols. Nevertheless, sticky substrates are universally used (19), and deposits are generally limited to a few monolayers of particles for a 2.5-μm particle. This limit amounts to no more than 7 μm of deposit, or, for 1.5-$\mu g/m^3$ aerosols (per stage), about 1000 $\mu g/cm^2$ of deposit.

In summary, chemical analysis of the collected particulate matter on a multiple-staged impactor poses serious difficulties:

1. There is only a severely limited amount of mass available for analysis, and efforts to increase size information through more stages simply makes the available mass even less. Attempts to collect more mass by longer runs are limited by particle-bounce effects.

2. Attempts to provide accurate particulate size information usually require adhesives on the collection surfaces. But these adhesives in turn make removal of particles for analysis difficult, contaminate the deposit, and provide background and contaminant problems for methods that analyze the substrates and deposit together. The problems of particle bounce are, of course, greatly reduced for submicrometer hygroscopic or organic aerosols, which usually include the important anthropogenic sulfates, nitrates, and organic compounds.

3. Impactor deposits are typically highly nonuniform; this non-uniformity reduces accuracy and precision for techniques that analyze the substrate and deposit together.

4. Many techniques used for compositional analysis of filters will not work for most impactors (e.g., gravimetric mass).

These factors combine to make impactors less precise and accurate than filters. Very few comparisons have been made between sizing impactors and those that have provided mixed results. The 1977 Environmental Protection Agency–Department of Energy Sampler Intercomparison included the Multi-Day Sampler, which performed well ($\pm 15\%$) for fine aerosols such as sulfur, lead, and zinc (15). The 1986 Carbonaceous Species tests at Glendora, California, included the DRUM sampler. It performed well for sulfur ($\pm 18\%$), as compared to the fine filter sampler (PM–2.5), but no other sizing impactor was available for comparison and no element other than sulfur was reported. DRUM versus filter comparisons were reported as part of the Southern California Air Quality Study of 1987 (2). Again, no other impactor was available for comparison, and the comparisons with filters were only fair ($r^2 \approx 0.7$; r, linear correlation coefficient).

Probably the first side-by-side comparison of multiple-stage impactors occurred as part of the Salt River Project's Grand Canyon Study of 1989–1990; the DRUM, LPI, and MOUDI were all used. This study had operational problems. Possibly because some of the samplers were rented out to the study and operated by third parties, a great deal of data were lost. Nevertheless, the overall behavior of the major aerosol species was usually reproduced in size and concentration. Although the DRUM was reported to have been less precise and accurate than the MOUDI or LPI, all correlations were far worse than similar correlations for filter samples, and slopes versus filters were well below unity (20). Moreover, a study was performed at Shenandoah National Park in 1991, during which filters and the MOUDI were operated versus three co-located DRUM samplers. Figure 3 shows the results over a 3-week period. Again, fair agreement is evident, but the data with $r^3 = 0.78$ must be compared to side-by-side filter samples at the same site, which achieved $r^3 = 0.96$. In an attempt to improve this situation, we developed a family of samplers with the rotating drum and slit configuration of the Multi-Day Impactor, which itself was a modification of the Lundgren Impactor. Flow was raised to 12 L/min, and a new analysis system was built that was dedicated to such difficult samples. Figure 4 shows the result of a side-by-side comparison of two IMPROVEd DRUM Samplers for particles with diameters between 2.5 and 0.34 μm. Although this was only a precision test, the results are better than other side-by-side tests of ambient sizing impactors; values of r^2 are as high as 0.96, and slopes are within 5% of unity.

The argument can be made that some lack of precision and accuracy is only to be expected, given the formidable difficulties in accurate collection and compositional analysis of aerosols by size. Thus, while efforts are being

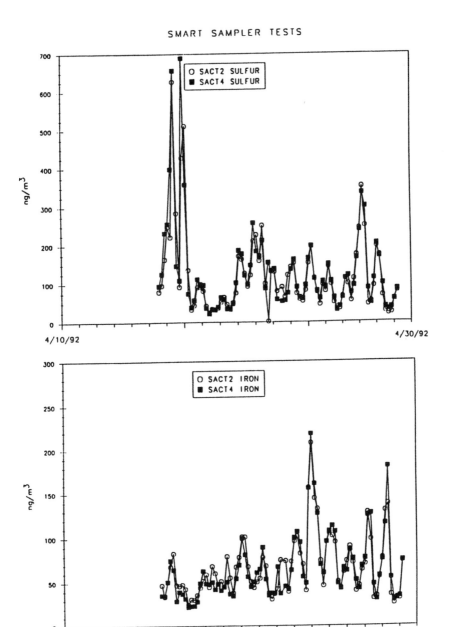

Figure 4. Comparisons of two IMPROVEd DRUM samplers (12 L/min, three size cuts below 2.5 μm, and slit orifices) running side by side at Sacramento, California. The size range shown is for a particle diameter of 0.34 to 2.5 μm. Iron is used as a tracer of soil particles. The mean r^2 for all stages less than 2.5 μm was 0.87 with a slope of 1.03.

expended to improve the analytical situation, a question must also be raised: Is the resulting information useful despite the limits to accuracy and precision?

Examples of the Utility of Size–Composition Profiles

With increased sensitivity, the size-segregating samplers have produced data that are most helpful in characterizing the size and composition of aerosols. The data from these samplers have been used to resolve a bimodal, and sometimes a trimodal, distribution of particles.

The first example involves the very large data set collected in California from 1973 to 1977 with the MD (20). Almost by accident, this impactor had a cut point at 0.5 μm. When an analysis was made of visibility degradation in California (2), only the sulfur in the 0.5-μm to 3.6-μm size range was found to correlate strongly with visibility. Thus, on statistical grounds alone, it was established that sulfur in the finest modes was not a major factor in California's visibility problems, whereas sulfur in the high end of the accumulation mode was disproportionately efficient in scattering light. If the aerosol data had been collected with a filter having a range from 2.5 to 0 μm, this conclusion could not have been drawn.

The physical reason for this statistical result was determined a decade later, after a nine-day study in Los Angeles in 1986 as part of the large Southern California Air Quality Study (SCAQS) (Figure 5) (24). Visibility was poor on the first five days but better on the last four days. Although most meteorological and air pollution parameters were relatively constant during this change, sulfur had a major size change from a relatively coarse aerosol, mass median diameter (MMD) 0.50 μm, to a finer aerosol, MMD 0.33 μm (21). The change in visibility as measured by a nephelometer was 0.184 km^{-1} (23), and the change in aerosol mass was 14.2 ± 1.0 μg/m^3. Almost all of this difference was ammonium sulfate (8.0 μg/m^3), with some water ammonium nitrate and organic matter. Almost all of the change in ammonium sulfate occurs in the optically efficient 0.34- to 1.15-μm-diameter size ranges (7.9 μg/m^3). These chemical changes resulted in a dramatic decrease in visibility (15.5 km to 8.8 km) despite only a modest 15% change in fine aerosol mass.

The third example (20) involves soils measured near an industrial area of California over two weeks (Figure 6). The data were taken on a DRUM impactor; stage 1 was coarse particles, 15 to 10 μm, and stage 2 was fine aerosols within the PM–10 cut, 10 to 5 μm. The aerosols in these ranges were almost entirely soils (Al, Si, K, Ca, Ti, Mn, Fe, Sr, and so forth), and iron is used as an example. Clearly, the coarse particles are anticorrelated with the finer particles despite the fact that both are soils. The explanation lies in wind suspension of the coarse particles; the highest wind velocities give the highest coarse particle loadings. This is the standard behavior seen

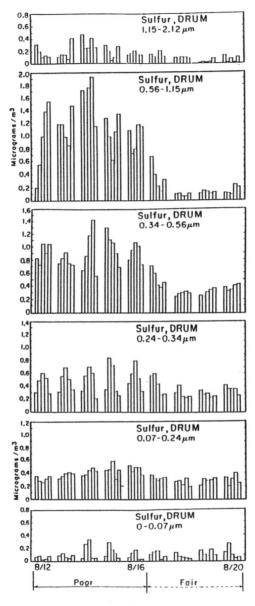

Figure 5. Sulfur size distribution in Glendora, California (east of Los Angeles), August 1986. The period of poor visibility during August 12 through 16 is best explained by the presence of sulfur mass in sizes above 0.34 μm in diameter. Surprisingly, fine mass and most of the other components changed little as the visibility improved. (Reproduced with permission from reference 24. Copyright 1990.)

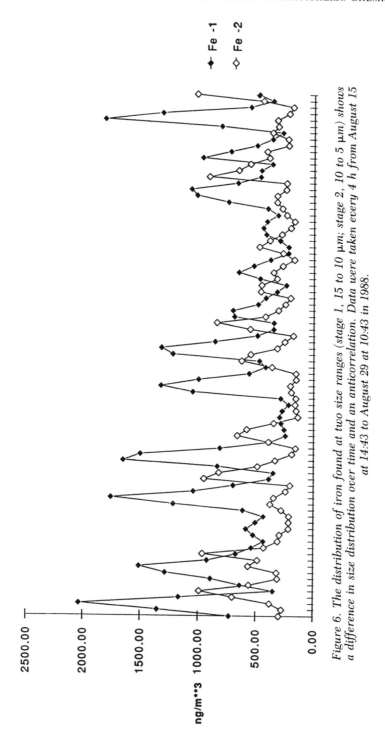

Figure 6. The distribution of iron found at two size ranges (stage 1, 15 to 10 μm; stage 2, 10 to 5 μm) shows a difference in size distribution over time and an anticorrelation. Data were taken every 4 h from August 15 at 14:43 to August 29 at 10:43 in 1988.

in arid sites around California. The surprise is the anticorrelation of the finer soils with wind velocity. This anticorrelation is typical for wind dilution of a constant source term. In fact, there was a constant level of activity at this site 24 h per day in the form of a great deal of truck traffic on unpaved roads. Thus, the finer aerosols are concluded to be anthropogenic. The presence of numerous other nonsoil elements in this finer mode supports this hypothesis.

A fourth example of data showing the particle distribution was a study that used the DRUM sampler at Grand Canyon National Park in 1984 (22). Recording the size distribution of sulfur was necessary in helping to understand the effects of sulfur on visibility degradation because there were two size modes: one near 0.3 μm and one around 0.1 μm. These modes were not present simultaneously but appeared somewhat anticorrelated (*see* August 14 in Figure 7).

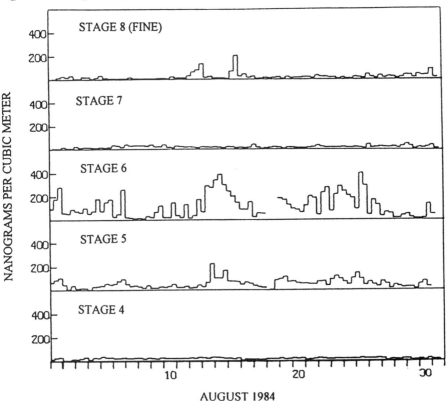

Figure 7. Sulfur size distributions at Grand Canyon National Park, August 1984. The finest aerosols are at the top (stage 8, around 0.1 μm), and the coarsest (8.5 to about 15 μm) are at the bottom. Most sulfur mass occurs between 0.34 and 0.56 μm (stage 6). The periods of August 14 are used for Table III. (Reproduced with permission from reference 15. Copyright 1988.)

Although the ultrafine particles were found at times to be a major component of the total amount of sulfur below 2.5 μm, the ultrafine particles are not predicted to be a major factor in visibility degradation because they are too fine to effectively scatter light. On the other hand, under the right conditions these same particles may grow in size to the accumulation mode, the mode most efficient in scattering light. As shown by this example and others, collecting one size of particles will not tell the whole story of the aerosol's influence. In addition, Table III compares the composition of the finest particles, ~0.1 μm, to that of the accumulation-mode particles, ~0.3 μm. There is a dramatic increase in primary combustion-derived particles in the finest mode.

Table III. Concentrations of Selected Elements at Grand Canyon National Park for an 8-h Period on August 16, 1984

Elements	Stage 8, 0.088–0.15 μm (ng/m³)	Stage 6, 0.15–0.60 μm (ng/m³)
Sodium	420	10
Silicon and aluminum	<8	<6
Sulfur	204	392
Chlorine	208	<5
Potassium	59	<3
Calcium	150	5
Titanium	<2	4
Vanadium	<2	3
Iron and nickel	<2	<2
Copper	100	1
Zinc	931	<2
Arsenic	13	<2
Bromine	<2	<2
Lead	63	<4

The rapid development of multistage sequential impactors designed for compositional analysis during the 1980s must not be allowed to disguise the fact that very few such devices exist as of 1992. We anticipate a major expansion in the number and types of such devices in the 1990s and much greater utilization of the impactors for studies in visibility, acidic dry deposition, and PM–10 diagnostic and control efforts.

Conclusions

Compositional analyses of size-segregated particles from ambient atmospheric aerosols are vital for understanding the sources and effects of these aerosols. Three challenges exist for analytical chemistry in the next decade:

1. Compositional analyses of particles in an aerosol, ideally in real time, should be possible without the necessity of isolating the particles.

2. Further refinements are needed in compositional analyses of filters; greater sensitivity, specifically of organic species, are needed.

3. Improvement in sensitivity and range of nondestructive analyses from impactor samples is needed, especially in the methods for detecting total particulate mass and organic species.

For the greatly increased use of size–composition aerosol analysis anticipated in the next decade, progress in all of these areas is essential.

References

1. Whitby, K. T. *Atmos. Environ.* **1978**, *12*, 135–159.
2. John, W.; Reischl, G. *J. Air Pollut. Control* **1980**, *30*, 872–876.
3. Dzubay, T. G.; Stevens, R. K.; Petersen, C. M. In *X-ray Fluorescence Analysis of Environmental Samples*; Dzubay, T. G., Ed.; Ann Arbor Science: Ann Arbor, MI, 1977; pp 95–105.
4. Cahill, T. A.; Ashbaugh, L. L.; Barone, J. B.; Eldred, R. A.; Feeney, P. J.; Flocchini, R. G.; Goodart, C.; Shadoan, D. J.; Wolfe, G. W. *J. Air Pollut. Control Assoc.* **1977**, *27*, 675–678.
5. Eldred, R. A.; Cahill, T. A.; Feeney, P. J. *Proceedings of the 4th International Conference on Particle Induced X-ray Emission and Its Analytical Applications* **1986**, *B22*, 289–295.
6. Cahill, T. A.; Eldred, R. A.; Feeney, P. J.; Beveridge, P. J.; Wilkinson, L. K. In *Visibility and Fine Particles*; Mathai, C. V., Ed.; Air and Waste Management Association: Estes Park, CO, 1989; pp 213–222.
7. Flocchini, R. G.; Cahill, T. A.; Pitchford, M. L.; Eldred, R. A.; Feeney, P. J.; Ashbaugh, L. L. *Atmos. Environ.* **1981**, *15*, 2017–2030.
8. Barone, J. B.; Ashbaugh, L. L.; Kusko, B. H.; Cahill, T. A. In *Atmospheric Aerosol: Source/Air Quality Relationships*; Macias, E. S.; Hopke, P. K., Eds.; ACS Symposium Series 167; American Chemical Society: Washington, DC, 1981; pp 327–346.
9. Eldred, R. A.; Cahill, T. A.; Feeney, P. J.; Malm, W. C. *Visibility Protection: Research and Policy Aspects*; Bhardwaja, P., Ed.; Air Pollution Control Association: Pittsburgh, PA, 1987; pp 386–396; Transactions TR–10.
10. Hering, S. V.; Flagan, R. C.; Friedlander, S. K. *Environ. Sci. Technol.* **1978**, *12*, 667–673.
11. Mercer, T. T. *Ann. Occup. Hyg.* **1964**, *7*, 115–124.
12. Van Grieken, R. G.; Johansson, J.; Winchester, J.; Odom, A. L. *Anal. Chem.* **1975**, *275*, 343.
13. Flocchini, R. G.; Cahill, T. A.; Shadoan, D. J.; Lange, S. J.; Eldred, R. A.; Feeney, P. J.; Wolfe, G. W.; Simmeroth, D. C.; Suder, J. K. *Environ. Sci. Technol.* **1976**, *10*, 76–82.
14. Cahill, T. A.; Goodart, C.; Nelson, J. W.; Eldred, R. A.; Nasstrom, J. S.; Feeney, P. J. In *Design and Evaluation of the DRUM Impactor*; Veziroglu, T. N., Ed.; Particulate Science and Technology, An International Journal, Hemisphere Publications: Washington, DC, 1985; Proceedings of International Symposium on Particulate and Multi-Phase Processes, Miami Beach, FL.
15. Raabe, O. G.; Braaten, D. A.; Axelbaum, R. L.; Teague, S. V.; Cahill, T. A. *J. Aerosol Sci.* **1988**, *19*(2), 183–195.
16. Berner, A.; Lurzer, C. *J. Phys. Chem.* **1980**, *84*, 2079, 2083.

17. Marple, V.; Rubow, K.; Anath, G.; Fissan, H. J. *Aerosol Sci.* **1986**, *17*, 489–494.
18. Cahill, T. A. *Nucl. Instrum. Methods* **1978**, *149*, 431–433.
19. Wesolowski, J. J.; John, W.; Devor, W.; Cahill, T. A.; Feeney, P. J.; Wolfe, C.; Flocchini, R. G. In *X-ray Fluorescence Analysis of Environmental Samples*; Dzubay, T. G., Ed.; Ann Arbor Science: Ann Arbor, MI, 1978; pp 121–130.
20. Flocchini, R. G.; Cahill, T. A.; Shadoan, D. J.; Lange, S. J.; Eldred, R. A.; Feeney, P. J.; Wolfe, G. W.; Simmeroth, D. C.; Suder, J. K. *Environ. Sci. Technol.* **1976**, *10*, 76–82.
21. Barone, J. B.; Cahill, T. A.; Eldred, R. A.; Flocchini, R. G.; Shadoan, D. J.; Dietz, T. M. *Atmos. Environ.* **1978**, *12*, 2213–2221.
22. Hering, S. V.; Appel, B. R.; Cheng, W.; Salaymeh, F.; Cadle, S. H.; Mulawa, P. A.; Cahill, T. A.; Eldred, R. A.; Surovik, M.; Fitz, D.; Howes, J. E.; Knapp, K. T.; Stockburger, L.; Turpin, B. J.; Huntzicker, J. J.; Zhang, X.; McMurry, P. H. *Aerosol Sci. Technol.* **1990**, *12*, 200–213.
23. Campbell, D.; Copeland, S.; Cahill, T. A.; Eldred, R. A.; Cahill, C.; Vesenka, J.; VanCuren, T. *Proc. Air Waste Mgmt. Assoc.*; 82nd Annual Meeting and Exhibition, 1989, Anaheim, CA; Air and Waste Management Association: Pittsburgh, PA, 1989; Paper 89–151.6; pp 1–14.
24. Cahill, T. A.; Surovik, M.; Wittmeyer, I. *Aerosol Sci. Technol.* **1990**, *12*, 149–160.

RECEIVED for review May 30, 1991. ACCEPTED revised manuscript August 19, 1992.

Measuring the Strong Acid Content of Atmospheric Aerosol Particles

Roger L. Tanner

Energy and Environmental Engineering Center, Desert Research Institute, Reno, NV 89506

Methods used to determine the strong acid content of aerosol particles in the ambient atmosphere are reviewed. These methods include those for generic determination of strong acid content and those in which the concentrations of individual strong acid species are determined. Difficulties in sampling these species due to their reactivity and occurrence under non-steady-state atmospheric conditions are discussed, and the methods currently used for resolving these difficulties are critically evaluated.

A FUNDAMENTAL QUESTION about the interpretation of acidic aerosol data is whether researchers can characterize past and current atmospheric concentrations and distributions (spatial and temporal) with sufficient accuracy for studies of their effects on ecosystems and human health. Part of the answer to this question can be provided by a review of the methods that have been used to measure the strong acid content of aerosol particles collected from the atmosphere. This chapter serves as such a review, and, in evaluating analytical procedures, it specifically assesses the ability of each procedure to overcome sampling artifacts, to distinguish between strong and weak acids, to properly partition strong acidity between gas-phase and aerosol-phase species, and to quantitate strong acidity at the levels at which it is found in the ambient atmosphere.

Definitions of Acids and Bases

Acids and bases are defined in accordance with Brønsted–Lowry theory in terms of their propensity to donate or accept hydrated protons in aqueous

0065–2393/93/0232–0229$06.00/0

solution, a definition that is readily extended to nonaqueous solvents. An acid is thus a substance with a tendency to dissociate into a hydrated proton in solution, and it is a strong acid if that tendency is quantitatively large compared to that of water; that is, at equilibrium its proton-donor reaction with water lies far to the right. Weak acids are, by contrast, only partially dissociated in aqueous solution, and the degree of dissociation is quantitatively expressed in terms of the equilibrium constant for its proton-donor reaction with water. A base in aqueous solution is defined analogously in terms of its proton-accepting capacity relative to H_2O; a strong base nearly completely dissociates in aqueous solution to form free hydroxide ions. Solvent dissociation is the process determining co-existing $[H^+]$ and $[OH^-]$ when a solution is acidic or basic, that is, when $[H^+] > [OH^-]$ or vice versa. The term $[H^+]$ signifies the molar concentration of hydrated protons in solution, $(H_2O)_n H^+$.

A logarithmic scale is used to express $[H^+]$ as pH, where pH $= -\log(a_{H^+}) = -\log \gamma[H^+]$; γ is the single-ion activity coefficient and a_{H^+} is the H^+ activity. Other terms have also been used to describe acid levels in environmental samples—titratable acidity, titratable strong acidity, and total acidity. Titratable acidity is the acid content in aqueous solution that is neutralizable by addition of strong base; this term has meaning only if the pH at the endpoint of the titration is defined. If an endpoint pH is chosen at the low end of observed pHs of cloud-water or rain (e.g., pH $= 2-4$), this quantity becomes total strong acidity in atmospheric samples. However, if a higher pH is chosen (e.g., pH $= 7$), this quantity is often referred to as total strong and weak acidity; at an even higher endpoint pH (pH $= 10$), it becomes total acidity, because now such conjugate acid species as ammonium are titrated. The degree of neutralization at a specified pH depends on concentration of the acidic species in solution as well as on the equilibrium constant.

Atmospheric Strong and Weak Acids

These definitions must be applied in a clear and unambiguous way to atmospheric samples. In this chapter, strong and weak acids are assumed to be present in the atmosphere in both condensed (solid particle and liquid droplet) and gaseous forms. In addition, fine particles ($<\sim 2 \mu m$ in diameter) as well as water-soluble gaseous species are removed in large part by scavenging into clouds and rain droplets. Strong acids, whether present as gases or in aerosol particles, are somewhat arbitrarily assumed to be completely dissociated ($>99\%$) into H^+ and an anion in atmospheric water samples with pH ≥ 4. This definition is equivalent to identifying strong acids as those with pK_a values $\leq \sim 2$. This definition will be used with the recognition that strong acid species may be undissociated in solid-particle or concentrated droplet phases in aerosols but dissociated when scavenged into the more dilute (by a factor of 1000) hydrometeor phase.

The most commonly occurring atmospheric strong acids are nitric and sulfuric acids, which are derived from the oxidation of nitrogen oxides and sulfur dioxide, and are present predominantly in the gaseous and particulate phases, respectively. Hydrochloric acid and oxalic acid (the latter qualifying as a strong acid based on its K_{a1} being less than 2) may also be present occasionally in the atmosphere (1, 2). H_3PO_4, methanesulfonic acid, and hydroxymethanesulfonic acid could also be present, but the acid forms of these species have not been specifically identified. These strong acid species are neutralized, partially or in toto, by atmospheric gaseous ammonia and to a lesser extent by soil-derived particulate matter; this neutralization produces the observed composition of sulfate- and nitrate-containing aerosols (with smaller amounts of other anions) along with the equilibrium-determined levels of gaseous nitric acid (3, 4). Acidic species that are present in atmospheric solid particles, condensed phases, or mixtures thereof may thus include sulfuric acid, ammonium bisulfate, and letovicite [$(NH_4)_3H(SO_4)_2$] mixed with neutral species (ammonium sulfate, ammonium nitrate, NaCl, $NaNO_3$, Na_2SO_4, and certain mixed ammonium sulfate and nitrate salts). Under higher humidity conditions, aqueous solutions containing H^+, NH_4^+, Na^+, HSO_4^-, SO_4^{2-}, NO_3^-, and Cl^- may also coexist with some of these solid–condensed phases, or the solid phases may be absent altogether.

Weak acid species in the atmosphere may include any or all of the following: inorganic species such as nitrous acid; dissolved S(IV); dihydrogen phosphates and hydrogen phosphates; hydrated transition metal species; organic species, including carboxylic acids and their hydroxy- and keto-acid relatives, phenols; and some pesticide compounds. Weak acid species, formic and acetic acids in particular, certainly contribute to the acidity of precipitation, especially in remote areas (5). An important and unresolved question is the extent to which these weakly acidic species contribute to the free acidity of aerosol particles. In this chapter the discussion is restricted to methods for determining strong acidity in aerosol particles and for distinguishing strong acid species from weak acids and acidic gases.

Measurement Methods

Descriptions of analytical methods for strong acid and acidic sulfate content of atmospheric aerosols have been reviewed (6–10). Methods for acidic aerosol determination are reviewed in this chapter according to the measurement principle: either filter collection and post-collection extraction, derivatization or thermal treatment, and analysis; or in situ collection (real-time or stepwise) and analysis.

Filter Collection

Thermal Volatilization.   Thermal volatilization schemes have been popular for speciation of acidic sulfate compounds in aerosols. Both filter-based and in situ approaches are used; the former are discussed in this section. Filter samples were heated and H_2SO_4 was collected by microdiffusion (11, 12) or determined directly by flame photometry after volatilization from poly(tetrafluoroethylene) (Teflon) filters (13, 14). In one method, H_2SO_4 was distinguished from other volatile sulfates (e.g., NH_4HSO_4 and $(NH_4)_2SO_4$), and from nonvolatile sulfates (e.g., Na_2SO_4) by heating consecutively to two different temperatures (15). Perimidinylammonium sulfate was formed from acidic sulfates and decomposed to SO_2 for West–Gaeke analysis in another approach (16).

These methodologies, developed in the late 1960s and early 1970s, were stimulating attempts to analyze acidic sulfate species in aerosols. However, because of serious recovery problems (15) and limited success in distinguishing the two major aerosol species (NH_4HSO_4 and $(NH_4)_2SO_4$) from each other (17), they have fallen into disfavor in the past decade. A few exceptions involving heated denuder collection and in situ analysis are discussed in the following sections.

Extraction with pH Measurement or H^+ Titration.   Filter-collected particle samples have been analyzed for net strong acid content by several related procedures, all starting with extraction of the sample into water or dilute mineral acid. The strong acid content of particles may be determined in principle, based on the definition, by a simple measurement of the pH of an aqueous extract. However, this procedure is nearly always subject to serious error when atmospheric buffering agents such as dissolved CO_2, weak carboxylic acids, or hydrated forms of transition metal ions (18, 19) are present, because free acidity (determined by the pH measurement) cannot then be equated to strong acid content (the analyte of interest) in these buffered solutions. Hence, most current procedures now prescribe filter extraction into weak mineral acid (e.g., 10^{-4} N $HClO_4$ or H_2SO_4), with measurement of the free acidity as the difference between $\exp(-pH_{obs})$ and the free acid concentration contained in the extractant (20, 21) (pH_{obs} is observed pH). The principal differences between these pH-measurement approaches are in the sampling apparatus, that is, the pretreatment of the atmospheric aerosol prior to collection on the filter. These procedural differences are discussed in relation to sampling artifacts.

Titration procedures for strong acid content of atmospheric aerosols that use an exponential display of data points (titration according to Gran, reference 22) were reported by Junge and Scheich (19), were perfected by Brosset and co-workers (23, 24), and have since been widely used by other groups (25, 26). Electrolytic generation of strong base in Gran titration procedures has been used by several groups (25, 27, 28). Again, dissolution of filter samples in ~0.1 mM mineral acid prior to Gran titration, with correction for the acid content of the extracting solution, allows for titration of 1-μmol amounts of strong acid (28), with precision and accuracy of the order of ±10% (26, 29).

Errors may occur in the Gran titration procedure if weakly acidic species with dissociation constants (expressed as pK_a) in the range of the extract pH are present. In particular, curvature or reduction (or both) of the slope of the Gran exponential plot results (24), because weak acid dissociation and titration of released free acidity take place during the portion of the titration used for end-point determination. Fortuitously, some of the common, weak carboxylic acids (e.g., formic and acetic) are not stable toward microbial decomposition when collected in aerosol samples from the atmosphere, so much of the historical data base on strong acid content of aerosols does not suffer from this positive error source, unless of course the microbial processes produce additional strong acids.

Specific Extraction of Atmospheric Acids. Several efforts to specifically extract atmospheric acids from aerosol samples have been reported, focusing on sulfuric acid. Benzaldehyde has been shown to specifically extract H_2SO_4 from dried acidic aerosol sulfate samples (15), with analysis for sulfate made in aqueous "back-extracts" of the H_2SO_4–benzaldehyde solutions. Isopropyl alcohol quantitatively extracts H_2SO_4 from aerosol samples on quartz filters (30) but also removes ammonium bisulfate from samples as well (15). Distribution coefficients of isopropyl alcohol with respect to mixed nitrate–sulfate aerosol phases are still not well known. In addition, difficulties have been reported in two separate investigations of the quantitative removal and selectivity of extraction techniques that use benzaldehyde (31, 32). Because free H_2SO_4 is not a common constituent of ambient aerosols, use of specific extractant methods has decreased in recent years in favor of generic strong acid determination. Techniques that specifically remove gaseous strong acids prior to aerosol collection on filters are discussed here.

Specific Extraction with Derivatization. A method has been reported (33) in which filter-collected H_2SO_4 is converted to dimethyl sulfate by reaction with diazomethane, with subsequent analysis by gas chromatography–flame photometric detection (GC–FPD). This method apparently does not specifically determine sulfuric acid in the presence of ammonium bisulfate and mixed acidic sulfate–nitrate salts (Tanner and Fajer, unpublished data, 1981). A method has been proposed for derivatization of collected H_2SO_4 by reaction with diethylamine in dry air, followed by reaction with CS_2 and cupric ion to form a colored complex for spectrophotometric determination (34). This method also suffers from a large ammonium bisulfate interference. A related method (35), in which H_2SO_4 and other aerosol strong acids are converted to ^{14}C-labeled bis(diethylammonium) sulfate and analogs, is a useful (and underutilized) technique for determination of low levels of strong acids in aerosols.

In Situ Procedures

Continuous Sampling with Periodic Determination. Sulfuric acid in atmospheric aerosols may be determined in situ by using a continuously sampling flame photometric detector (FPD) (36–38). The technique uses a diffusion denuder inlet tube for SO_2 removal prior to FPD determination of aerosol, in the same way that an FPD instrument has been used for continuous aerosol sulfur analysis by several groups (39–41). However, in this adaptation the temperature of the denuder tube or a zone just upstream from it is cycled between room temperature and about 120 °C. At ambient temperatures, sulfuric acid remains in the aerosol phase, but at about 120 °C it is volatilized and removed to the walls of the denuder tube. The difference between the signal at ambient temperatures and 120 °C is proportional to $[H_2SO_4]$. The minimum cycle time, and thus the time resolution, is about 6–8 min. Sensitivity of the instrument is barely adequate for high ambient levels of acidic sulfate, and the limit of detection in this difference mode is insufficient for the usual levels of atmospheric H_2SO_4.

One adaptation of this approach uses a further temperature cycle to 220 °C, which volatilizes ammonium sulfate and bisulfate in aerosols but not nonvolatile sulfates (e.g., Na_2SO_4) (37, 40). Ammonium bisulfate and ammonium sulfate are not differentiated by this approach. Another adaptation collects the aerosol sulfuric acid on a heated denuder tube (38) for ~15 min, then removes it thermally for FPD analysis while collecting another sample on a second denuder tube. Because the sample is preconcentrated in this approach, the more sensitive version of the FPD is not required.

Acidic sulfates, including sulfuric acid aerosol, may be differentiated by using the humidograph technique of Charlson et al. (42) with the improved speciation of more recently developed thermidograph variations (43, 44). The latter technique involves heating the air stream containing the aerosol from 20 to 380 °C in 5-min cycles, rapidly cooling it to the dry bulb temperature, and measuring the light scattering at 65 to 70% relative humidity (RH) with a nephelometer. By comparing actual thermidograms with those of test aerosols, the fractional acidity (neq H^+/neq SO_4^{2-}; neq denotes nano-equivalents) can be measured and the approximate level of H_2SO_4 determined. More recent work (45) demonstrates the utility of this technique for determining even background levels of acidic sulfate (<10 neq/m^3).

One other in situ technique can be used to determine fractional acidity in atmospheric aerosols by means of Fourier transform infrared (FTIR) spectroscopy (46). Originally, impactor samples were collected and were pressed into a KBr matrix, and then the IR spectrum was taken by attenuated total reflectance (ATR) FTIR spectroscopy to determine relative acidity, based on differences in absorption bands for sulfate and bisulfate species. Aerosols with $[H^+]/[SO_4^{2-}]$ ratios greater than 1 could also be qualitatively identified. More recent innovations in the FTIR technique (47, 48) have made possible the quantitation of sulfuric acid; further, with the use of ATR plates as substrates in impactive collection of atmospheric aerosols, the technique can be used for in situ determination of acidic sulfates and other aerosol constituents with distinguishable IR spectra (e.g., nitrates and silicates). A modification of this technique has also been applied to cascade impactor samples by Brown et al. (49), albeit not in an in situ mode. The former technique deserves further study and might serve as a standard method for acidic aerosol species.

Continuous Sampling and Determination.   There are no truly continuous techniques for the direct determination of sulfuric acid or other strong acid species in atmospheric aerosols. The closest candidate method is a further modification of the sensitivity-enhanced, flame photometric detector, in which two detectors are used, one with a room-temperature denuder and one with a denuder tube heated to about 120 °C. Sulfuric acid is potentially determined as the difference between the two channels. In fact, a device based on this approach did not perform well in ambient air sampling (Tanner and Springston, unpublished data, 1990). Even with the SF_6-doped H_2 fuel gas for enhanced sensitivity, the limit of detection is unsuitably high ($5 \mu g/m^3$ or greater) because of the difficulty in calibrating the two separate FPD channels with aerosol sulfates.

Pitfalls of Sampling for Acidic Aerosols

General Observations.   Measurement techniques for traces of strong acids in atmospheric aerosols have been plagued with difficulties, often referred to as sampling anomalies or artifacts. Often, these difficulties have derived from the use of materials on collection surfaces (e.g., filters and impactor slides) that were designed for other purposes. These difficulties and means of circumventing them are discussed in this section. Artifacts are of two general types: (1) reversible or irreversible sorption losses to filter or impactor surfaces in integrative sampling or to the sampling lines (e.g., nitric acid on Teflon lines) in continuous, in situ techniques; and (2) equilibrium-driven loss or gain of species due to particle–particle reactions on the collection surface or non-steady-state conditions in the sampled atmosphere during the time period of sample collection. These types of sampling artifacts are summarized in terms of surface, gas–aerosol, and aerosol–aerosol interactions.

Surface Interactions.   Loss of strong acid content of atmospheric aerosols was observed and attributed to reaction with basic sites in the glass or cellulose filter matrices commonly used for high-volume sampling of atmospheric aerosols (46, 50). These filter materials, and glass fiber filters of all types in particular, are unsuitable for collection of acidic aerosol particles for subsequent extraction and analysis. Even if they are pretreated with acid and fired to a high temperature, subsequent rinsing exposes additional free basic sites that neutralize acidic particles (28). High-purity quartz filters can be pretreated with acid to remove basic sites prior to sampling, and Teflon filter media are generally inert to acidic particles. Thus, Teflon and treated quartz filter media have generally replaced glass or cellulose filters for sampling acidic aerosols. In addition, their use eliminates a positive source of error in sulfate measurements due to the base-catalyzed conversion of sorbed, gaseous SO_2 to sulfuric acid, analyzed as sulfate after extraction (50–52).

Gas–Aerosol Interactions. Re-equilibration involving gas and particulate-phase species may lead to errors in sampling and analysis of aerosol strong acids and other aerosol species, both under steady-state and non-steady-state conditions. In particular, transport or exposure of condensed-phase species on particles to environments with lower gas-phase concentrations may lead to loss of particulate species in order to reestablish pertinent equilibria. The example of greatest significance for atmospheric acid determinations is the equilibrium between volatile ammonium salts and gaseous acid and ammonia levels. Although the simple evaporation of neutral salts from collected particles should not directly introduce errors in strong acid determinations, the more subtle effects of equilibration processes on accurate determination of aerosol strong acid levels should be considered.

Several studies have demonstrated that observed, atmospheric levels of gaseous ammonia and nitric acid (or HCl) are in order-of-magnitude agreement with calculated equilibrium values under ambient temperature and relative humidities, based on thermodynamic properties, whenever a solid ammonium nitrate–chloride aerosol phase (or mixed sulfate–nitrate–chloride phase) is present (53–57). A thermodynamic model has been developed and refined for aerosol formation and gas–aerosol partitioning of these phases (58, 59). These and other studies dealing with the effects of using averaged aerosol compositions, with long sampling time compared with the mean time for changes in thermodynamically significant variables, have been reviewed by Tanner and Harrison (60). The likelihood of kinetic limitations in the achievement of gas–aerosol equilibria has also been evaluated as a function of aerosol particle size and temperature, and the effect of such limitations on strategies for sampling aerosol strong acids needs to be considered.

The principal implication of these studies for the accurate measurement of strong acids in gaseous and aerosol phases is that sampling must be performed in a way that does not disturb the equilibrium significantly in the process. This necessity has led to the development of the diffusion denuder sampler (61), which measures nitric acid after its removal to the walls of a tube coated with a material that reacts irreversibly with HNO_3. The residence time in the tube is short enough (0.2–2 s) that particles are not measurably removed by diffusion, nor do they evaporate significantly therein to reestablish gas–aerosol equilibrium.

Is this re-equilibration phenomenon important for the measurement of aerosol strong acid content, as it is for gaseous nitric acid? Generally, nitric acid is taken up into aerosol particles (solid or liquid droplets) only if the particles have been nearly completely neutralized by ambient ammonia, because the nitrate–nitric acid equilibrium favors the gas phase in the presence of significant particulate strong acid (62). Most concern, however, has been expressed concerning sampling of acidic aerosols in the presence of ambient ammonia.

The existence of thermodynamic equilibrium between gaseous NH_3 and HNO_3 and aerosol particles in ambient air should preclude the uptake of ammonia during sampling except under two circumstances. The first condition involves post-collection neutralization of fine, acidic particles by impacting, coarse, basic particles; populations of these particles were not steady-state in the atmosphere. This condition is discussed in the following section. The second circumstance is when sampling must proceed for a period of many hours in order to acquire enough material for analysis. In this case the ambient conditions may change during sampling; for example, ambient $[NH_3]$ may increase, and already collected acidic aerosols may be neutralized by exposure to air of higher ammonia levels than those extant during sampling. If the ambient $[NH_3]$ decreases but still exceeds gaseous $[HNO_3]$, net ammonia might be released from collected ammonium acid salts, increasing the $[H^+]/[anion]$ ratio (and the apparent $[H^+]$) in the remaining deposit. Experimental confirmation of this phenomenon has apparently not been reported.

One procedure that is widely used to circumvent these complications is to remove ambient ammonia from the sampled air without removing particles by inserting one of several types of diffusion denuders upstream from the filter(s). In fact, in a recent Environmental Protection Agency (EPA)-sponsored intercomparison of methods for determination of strong acid content of aerosols, all but one protocol utilized an ammonia denuder (63), and all used an impactor or cyclone to remove coarse particles. The presence of this denuder clearly prevents neutralization of acidic aerosols by ammonia but also disturbs the gas–aerosol equilibrium between sulfate–nitrate aerosols and gaseous species. Ammonia and nitric acid are released from the depositing particles (64, 65) and must be collected downstream if accurate particulate ammonium and nitrate determinations are to be made. If equal amounts of ammonia and nitric acid are released, then the absolute $[H^+]$ (neq/m^3) will not be altered. No specific evidence is available in the literature to demonstrate alteration of the observed $[H^+]$ as the result of re-equilibration, but this area deserves further study.

Particle–Particle Interactions.   Loss of strong acid content of aerosol particles can also occur because of reactions between co-collected acidic and basic particles impacted together on the collection surface. This phenomenon most frequently occurs as the result of interaction of coarse (>2.5 μm diameter), alkaline, soil-derived particles with fine (<2.5 μm diameter) acidic sulfate particles (66). Particle–particle interactions with net neutralization can be reduced in many cases by sampling with a virtual impactor or a cyclone to remove coarse particles, although this procedure does not prevent the effect if external mixtures of fine particles of different acid contents are sampled. In situ methods with shorter sampling times can be used such that these topochemical reactions are less likely to occur.

Summary and Conclusions

Modified filter sampling methods that are available will measure ambient levels of strong acid in ambient aerosol samples, and these methods do so with acceptable precision and accuracy [as indicated by the balance between measured anions and cations (56, 57)] in the absence of significant levels of particulate weak acids. Additional intercomparisons involving intrinsically different techniques for particulate strong acidity [e.g., IR spectroscopy (48), thermal speciation (38, 45), and filter methods (28)] are needed. Further information on the occurrence of various weak acids in airborne particles is needed, along with further studies of techniques for their specific determination in atmospheric aerosol samples.

If long sampling times are required to collect sufficient sample for analysis, it is desirable to remove ambient ammonia (by diffusional removal to the walls of a coated denuder tube) because increases in the concentration of this gas during sampling may result in post-collection neutralization of acidic aerosol particles. No evidence of changes in absolute [H$^+$] caused by this procedure is reported, but further research is needed in this area.

Sulfuric acid may be determined in the presence of other acidic aerosol constituents, but no method for the separation and determination of bisulfate, the most common acidic species in atmospheric aerosols, has been successfully developed.

Removal of coarse particles from the air stream is desirable for determination of fine particulate acidity because ambient coarse particles may contain basic substances.

Further study is needed of the phenomenon of kinetic limitations to the neutralization of acidic aerosols. Simultaneous occurrences of acidic aerosols at gaseous [NH$_3$] well above the equilibrium values have been reported (56, 67), and it is still unclear whether kinetic limits to microscale neutralization or boundary layer mixing (macroscale) kinetics (or both) are responsible for these limitations. An understanding of the extent of human exposure to acidic aerosols, as well as of the availability of acidic aerosols for wet scavenging into hydrometeors, may depend on the results of further studies of neutralization kinetics in ambient air.

References

1. Lightowlers, P. J.; Cape, J. N. *Atmos. Environ.* **1988**, *22*, 7–15.
2. Norton, R. B.; Roberts, J. M.; Huebert, B. J. *Geophys. Res. Lett.* **1983**, *10*, 517–520.
3. Brosset, C.; Andreasson, K.; Ferm, M. *Atmos. Environ.* **1975**, 9, 631–642.
4. *Nitrogen Oxides*; National Research Council, National Academy of Sciences: Washington, DC, 1977; 333 pp.
5. Keene, W. C.; Galloway, J. N.; Holden, J. D., Jr. *J. Geophys. Res.* **1983**, *88*, 5122–5130.
6. Tanner, R. L.; Forrest, J.; Newman, L. In *Sulfur in the Environment, Part I, The Atmospheric Cycle*; Nriagu, J. O., Ed.; Wiley: New York, 1978; pp 371–452.
7. *Air Quality Criteria for Particulate Matter and Sulfur Oxides*; U.S. Environmental Protection Agency Center for Environmental Research Information: Cincinnati, OH, 1982; Report EPA–600/8–82–029b; Vol. II, pp 3-62–3-75.
8. Lioy, P. J.; Waldman, J. M. *EHP, Environ. Health Perspect.* **1989**, *79*, 15–34.
9. Tanner, R. L. In *Methods of Air Sampling and Analysis*, 3rd ed.; Lodge, J. P., Jr., Ed.; Lewis Publishers: Chelsea, MI, 1989; pp 703–714.
10. *An Acid Aerosols Issue Paper: Health Effects and Aerometrics*; U.S. Environmental Protection Agency Center for Environmental Research Information: Cincinnati, OH, 1989; Report No. EPA/600/8–88/00SF.
11. Scarengelli, F. P.; Rehme, K. A. *Anal. Chem.* **1969**, *41*, 707–713.
12. Dubois, L.; Baker, C. J.; Teichman, T.; Zdrojewski, A.; Monkman, J. L. *Mikrochim. Acta* **1969**, 269–279.
13. Richards, L. W.; Mudgett, P. S. U.S. Patent 3 833 972, 1974.
14. Richards, L. W.; Johnson, K. R.; Shepard, L. S. *Sulfate Aerosol Study*; Rockwell International: Newbury Park, CA, 1978; Report AMC8000.13FR.
15. Leahy, D.; Siegel, R.; Klotz, P.; Newman, L. *Atmos. Environ.* **1975**, 9, 219–229.
16. Maddalone, R. F.; Shendrikar, A. D.; West, P. W. *Mikrochim. Acta* **1974**, 391–401.
17. Thomas, R. L.; Dharmarajan, V.; Lundquist, G. L.; West, P. W. *Anal. Chem.* **1976**, *48*, 639–642.
18. Commins, B. T. *Analyst (London)* **1963**, *88*, 364–367.
19. Junge, C.; Scheich, G. *Atmos. Environ.* **1971**, *5*, 165–175.
20. Koutrakis, P.; Wolfson, J. M.; Spengler, J. D. *Atmos. Environ.* **1988**, *22*, 157–162.
21. Waldman, J. D.; Lioy, P. J.; Thurston, G. D.; Lippmann, M. *Atmos. Environ.* in press.
22. Gran, G. *Analyst (London)* **1952**, *77*, 661–671.
23. Askne, C.; Brosset, C.; Ferm, M. *Determination of the Proton-Donating Property of Airborne Particles*; Swedish Water and Air Pollution Research Laboratory: Goteborg, Sweden, 1973; IVL Report B157.
24. Brosset, C.; Ferm, M. *Atmos. Environ.* **1978**, *12*, 909–916.
25. Liberti, A.; Possanzini, M.; Vicedomini, M. *Analyst (London)* **1972**, *97*, 352–356.
26. Stevens, R. K.; Dzubay, T. G.; Russworm, G.; Rickel, D. *Atmos. Environ.* **1978**, *22*, 55–68.

27. Krupa, S. V.; Coscio, M. R., Jr.; Wood, F. A. *J. Air Pollut. Control Assoc.* **1976,** *26,* 221–223.
28. Tanner, R. L.; Cederwall, R.; Garber, R.; Leahy, D.; Marlow, W.; Meyers, R.; Phillips, M.; Newman, L. *Atmos. Environ.* **1977,** *11,* 955–966.
29. Phillips, M. F.; Gaffney, J. S.; Goodrich, R. W.; Tanner, R. L. *Computer-Assisted Gran Titration Procedure for Strong Acid Determination;* Brookhaven National Laboratory: Upton, NY, 1984; Report BNL 35734.
30. Barton, S. C.; McAdie, W. G. In *Proc. Int. Clean Air Congr., 2nd;* Academic: New York, 1971; pp 379–382.
31. Eatough, D. J.; Izatt, S.; Ryder, J.; Hansen, L. D. *Environ. Sci. Technol.* **1978,** *12,* 1276–1279.
32. Appel, B. R.; Wall, S. M.; Haik, M.; Kothy, E. L.; Tokiwa, Y. *Atmos. Environ.* **1980,** *14,* 559–563.
33. Penzhorn, R. D.; Filby, W. G. *Staub-Reinhalt. Luft* **1976,** *36,* 205–207.
34. Huygen, C. *Atmos. Environ.* **1975,** *9,* 315–319.
35. Dzubay, T. G.; Snyder, G. K.; Reuter, D. J.; Stevens, R. K. *Atmos. Environ.* **1979,** *13,* 1209–1212.
36. Tanner, R. L.; D'Ottavio, T.; Garber, R.; Newman, L. *Atmos. Environ.* **1980,** *14,* 121–127.
37. Allen, G. A.; Turner, W. A.; Wolfson, J. M.; Spengler, J. D. In *U.S. Environ. Prot. Agency, Res. Dev. [Rep.] EPA* **1984,** EPA/600/9–84/019.
38. Slanina, J.; Schoonbeek, C. A. M.; Klockow, D.; Niessner, R. *Anal. Chem.* **1985,** *57,* 1955–1960.
39. Huntzicker, J. J.; Hoffman, R. S.; Ling, G. S. *Atmos. Environ.* **1978,** *12,* 83–88.
40. Cobourn, W. G.; Husar, R. B.; Husar, J. D. *Atmos. Environ.* **1978,** *12,* 89–98.
41. Camp, D. C.; Stevens, R. K.; Cobourn, W. G.; Husar, R. B.; Collins, J. F.; Huntzicker, J. J.; Husar, J. D.; Jaklevic, J. M.; McKenzie, R. L.; Tanner, R. L.; Tesch, J. W. *Atmos. Environ.* **1982,** *16,* 911–916.
42. Charlson, R. J.; Vanderpohl, A. H.; Covert, D. S.; Waggoner, A. P.; Ahlquist, N. C. *Atmos. Environ.* **1974,** *8,* 1257–1268.
43. Larson, T. V.; Ahlquist, N. C.; Weiss, R. E.; Covert, D. S.; Waggoner, A. P. *Atmos. Environ.* **1982,** *16,* 1587–1590.
44. Rood, M. J.; Larson, T. V.; Covert, D. S.; Ahlquist, N. C. *Atmos. Environ.* **1985,** *19,* 1181–1190.
45. Covert, D. *J. Geophys. Res.* **1988,** *93,* 8455–8458.
46. Cunningham, P. T.; Johnson, S. A. *Science (Washington, D.C.)* **1976,** *191,* 77–79.
47. Johnson, S. A.; Kumar, R. In *Acid Aerosol Measurement Workshop;* Tropp, R. J., Ed.; U.S. Environmental Protection Agency Center for Environmental Research Information: Cincinnati, OH, 1989; Report EPA/600/9–89/056; pp 36–38.
48. Johnson, S. A.; Kumar, R. *J. Geophys. Res.* **1990,** *97,* accepted.
49. Brown, S.; Dangler, M. C.; Burke, S. R.; Hering, S. V.; Allen, D. *Aerosol Sci. Technol.* **1990,** *12,* 172–181.
50. Coutant, R. W. *Environ. Sci. Technol.* **1977,** *11,* 873–878.
51. Pierson, W. R.; Hammerle, R. H.; Brachaczek, W. W. *Anal. Chem.* **1976,** *48,* 1808–1811.
52. Pierson, W. R.; Brachaczek, W. W.; Korniski, T. J.; Truex, T. J.; Butler, J. W. *J. Air Pollut. Control Assoc.* **1980,** *30,* 30–34.
53. Stelson, A. W.; Friedlander, S. K.; Seinfeld, J. H. *Atmos. Environ.* **1979,** *13,* 369–371.

54. Doyle, G. J.; Tuazon, E. C.; Graham, R. A.; Mischke, T. M.; Winer, A. M.; Pitts, J. N., Jr. *Environ. Sci. Technol.* **1979**, *13*, 1416–1419.
55. Harrison, R. M.; Pio, C. A. *Tellus* **1983**, *35B*, 155–159.
56. Tanner, R. L. *Atmos. Environ.* **1982**, *16*, 2935–2942.
57. Erisman, J. W.; Vermetten, A. W. M.; Asman, W. A. H.; Waijers-IJpelaan, A.; Slanina, J. *Atmos. Environ.* **1988**, *22*, 1153–1160.
58. Bassett, M.; Seinfeld, J. H. *Atmos. Environ.* **1983**, *17*, 2237–2252.
59. Pilinis, C.; Seinfeld, J. H. *Atmos. Environ.* **1987**, *21*, 2453–2466.
60. Tanner, R. L.; Harrison, R. M. In *Environmental Particles*; Bouffle, J.; van Leeuwen, H. P., Eds.; Lewis Publishers: Chelsea, MI, 1992; Vol. 1, Chapter 2.
61. Shaw, R. W., Jr.; Stevens, R. K.; Bowermaster, J. W. *Atmos. Environ.* **1982**, *16*, 845–853.
62. Tang, I. *Atmos. Environ.* **1980**, *14*, 819–828.
63. Ellestad, T. G.; Barnes, H. M.; Kamens, R. M.; McDow, S. R.; Sickles, J. E., II; Hodson, L. L.; Waldman, J. M.; Randtke, S. J.; Lane, D. D.; Springston, S. R.; Koutrakis, P.; Thurston, G. D. In *Measurement of Toxic and Related Air Pollutants*; Air and Waste Management Association: Pittsburgh, PA, 1991; Vol. 1, pp 122–127.
64. Appel, B. R.; Wall, S. M.; Tokiwa, Y.; Haik, M. *Atmos. Environ.* **1980**, *14*, 549–554.
65. Forrest, J.; Spandau, D. J.; Tanner, R. L.; Newman, L. *Atmos. Environ.* **1982**, *16*, 1473–1485.
66. Tanner, R. L.; Garber, R.; Marlow, W.; Leaderer, B. P.; Leyko, M. A. *Ann. N.Y. Acad. Sci.* **1979**, *322*, 99–114.
67. Harrison, R. M.; Sturges, W. T.; Kitto, A. M. N.; Li, Y. *Atmos. Environ.* **1990**, *24A*, 1883–1888.

RECEIVED for review March 20, 1991. ACCEPTED revised manuscript August 10, 1992.

Measurement Challenges of Nitrogen Species in the Atmosphere

David D. Parrish[1] and Martin P. Buhr[1,2]

[1]Aeronomy Laboratory, National Oceanic and Atmospheric Administration, Boulder, CO 80303
[2]Cooperative Institute for Research in Environmental Sciences, University of Colorado, Boulder, CO 80309

Understanding the chemistry and physics of atmospheric nitrogen species presents several challenges for analytical chemists; these challenges are discussed in the context of an overview of measurement techniques and recent results from field studies. In the troposphere, reliable in situ techniques to measure HNO_3, organic nitrates, NO_3, N_2O_5, and HONO are required, and fast response (1 to 10 Hz) techniques are needed to measure the surface fluxes of N_2O, NH_3, NO_2, and HNO_3 by micrometeorological techniques. In the stratosphere, fast response (about 1 Hz) instruments are required for in situ measurements of NO_2, N_2O_5, HNO_3, $ClONO_2$, and HO_2NO_2 from aircraft. In the troposphere and stratosphere, instruments to characterize aerosols must be developed. These instruments must be integrated into packages for surface and aircraft studies that simultaneously measure a wide range of atmospheric species, and these packages must be deployed in field studies to elucidate atmospheric processes and to define the spatial and temporal distributions of the atmospheric nitrogen species.

SOCIETY IS FACING SEVERAL CRUCIAL ISSUES involving atmospheric chemistry. Species containing nitrogen are major players in each. In the troposphere, nitrogen species are catalysts in the photochemical cycles that form ozone, a major urban and rural pollutant, as well as other oxidants (references 1 and 2, and references cited therein), and they are involved in acid precipitation, both as one of the two major acids (nitric acid) and as a base (ammonia) (3, 4). In the stratosphere, where ozone acts as a shield for the earth from the sun's harsh ultraviolet radiation, nitrogen species play several roles in the ozone destruction cycles that threaten to reduce that shield (reference 5 and references cited therein). Ozone, as well as one member of the nitrogen family, N_2O, are "greenhouse" gases, which threaten to induce climate change through atmospheric warming (reference 6 and references cited therein). Thus, the distributions of the nitrogen species must be well known to understand these important issues.

The nitrogen species enter the atmosphere from a variety of natural and anthropogenic sources (7). The largest sources are concentrated in urban and industrialized areas. The levels of the species in the atmosphere vary from hundreds of parts per billion by volume (ppbv, that is, 10^{-9} mole fraction) in these source regions to below one part per trillion by volume (pptrv, 10^{-12} mole fraction) in remote areas. Even at the pptrv level, these species can play significant roles in atmospheric chemistry, and measurements of species at the sub-pptrv level can yield useful information concerning atmospheric photochemistry.

Atmospheric nitrogen species—both characterized and uncharacterized—include nearly all oxidation states of nitrogen and encompass a large number of distinct molecules. With the crucial roles these species play, the low concentrations of interest, the wide dynamic range of concentrations encountered, and the wide variety of species included in this family, analytical chemists face many challenges in the development of instrumentation for characterizing the atmospheric nitrogen family.

The development of instrumentation is only the first step. The deployment of the instrumentation in pioneering field studies is a second challenge. Atmospheric chemistry is a new enough field, and the instrumental challenges are great enough, that progress is still limited by data. Every new advance in instrumentation deployed in a well-planned field study can bring exciting new insights into the controlling processes of atmospheric chemistry.

The concentration of each nitrogen species exhibits systematic variations with latitude, longitude, and altitude. The concentration at each location in the atmosphere will usually exhibit diurnal and seasonal cycles as well as long-term trends. Superimposed on these systematic variations are more irregular changes that reflect the history of the particular air parcel transported to the location. Thus, there is the mundane but daunting challenge of characterizing these variations for each of the species by field studies incorporating long-term monitoring. Fortunately there are local, regional, and global computer models in various stages of development that seek to predict these variations. Ultimately it should be possible to rely on the results of these models to minimize the routine monitoring requirements, but currently field measurements are needed to validate the results of the computer models.

A particularly fruitful interaction exists between the experimentalist and the modeler. The experimentalist makes pioneering field measurements that guide the modeler in selecting the important chemical species and reactions to include in the developing model, and the sensitivity of the model results to particular parameters guides the experimentalist in planning fruitful field and laboratory kinetic studies. Several examples of computer model results and how this interaction operates are discussed.

This chapter focuses on several specific current challenges in the measurement of the nitrogen species. Examples of past results are included to give an indication of the insights into atmospheric processes that are potentially realizable from field studies. Many laboratories around the world are involved in these studies. The examples are drawn disproportionately from our laboratory simply for convenience. Although other examples can be cited, the ones included here are to a great extent representative of the field's results.

An introduction to the species composing the atmospheric nitrogen family is followed by a discussion of the specific goals of the measurements. A survey of the general approaches to these measurements and a description of measurement platforms completes the introductory material. An outline of the general challenges that are encountered in making such measurements and descriptions of specific challenges that are particularly important to the atmospheric chemistry community at the present time complete the chapter.

Atmospheric Nitrogen Species

The primary sources that are responsible for the presence of this family of compounds in the atmosphere emit NH_3, N_2O, and NO to the troposphere, the lowest level of the atmosphere, which extends to approximately 10 km from the earth's surface. NH_3 seems to undergo very little chemistry in the atmosphere except for the formation of aerosols, including ammonium nitrate and sulfates. NH_3 and the aerosols are highly soluble and are thus rapidly removed by precipitation and deposition to surfaces. N_2O is unreactive in the troposphere. On a time scale of decades it is transported to the stratosphere, the next higher atmospheric layer, which extends to about 50 km. Here N_2O either is photodissociated or reacts with excited oxygen atoms, O (^1D). The final products from these processes are primarily unreactive N_2 and O_2, but about 10% NO is also produced. The product NO is the principal source of reactive oxidized nitrogen species in the stratosphere.

NO initiates the rapid photochemistry both in the troposphere and in the stratosphere that produces the majority of the family members. As the nitrogen is oxidized further, the manifold of species shown in Figure 1 is formed. These include the listed inorganic species and, in regions where nonmethane hydrocarbons are also present, organic nitrates. The most important organic nitrate is PAN, peroxyacetyl nitrate, with the formula $CH_3C(O)O_2NO_2$; other peroxyacyl nitrates and alkyl nitrates are also known. There is evidence that other nitrogen-containing organic species, and per-

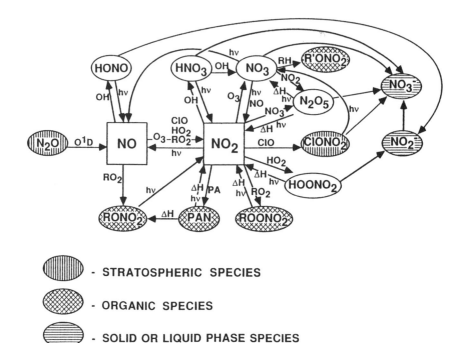

Figure 1. Reactive oxidized nitrogen species present in the atmosphere. R represents single- or multifunctional organic groups. (Reproduced with permission from reference 57. Copyright 1990 Pergamon Press.)

haps other inorganic species, are presently undetected; these species complete the family.

Two terms are commonly used in the field to collectively refer to these species. NO_x represents the sum of NO plus NO_2. These two species are combined in one term because they are interconverted in the sunlit atmosphere on a time scale of approximately 1 min, so their sum is a more conserved quantity than is either separately. However, no technique currently exists for directly measuring the sum of the NO and NO_2 concentrations; each must be determined separately. NO_y, which can be referred to as total reactive oxidized nitrogen, represents the sum of the species that have nitrogen in an oxidation state of $+2$ or higher. However, techniques have been developed that are believed to measure the complete family concentration of at least the gaseous species in a single determination.

Goals of the Measurements

In essence, measurements of these nitrogen species are required to understand the several biogeochemical cycles that transport material through the atmosphere. Several specific goals can be identified. First, as discussed in the introduction, these species are so intimately involved in atmospheric photochemistry that the spatial and temporal distributions of their concentrations must be known to begin to develop a clear picture of the atmospheric processes. Second, it is also important to quantify the fluxes of these species between the surface and the atmosphere and within the atmosphere. If the atmospheric concentrations and rates of transformation processes are known, then they can be combined with the fluxes to develop a closed budget for a given atmospheric species. Determination of the fluxes requires measurement of the concentrations of the species and of concurrent meteorological parameters. Third, and perhaps most interesting, if these species are measured in well-designed studies, photochemical relationships can be examined that provide direct tests of our understanding of the processes of atmospheric chemistry.

Two examples serve to illustrate these photochemical relationships. One member of the family of atmospheric peroxy radicals is the peroxy acetyl radical, $CH_3C(O)O_2$. In at least the warm portion of the troposphere, PAN is near thermal equilibrium with the peroxy acetyl radical and NO_2. The equilibrium constant for this reaction has been measured in laboratory studies. Therefore, if concentrations of both PAN and NO_2 are measured, the concentration of these radicals can be calculated from the equilibrium constant and the ratio of the two nitrogen species as shown in Figure 2. The

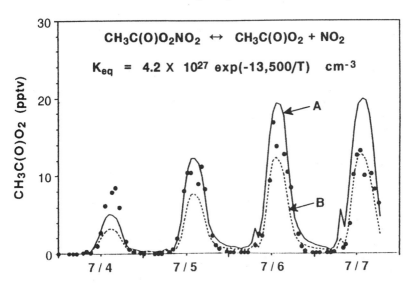

Figure 2. Peroxy acetyl radical equilibrium chemistry and concentrations derived for a 4-day period during a field study in rural Pennsylvania. The two curves give model predicted concentrations for two different scenarios. (Adapted with permission from reference 8. Copyright 1991 American Geophysical Union.)

data points show the inferred peroxy acetyl radical concentrations, and the lines are a photochemical model calculation for two different scenarios (8). Thus, more information concerning the atmosphere has been gained than simply the concentrations of the two nitrogen species. Furthermore, a critical test of the model results is possible. In addition, the behavior of the inferred concentration of the radicals can be evaluated for reasonableness. Because the radicals are produced from relatively rapid reactions initiated by photolysis, their concentration is expected to drop at night. The results follow this expectation and give the investigators confidence in the validity of their measurements.

Another example of a photochemical relationship is in the alkyl nitrate, $RONO_2$, chemistry. The alkyl nitrates are formed in the atmosphere during the oxidation of hydrocarbons (outlined in Figure 3). The rate-determining step of the oxidation is attack by a hydroxyl radical, which extracts a hydrogen atom from the hydrocarbon to yield the alkyl radical. The alkyl radical quickly combines with O_2 to give the peroxy radical. This radical then reacts with NO (usually by transferring an oxygen atom to oxidize it to NO_2) and leaves the oxy radical, which then goes on to form an aldehyde or ketone. However, there is some probability, α, that the peroxy radical simply combines with the NO to give the alkyl nitrate. The probability of forming an alkyl nitrate varies from about 0.08 for a C_2 peroxy radical to about 0.30 for a C_8 peroxy radical (9). The alkyl nitrate is removed from the atmosphere by photodissociation (at the light-dependent photolysis rate J) or oxidation by a hydroxyl radical. These removal reactions are slow, so this reaction sequence can be

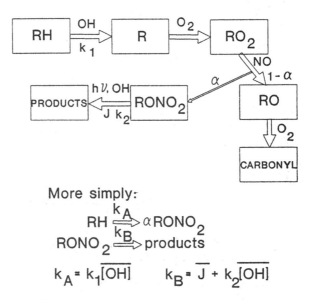

Figure 3. Alkyl nitrate chemistry and kinetics.

simplified to two sequential pseudo-first-order reactions. The rate constants, k_A and k_B, are related to the rate constants for the hydroxyl radical reactions and to the diurnal averages of the hydroxyl radical concentration and photolysis rate. Two sequential first-order reactions constitute a common kinetics example handled in undergraduate physical chemistry, and the time evolution of the ratio of the alkyl nitrate to the parent alkane can be integrated to yield

$$[\text{RONO}_2]/[\text{RH}] = [\alpha k_A/(k_B - k_A)][1 - e^{-(k_B - k_A)t}] \qquad (1)$$

The ratio of two isomers of pentyl nitrate to pentane is compared with the ratio of butyl nitrate to butane in Figure 4. These two ratios are correlated. The solid line is the behavior expected from equation 1 with the

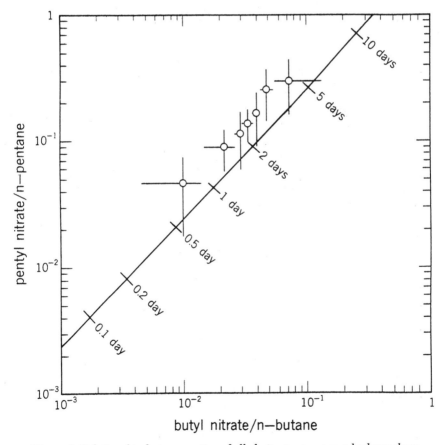

Figure 4. Relationship between ratios of alkyl nitrates to parent hydrocarbons measured in a field study in rural Pennsylvania.

substitution of laboratory-determined rate constants. Indicated on the line are the times required for the ratio to be reached for an assumed diurnally averaged hydroxyl radical concentration of 10^6 cm^{-3}. The measured pentyl nitrate to pentane ratio is somewhat higher than expected. This deviation indicates that more remains to be learned about this system. Possible causes of the deviation include measurement error, uncertainty in the kinetic data, and the lack of completeness in the description of the atmospheric photochemistry outlined in Figure 3.

Measurement Approaches

Measurements can be made both in situ, where a particular air parcel is taken into an instrument and analyzed by an appropriate technique, or by remote techniques. The latter include long-path absorption and emission measurements, which yield path-averaged concentrations, and potentially LIDAR (light detection and ranging) techniques, which give concentration as a function of distance along a propagating laser pulse. Both in situ and remote approaches have characteristic advantages and disadvantages as well as certain engineering requirements. Both approaches are referred to in this chapter, but the challenges of in situ techniques are emphasized. The different approaches and platforms are to a large extent complementary; each has its own advantages.

A critical concern of all approaches is measurement specificity. The atmosphere is a complex mixture of gases and aerosols with a multitude of opportunities for interferences and artifacts. One means of approaching definite specificity is through spectroscopic techniques. Such techniques are most readily applied to the stratosphere, where the ambient pressure is low enough to avoid serious pressure broadening of spectroscopic features. Stratospheric applications are briefly surveyed in the Measurement Platforms section.

Spectroscopic applications in the troposphere must address the pressure-broadening problem. Three approaches have been successfully used to measure concentrations of many of the nitrogen species. In differential optical absorption spectrometry (DOAS) and Fourier transform infrared spectrometry (FTIR), the problem is simply accepted as unavoidable, and in favorable situations the broadened ambient spectrum of interest can be carefully separated from other interfering spectral features (*see* references 10 and 11 for an example of each technique). An advantage of these two methods is that one or more nitrogen species plus other molecules can be measured simultaneously. Alternatively, in tunable diode laser absorption spectrometry (TDLAS) the ambient air is sampled into an absorption cell held at low pressure in order to reduce the pressure broadening to acceptable levels (*see* reference 12 and references therein). In addition, multiphoton, laser-induced fluorescence (LIF) techniques can be used to avoid the pressure-broadening problem through the use of diatomic molecules as the fluorescing entity; with this method the spectral features are adequately well defined for specificity.

FTIR, TDLAS, and LIF are in situ techniques, whereas DOAS is a long-path method that gives only a path-integrated result. NO, NO_2, NO_3, and HONO have been successfully measured with DOAS even in rural and remote regions. PAN, HNO_3, and NH_3 have been measured with FTIR in urban areas, but its sensitivity at present is not adequate for levels below a few parts per billion by volume. NO, NO_2, PAN, HNO_3, and NH_3 have been measured with TDLAS down to sub-ppbv levels. A review with references for applications of these three methods is available (13). The LIF method has been more recently developed, has been applied to the measurement of NO, NO_2, NH_3, and HONO (*see* reference 14 for an example), and offers sensitivity down to the parts per trillion by volume level.

The LIF technique is well illustrated by the example outlined in Figure 5 and diagramed in Figure 6: a vacuum UV, photofragmentation, laser-induced fluorescence system for measuring ammonia (15). The air sample is passed through a photolysis cell where an excimer laser produces an intense pulse of vacuum UV light. Two photons are absorbed by the ammonia molecule to produce the NH radical in a metastable electronic state. A particular vibrational level of this NH radical is excited with a dye laser pulse to a higher electronic state, again in a particular vibrational level. In this level the NH fluoresces to the ground state at a third wavelength. The prompt fluorescence induced from a variety of sources by the vacuum UV pulse decays to negligible levels while the metastable state persists before the dye laser fires. The fluorescence occurs to the blue of the exciting radiation (at higher energy), so it is easily distinguishable.

Application of LIDAR techniques to the measurement of nitrogen species has been very limited. Sensitivity seems to be the limiting factor. For NO_2 a detection limit of about 100 ppbv in the lower troposphere is reported for a resolution of 50 m along the laser path (16).

In situ measurements based on various nonspectroscopic analytical chemistry techniques have been successful in achieving the low detection limits required. More or less traditional analytical chemistry, requiring ex-

$$NH_3 + 2h\upsilon_1 \longrightarrow NH\,(b^1\Sigma^+) + products$$

$$NH\,(b^1\Sigma^+, v''{=}0) + h\upsilon_2 \longrightarrow NH\,(c^1\pi, v'{=}0)$$

$$NH\,(c^1\pi, v'{=}0) \longrightarrow NH\,(a^1\Delta, v''{=}0) + h\upsilon_3$$

$$\lambda_1 = 193\,nm$$

$$\lambda_2 = 452\,nm$$

$$\lambda_3 = 325\,nm$$

Figure 5. Processes involved in vacuum UV photofragmentation–laser induced fluorescence detection of ammonia.

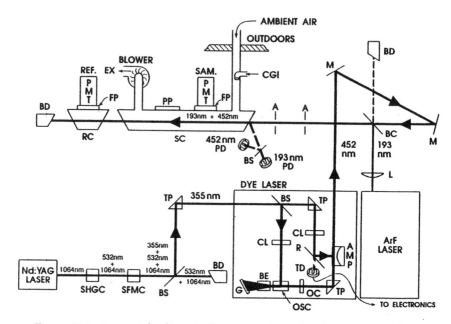

Figure 6. Instrumental schematic for vacuum UV photofragmentation–laser induced fluorescence measurement of ammonia: SHGC, second harmonic generation crystal; SFMC, sum frequency mixing crystal; BS, beam splitter; BD, beam dump; TP, turning prism; CL, cylindrical lens; R, reflector; TD, trigger diode; OSC, oscillator cell; AMP, amplifier cell; BE, beam expander; G, grating; OC, output coupler; M, mirror; BC, beam combiner; L, lens; A, aperture; PD, photodiode; SC, sample cell; RC, reference cell; FP, filter pack; SAM.PMT, sample cell photomultiplier; REF.PMT, reference cell photomultiplier; PP, additional photomultiplier port; EX, exhaust; and CGI, calibration gas inlet to flow line. (Reproduced with permission from reference 15. Copyright 1990 Optical Society of America.)

tensive manual operation and procedures, still plays an important role, although typically there is a critical compromise between adequate detection limit and long exposure time. An example of this approach is the filter-pack measurement of nitric acid and nitrate aerosols. The filters in an automatic sequencing system are exposed for from 1 to 4 h and are changed once per day. The filters are taken to the laboratory, extracted, and analyzed by ion chromatography. An important aspect of the measurement program is a careful monitoring of field blanks. This approach is certainly labor-intensive, but with careful work low detection limits can be obtained. For example, a detection limit of 2 pptrv of HNO_3 with a 2-h exposure is possible (17).

More modern instrumental analysis has been used to improve sensitivity and response time. Shown in Figure 7 is a schematic diagram of an instrument that has been developed to measure NO and NO_2 (18). The detection scheme is based on the chemiluminescence produced when NO in the ambient air reacts with ~1% ozone added as a reagent to the sampled airstream.

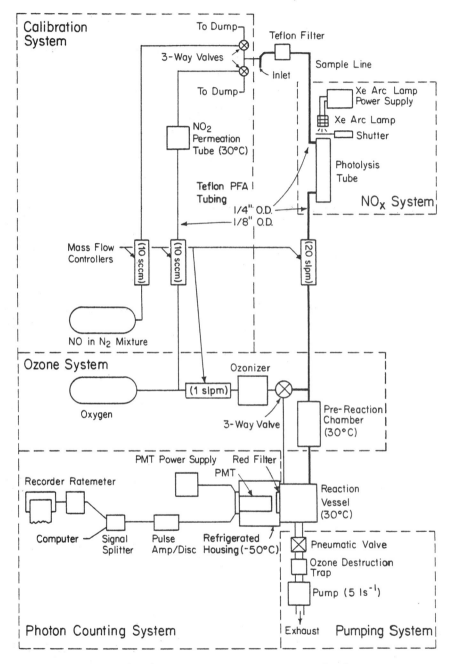

Figure 7. Schematic of O_3 chemiluminescence instrument for the measurement of NO and NO_2. (Reproduced with permission from reference 18. Copyright 1990 American Geophysical Union.)

The ambient NO concentration is proportional to the measured light signal. NO alone is measured when the arc-lamp shutter is closed, and NO_2 can be determined from the increase in the signal when the shutter is opened. This instrument has a detection limit of 2 and 10 pptrv for NO and NO_2, respectively, for a 10-s averaging time.

With these nonspectroscopic approaches there is no guarantee of specificity, and even for the spectroscopic techniques, problems may exist. An essential component of technique development is the evaluation of potential sampling artifacts and interferences. This evaluation well may require more time and effort than the instrument development itself.

Measurement Platforms

In Situ Measurements: Surface Sites.   For the determination of temporal cycles and trends, surface sites where long-term measurements can be carried out are ideal. Depending on the meteorological conditions, many air masses with different characteristics and histories will generally be transported to the site. Automatic instrument operation is advantageous so that the data can be collected continually over weeks, months, or longer without constant operator control. In this application the requirements for instrument response time are usually easily met; a 1-min or longer time average is generally adequate for surface measurements unless flux determinations are desired (*see* discussion in the following sections).

Concentrations at surface sites are often affected by the proximity of the surface. The surface may emit the species, as is the case for NO, or the species may be deposited at the surface, as is the case for HNO_3. In either case, the atmospheric concentration measured within several meters of the surface may be significantly perturbed from the average for the troposphere as a whole or even for the boundary layer (the lowest layer of the troposphere in direct thermal contact with the surface; its thickness varies from 10^2 m or less at night to 10^3 m or more during the day, at least over land). Therefore, the surface complicates the interpretation of concentration measurements. However, proper measurements near the surface can be used to determine surface fluxes of the species.

In Situ Measurements: Aircraft.   In situ measurements can also be made from aircraft platforms, but the engineering requirements are much more stringent. These challenges include design of compact, low-weight, low-power-consumption instruments; achievement of quick start-up times and automatic operation to the fullest extent possible; and, because the aircraft is moving rapidly through the air, a fast response time (approximately 1 Hz or greater) to get good spatial information. An example is the NO–NO_y instrument that has been flown on the ER–2 aircraft that the National

Aeronautics and Space Administration (NASA) operates. This aircraft is a converted U–2 spy plane that was used in, among other studies, the Antarctic mission to investigate the ozone hole (*19*). This instrument weighs 170 kg, consumes about 1 kW, and has a 1-Hz response. At takeoff the pilot actuates one switch, which causes the instrument to automatically begin operation and collect and store the measurements for later retrieval. The deployment of such instrumentation on tropospheric aircraft is perhaps more relaxed, at least in size, power, and weight constraints. For example, the NASA Lockheed Electra aircraft was equipped with a wide variety of instrumentation for the measurement of NO, NO_2, PAN, HNO_3, and NO_y (as well as other trace species) during the Global Troposphere Experiment/Chemical Instrumentation Test and Evaluation (GTE/CITE–2) study (*20*).

An aircraft has three obvious advantages over a surface site. First, the aircraft can measure vertical and horizontal profiles of concentrations; such measurements are not possible at the surface. However, because the typical aircraft travels horizontally much faster than it ascends or descends, it may be difficult to deconvolute vertical from horizontal variations. Second, the aircraft allows the investigator to choose the general type of air parcels to study rather than simply allowing sampling of whatever parcels are brought to a site (as on the surface). Third, the aircraft can measure concentrations free from the surface influences discussed in the preceding sections.

With an aircraft platform the data set that is collected is likely more limited in time than is possible on the surface. However, this limitation is at least partially compensated by the aircraft's ability to rapidly sample many different air parcels, whereas at a surface site different parcels must be transported to it. However, diurnal and seasonal cycles may be more difficult to measure from an aircraft than from the surface.

In Situ Measurements: Balloons.   Balloons currently provide the only in situ platform that allows access to the upper part of the stratosphere (above 20 km). The engineering requirements are similar to those for aircraft except for a more relaxed time response. Regional coverage from balloons is difficult, particularly because the launching facilities for the large stratospheric balloons are very limited and generally localized in the midlatitudes. However, vertical profiles without horizontal distortion are the natural data collection mode. Measurement contamination due to emissions from the balloon is a potential problem.

A balloon-borne, open path, tunable diode laser spectrometer (*21*) provides a particularly elegant technique that combines the advantages of in situ and remote sampling. The radiation absorption path is defined by the laser on the balloon gondola and a retro reflector suspended up to 500 m below. Thus, a well-defined parcel of air is analyzed and the effects of sampling inlets are avoided. NO, NO_2, HNO_3, N_2O, and O_3 have been measured simultaneously (*22*).

Remote Measurements.   Long-path differential absorption measurements of several nitrogen species have been made from the surface, from aircraft, from balloons, and from satellites. These measurements are all characterized by light that passes from a source through some atmospheric path to a detector. Surface-based measurements have used both artificial light sources within the troposphere and the natural sources of the sun and the moon, either direct rays or sky-scattered radiation. The artificial light sources can shine directly to the detector from a distance or can be reflected to the detector from a mirror some distance away so that the detector and light source can be colocated. One example of such an approach has been effected in Colorado, where a visible–UV light source (a xenon arc lamp) is focused into a beam and passed 10.3 km across a valley to a retro-reflector array. Thus, the absorption measurement is carried out over nearly a 21-km path length.

Atmospheric column absorption measurements from the surface are possible for NO, NO_2, NO_3, HNO_3, and $ClONO_2$ (23–25). These measurements have defined the seasonal cycle (25) as well as much more rapid variations (26) in the stratospheric levels of NO_3. Under favorable conditions these measurements can yield information concerning the vertical profiles of the measured species (24, 26). Such techniques can also be used from aircraft platforms (27, 28).

Long-path absorption measurements from satellites are also possible. For example, Figure 8 shows some results from the Atmospheric Trace Molecule Spectroscopy (ATMOS) experiment (29) that was carried out on Spacelab 3. The light source is solar infrared radiation, which passes through progressively deeper layers of the atmosphere as the satellite moves into and out of the earth's shadow. The investigators could extract the average vertical profiles in the stratosphere of the several nitrogen species included in Figure 8 plus N_2O (30) and N_2O_5 (31). Measurements of oxidized nitrogen species in the stratosphere have also been made from the Stratospheric Aerosol and Gas Experiment (SAGE), Limb Infrared Monitor of the Stratosphere (LIMS), and Solar Mesosphere Explorer (SME) satellite experiments; reference 32 gives a review with references to the satellite methods.

Infrared emission spectra have been used to extract the concentrations of HNO_3 (33), N_2O_5 (34, 35), and N_2O (36) from balloon-borne, cryogenically cooled interferometers.

General Challenges

The challenges that are presently faced in the field can be divided into five general categories:

1. Engineering challenges specific to the particular measurement approach, platform, and application. They have been intro-

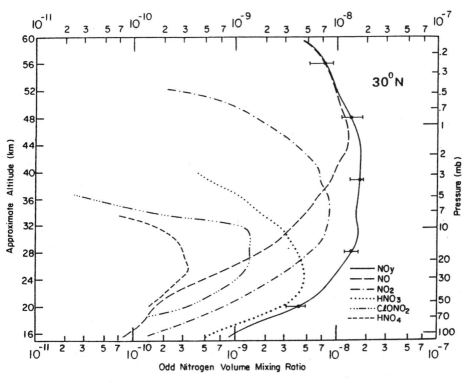

Figure 8. Measurements of odd nitrogen compounds in the stratosphere by the ATMOS experiment on Spacelab 3. (Reproduced with permission from reference 29. Copyright 1988 American Geophysical Union.)

duced in the preceding sections, and, although they are of course very important, they are probably different for each experiment and thus, with the exception of required response times, are not discussed further.

2. Demonstration that techniques now in use are specific, sensitive, accurate, and precise enough for the measurements envisioned.

3. Extension of measurement capabilities. Techniques must be developed for species that at present cannot be measured. Additionally, for species presently measured, new techniques, as well as adaptations of present techniques to other measurement platforms, are required.

4. The simultaneous measurement of as many species as possible. As was shown in the photochemical relationship examples, and as is shown in a few examples below, such simultaneous

measurements yield much additional insight into atmospheric chemistry.

5. The planning and completion of well-designed field studies that will yield a maximum of information.

These general challenges are addressed by examples of specific challenges that exemplify them.

Specific Challenges

Rigorous Field Intercomparisons.   To quantitate measurement uncertainties, the atmospheric chemistry community has developed a procedure for carrying out rigorous field intercomparisons. The features of the most instructive of the intercomparisons that have been completed include

- involvement of several different techniques together at the same site measuring the same species under typical operating conditions,

- supervision of the intercomparison by an independent referee,

- analysis of each investigator's results in a blind manner from the other investigators to a publication-ready status,

- not only simultaneous ambient measurements, but also sampling of prepared mixtures of standards as well as potential interferences in air in order to facilitate the interpretation of the results that are collected, and

- publication of the results in a refereed journal.

As an example, a recent intercomparison (37) included three NO_2 measurement techniques: a TDLAS-based system and two chemical-based systems— the photolysis–ozone chemiluminescence system diagramed in Figure 7 and an instrument based on NO_2 plus luminol chemiluminescence. Above 2 ppbv the three instruments gave similar results, but at sub-ppbv the results from the three techniques became dissimilar. Tests on the prepared mixtures showed that the luminol results were affected by expected interferences from O_3 and PAN. No interferences were found in the TDLAS system, but near the detection limit the data analysis procedures calculated levels of NO_2 that were too high. The outcome of this intercomparison was close to the ideal: the sensitivity, specificity, accuracy, and precision of each instrument were objectively analyzed; previous data sets taken by different systems can now be reliably evaluated; and each investigator was able to perceive areas in which the technique could be improved.

Completed intercomparisons that have reached definitive conclusions have involved NO (reference 38 and references cited therein), NO_y (39), and NH_3 (40) in addition to others for NO_2 (reference 41 and references cited therein). For these species the atmospheric chemistry community can objectively evaluate the current measurements of these species, at least in the troposphere. The intercomparisons have demonstrated that measurements of NO_2 involving surface conversion of NO_2 to NO have significant interferences that are due to the conversion of nitrogen species other than NO_2.

Measurement techniques for two other nitrogen species have been intercompared, with instructive but less definitive results. Two nearly identical systems for PAN measurements were intercompared as part of the NASA CITE 2 program (*42*). Generally, the agreement between the instruments for measurements in the remote troposphere was within the expected limits of accuracy and precision. However, the results indicated that there were difficulties in the calibration of this species and that at least for some periods there were significant disagreements between the results of the two systems. An additional intercomparison of methods to measure PAN would be desirable, including at least one fundamentally different technique if it can be developed. Additionally, an intercomparison of existing methods used to calibrate PAN instruments would be worthwhile, as would development of new calibration methods, given the demonstrated uncertainties in this critical procedure.

Several formal and informal intercomparisons of nitric acid measurement techniques have been carried out (*43–46*); these intercomparisons involve a multitude of techniques. The in situ measurement of this species has proven difficult because it very rapidly absorbs on any inlet surfaces and because it is involved in reversible solid–vapor equilibria with aerosol nitrate species. These equilibria can be disturbed by the sampling process; these disturbances lead to negative or positive errors in the determination of the ambient vapor-phase concentration. The intercomparisons found differences of the order of a factor of 2 generally, and up to at least a factor of 5 at levels below 0.2 ppbv. These studies clearly indicate that the intercompared techniques do not allow the unequivocal determination of nitric acid in the atmosphere. A laser-photolysis, fragment-fluorescence method (*47*) and an active chemical ionization, mass spectrometric technique (*48*) were recently reported for this species. These approaches may provide more definite specificity for HNO_3. Challenges clearly remain in the measurement of this species.

Development of Techniques for Currently Unmeasured Species in the Troposphere.

There is evidence that there are additional members of the nitrogen family that have not been measured. With the possible exceptions of HNO_3 and particulate NO_3^-, reasonably reliable techniques are currently available for in situ measurement of the concentrations of the major contributors to the NO_y family in the rural troposphere. In addition, a measurement of the total family concentration is available. Therefore the total of the concentrations of the individually measured species can be compared with the measured family concentration. This comparison can then provide a check for the contribution of any unmeasured species. In this comparison the problems in the measurement of HNO_3 and particulate NO_3^- contribute relatively small uncertainties because their sum enters the calculation; because the sum is used, the effect of the equilibria shifts is eliminated.

As an example of an evaluation of the NO_y budget, Figure 9 shows the measurement results from three field studies. The ratio of the sum of the five independently measured species to the measured NO_y level as a function of NO_x level is shown. In each of these studies, for levels of NO_x below about 2 ppbv the measured species did not account for the total measured NO_y. Important contributions to NO_y, approaching 30 to 40% under relatively clean conditions, must come from currently unmeasured species at these two sites. The closed circles show the results of calculations from a model excluding organic nitrates other than PAN (M. Trainer, personal communication). The difference between the calculated NO_y concentration and the calculated sum of the concentrations of the five individual species was composed of a variety of organic nitrates. Predicted species include acyl nitrates in addition to PAN, alkyl nitrates ($RONO_2$), and difunctional organic nitrates of the form $R'ONO_2$, where R' included either a carbonyl or hydroxyl moiety.

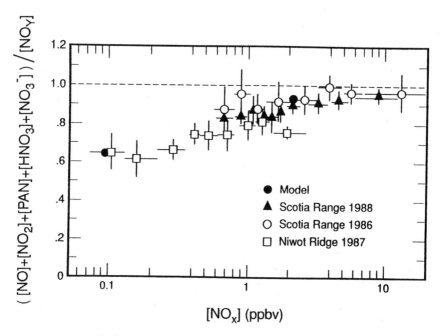

Figure 9. Results from three field studies for the ratio of the sum of NO_y species concentration to total NO_x concentration as a function of NO_x concentration. The model result at the higher NO_x levels is for the conditions at Scotia Range, Pennsylvania, and that for the lower NO_x levels is for the conditions at Niwot Ridge, Colorado (M. Trainer, personal communication).

A number of other theoretical and laboratory studies have predicted the existence of alkyl and multifunctional organic nitrates in the atmosphere (9, 49–56). The atmospheric chemistry of the organic nitrates was recently reviewed (57).

Some progress has been made in measuring these other organic nitrates, including peroxypropionyl nitrate (PPN), the three-carbon analog of PAN, peroxybenzoyl nitrate (PBzN), and the C_1–C_8 alkyl nitrates ($RONO_2$). PPN has been detected in the same chromatographic measurements that were used to measure PAN by a number of investigators (58–60). Peroxybenzoyl nitrate has been measured by first collecting a sample in a bubbler containing methanol–NaOH and then using solvent extraction on the resulting methyl benzoate, which was quantified with gas chromatography–flame ionization detection (GC–FID) (57, 61). The alkyl nitrates have been measured by a number of laboratories by using either packed or capillary column gas chromatography coupled with either an electron capture detector (ECD) (62–63) or with a nitrogen-specific NO_y detector (64). The measurements that have been made of PPN and the alkyl nitrates, in conjunction with NO_y measurements, have shown that the PPN and $RONO_2$ contributions to NO_y are not great enough to balance the NO_y budget. Therefore, attempts must be made to measure the other organic nitrate species predicted by model calculations.

Measurements of these relatively minor species will not only complete the budget of NO_y but will also indicate if our understanding of the hydrocarbon oxidation schemes in the atmosphere is complete. The organic nitrates that completed the NO_y budget in the example in Figure 9 arose primarily from the oxidation of the naturally emitted hydrocarbon, isoprene (2-methylbutadiene). To demonstrate the oxidation mechanisms believed to be involved in the production of multifunctional organic nitrates, a partial OH oxidation sequence for isoprene is discussed. The reaction pathways described are modeled closely to those described in reference 52 for propene. The first step in this oxidation is addition of the hydroxyl radical across a double bond. Subsequent addition of O_2 results in the formation of a peroxy radical. With the two double bonds present in isoprene, there are four possible isomers, as shown in reactions 2–5:

$$CH_2C(CH_3)CHCH_2 + OH$$

$$\xrightarrow{O_2} (0.35)\ HOCH_2C(CH_3)(OO)CHCH_2 \tag{2}$$

$$\xrightarrow{O_2} (0.15)\ OOCH_2C(CH_3)(OH)CHCH_2 \tag{3}$$

$$\xrightarrow{O_2} (0.35)\ CH_2C(CH_3)CH(OO)CH_2OH \tag{4}$$

$$\xrightarrow{O_2} (0.15)\ CH_2C(CH_3)CH(OH)CH_2OO \tag{5}$$

The branching ratios shown are estimates based on reference 65. The further oxidation of each of the isomers is similar, so only the first isomer will be examined in detail. In continental areas with at least tens of pptrv's of NO, the peroxy radical produced in the first step will react with NO. This reaction can result in either oxidation of NO to NO_2 or production of an organic nitrate (bold type indicates organic nitrates that are assumed to be stable):

$$HOCH_2C(CH_3)(OO)CHCH_2 + NO$$

$$\longrightarrow (0.7)\ NO_2 +\ HOCH_2C(CH_3)(O)CHCH_2 \qquad (6)$$

$$\longrightarrow (0.3)\ \mathbf{HOCH_2C(CH_3)(ONO_2)CHCH_2} \qquad (7)$$

The branching ratio shown here for the production of the organic nitrate is based on reference 65. Both of the organic products still contain one double bond, so further oxidation is likely. The alkoxy radical produced in the first step rapidly decomposes; methylvinylketone and formaldehyde are produced, as well as a hydroperoxy radical:

$$HOCH_2C(CH_3)(O)CHCH_2$$

$$\xrightarrow{O_2,\ fast} CH_2CHC(O)CH_3 +\ HCHO +\ HO_2 \qquad (8)$$

The organic nitrate produced in reaction 7 may undergo further oxidation by the hydroxyl radical via addition across the remaining double bond as shown in reactions 9 and 10. Again, in air the resulting radicals rapidly add O_2 to form peroxy radicals:

$$\mathbf{HOCH_2C(CH_3)(ONO_2)CHCH_2} + OH$$

$$\xrightarrow{O_2} (0.65)\ HOCH_2C(CH_3)(ONO_2)CH(OO)CH_2OH \qquad (9)$$

$$\xrightarrow{O_2} (0.35)\ HOCH_2C(CH_3)(ONO_2)CH(OH)CH_2OO \qquad (10)$$

In a manner similar to the reaction pathway shown in reactions 6 and 7, the peroxy radicals generated in reactions 9 and 10 will react with NO to yield either NO_2 and an alkoxy radical or a dinitrate:

$$HOCH_2C(CH_3)(ONO_2)CH(OO)CH_2OH + NO$$

$$\longrightarrow HOCH_2C(CH_3)(ONO_2)CH(O)CH_2OH +\ NO_2 \qquad (11)$$

$$\longrightarrow \mathbf{HOCH_2C(CH_3)(ONO_2)CH(ONO_2)CH_2OH} \qquad (12)$$

The stability of the dinitrate produced in reaction 12 is not known. The alkoxy radical produced in reaction 11 probably decomposes and then forms the organic nitrate shown in reaction 13, formaldehyde, and a hydroperoxy radical. Alternatively, the decomposition of this alkoxy radical can result in hydroxy acetaldehyde and an organic nitrate peroxy radical as shown in reaction 14.

$HOCH_2C(CH_3)(ONO_2)CH(O)CH_2OH$

$$\xrightarrow{O_2} \mathbf{HOCH_2C(CH_3)(ONO_2)CHO} + \mathbf{HCHO} + \mathbf{HO_2} \qquad (13)$$

$$\xrightarrow{O_2} HOCH_2C(CH_3)(ONO_2)OO + OHCH_2CHO \qquad (14)$$

The fate of the organic nitrate peroxy radical produced in reaction 14 is probably oxidation of NO to NO_2 and then decomposition, yielding acetyl nitrate, formaldehyde, and a hydroperoxy radical as shown in reactions 15 and 16.

$HOCH_2C(CH_3)(ONO_2)OO + NO$

$$\longrightarrow HOCH_2C(CH_3)(ONO_2)O + NO_2 \qquad (15)$$

$HOCH_2C(CH_3)(ONO_2)O$

$$\xrightarrow{O_2} HCHO + HO_2 + \mathbf{CH_3C(O)ONO_2} \qquad (16)$$

The further decomposition of acetyl nitrate in the atmosphere has not been studied. The oxidation of isoprene by the hydroxyl radical proceeds via repeated steps of OH addition across the double bond, followed by addition of O_2 to form a peroxy radical. The peroxy radical then either oxidizes NO to NO_2 or adds NO to form an organic nitrate. The alkoxy radical produced in the former step underwent decomposition to form both stable and reactive products. A number of possible pathways exist for forming presumably stable organic nitrates (bold in reactions 7 through 16).

In addition to being oxidized by the hydroxyl radical, alkenes may react with the NO_3 radical as has been described by several investigators (52, 56, 66). Listed in Table I are some of the organic nitrates that have been predicted to be produced via reaction of OH and NO_3 with isoprene and propene. Analogous compounds would be expected from other simple alkenes and from terpenes such as α- and β-pinene. Other possible organic nitrates may be produced via the oxidation of aromatic compounds (53, 54) and the oxidation of carbonaceous aerosols (67). Quantitative determination of these species has not been made in the ambient atmosphere.

The polar nature of many of these species may dictate the use of either solvent extraction followed by gas or liquid chromatography or supercritical extraction–chromatography (SFE–SFC) in order to make effective ambient measurements. When the measurements of these organic species are avail-

Table I. Possible Unmeasured Atmospheric Organic Nitrates

Organic Nitrate Formula	Parent Hydrocarbon
$HOCH_2C(CH_3)(ONO_2)CHCH_2$	$CH_2C(CH_3)CHCH_2$
$HOCH_2C(CH_3)(ONO_2)CH(ONO_2)CH_2OH$	$CH_2C(CH_3)CHCH_2$
$HOCH_2C(CH_3)(ONO_2)CHO$	$CH_2C(CH_3)CHCH_2$
$CH_3C(O)ONO_2$	$CH_2C(CH_3)CHCH_2$
$CH_3CH(O)CH_2ONO_2$[a]	C_3H_6
$CH_3CH(ONO_2)CH_2ONO_2$[a]	C_3H_6
$CH_3CH(OH)CH_2ONO_2$[b]	C_3H_6
$CH_3CH(ONO_2)CH_2OH$[b]	C_3H_6

[a]Reference 52; organic nitrate produced via addition of an NO_3 radical.
[b]Reference 52; organic nitrate produced via OH oxidation of propene.

able it should be possible to determine if any other species make significant contributions to NO_y.

Completion of Field Studies in the Troposphere with Better Spatial Coverage.   All of the examples of tropospheric measurements in this chapter were made at surface sites with the instrument inlets within 4 to 10 m of the ground. With the important influences of the surface on the near-surface concentrations of some atmospheric species and the lack of spatial information from surface studies, it is important to coordinate the surface measurements with aircraft (or balloon) measurements. Some aircraft measurements of nitrogen species in the troposphere have been made, but it is important to ensure that the measurement techniques on the surface and in the air give equivalent results. This assurance can only be accomplished through careful intercomparison of instruments utilized in each platform.

Figure 10 shows an example of needed data that cannot be obtained from surface studies; the model predicted nitric acid concentrations are shown as a function of altitude in the troposphere up to 2000 m immediately before sunrise. Three scenarios are given: no nighttime nitric acid source; inclusion of nitric acid from nitrate radicals reacting with carbonyl compounds and hydroperoxy radicals; and the addition of N_2O_5 reacting on aerosols to produce nitric acid. At the surface, deposition from the shallow, near-surface layer under a low-lying nocturnal inversion reduces the nitric acid concentration to near zero at night. Measurements and calculations for all three scenarios agree on this characteristic, so surface measurements cannot give any information concerning these postulated nighttime reactions. However, if it were possible to measure a vertical profile of nitric acid, the validity of each scenario could be assessed. Measuring vertical profiles of the nitrate radicals and N_2O_5 would be even more informative, but this measurement would require the development of new techniques as discussed in the next section.

Once developed, pilotless aircraft with long flight-time capabilities will be an exciting new platform for extending spatial coverage of measurements,

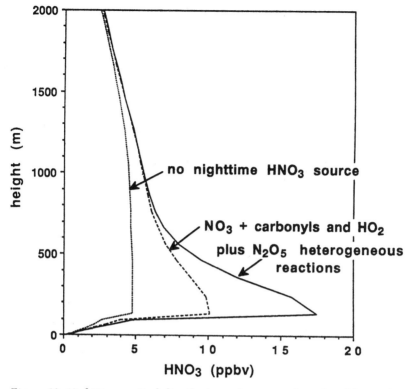

Figure 10. Nighttime vertical distributions of nitric acid predicted by model calculations for three scenarios. (Adapted with permission from reference 8. Copyright 1991 American Geophysical Union.)

both in the upper troposphere and the lower stratosphere. For example, the Boeing Condor (68) can operate at up to 20 km, can remain aloft for several days, and can carry a payload of 800 kg. Utilization of such an aircraft would allow measurements by a sophisticated suite of instrumentation at any point from the middle troposphere to the lower stratosphere.

Development of In Situ Techniques for Additional Inorganic NO$_y$ Species. Of the tropospheric inorganic species in Figure 1, NO$_3$, N$_2$O$_5$, and HONO lack reasonably well established, in situ measurement techniques that are routine and provide time resolution on the order of minutes. These three species are all believed to play important roles in the atmosphere, even though their concentrations are expected to be no higher than the low-pptrv to sub-ppbv range outside urban areas, but a paucity of measurements has prevented the full verification of these roles.

The roles that these species play are a strong function of their photolytic and thermal stability. Full sunlight photodissociates NO$_3$ and HONO on time scales of seconds and tens of minutes, respectively (17). N$_2$O$_5$ thermally dissociates to NO_2 and NO_3 on a time scale of minutes at temperatures of the lower troposphere. Thus, the concentrations of all three species are expected to reach significant levels in the troposphere only at night, and their roles are expected to be significant only during this time. As illustrated in Figure 10, nighttime production of HNO_3 from both NO_3 and N_2O_5 is possible. In addition, NO_3 reactions with organic compounds are believed to be an important source of peroxy and hydroxyl radicals at night (69). The properties, atmospheric roles, and measurements of NO_3 and N_2O_5 have been extensively reviewed recently (70). If significant sources are present, HONO concentrations will increase at night (or perhaps during other low-sunlight situations) and then provide a very important source of OH radicals (71) after sunrise (or when the sunlight intensity increases). The source of HONO shown in Figure 1 is combination of OH with NO, which will be significant only in sunlit conditions, but nonphotolytic sources have been suggested: direct emission in automobile exhaust (72) and biomass burning (73), conversion of NO_x on wet aerosols (reference 74 and references cited therein), and heterogeneous hydrolysis of NO_2 on surfaces of physical objects (75).

Measurement techniques for the troposphere for these three species require significant further work. As mentioned in preceding sections, NO_3 and HONO (but not N_2O_5) have been measured by DOAS, but in situ techniques are much less well developed. NO_3 has been measured by matrix isolation electron spin resonance (ESR) (see references and discussion in reference 70), but the time resolution is of the order of an hour, and the method is very labor-intensive. No measurements at all of N_2O_5 have been reported for the troposphere. Denuder tube techniques (references 76 and 77 and references cited therein) have been applied to measure HONO, but hydrolysis of NO_2 on the denuder collection surfaces, and perhaps other processes, evidently produce artifact signals. An intercomparison (78) between a denuder system and a DOAS technique indicated poor agreement at sub-ppbv levels, with the denuder system giving higher levels. An LIF method is available for HONO (79), but it suffers from an interference associated with ambient ozone. It would be very useful to develop sensitive, routine, in situ methods for the measurement of all three species to verify their postulated roles and tropospheric levels, and it would be particularly useful if these techniques were portable enough to deploy on aircraft with a wide suite of other measurements, because they are likely to exhibit significant vertical gradients.

Micrometeorological Flux Measurements.   It is important to quantify the flux of the nitrogen species to or from the atmosphere due to surface emission or deposition. Measurements of fluxes of NO and N_2O have been made by enclosure techniques, but the enclosures placed on a surface must be suspected of disturbing the flux that they are designed to measure. Fluxes also have been determined from measurements of vertical concentration gradients; however, this technique is of limited accuracy and applicability.

A general technique for determining fluxes is to measure the correlation of fluctuations of concentration with those of vertical wind speed, which are due to vertical eddies in the atmosphere (*see* reference 80 and references cited therein). If a species is emitted from the surface, for example, it is somewhat more concentrated in air parcels moving upward than in those moving downward. To effect this approach requires sensors with at least 1 Hz, and preferably 10 Hz, time response. Presently, such instrumentation is unavailable, except for NO and NO_y measurements. Developing such techniques for N_2O, NO_2 (or NO_x), NH_3, and HNO_3 would be desirable.

Development of Techniques for the Stratosphere.   Simultaneous measurement of as many of the nitrogen species as possible with high spatial resolution is a powerful technique for the stratosphere, just as it is for the troposphere. For example, Figure 11 shows data that were collected in the polar stratospheric ozone studies. The figure shows the relationships between NO_y and N_2O in the stratosphere, both inside and outside the polar vortices. Generally, the longer the air has been in the stratosphere, the lower the N_2O and the higher the NO_y levels. This behavior occurs because N_2O reacts in the stratosphere to yield NO with about a 7% net efficiency. This reaction is the source of nearly all the NO_y in the stratosphere. Outside the polar vortex regions NO_y is negatively correlated with N_2O, and the 7% net yield of NO_y is well matched by the slope of the NO_y versus N_2O regression line.

In Figure 11 the N_2O levels are shown to have continually decreased as the aircraft passed into each polar vortex, because air parcels that had spent progressively longer periods of time in the stratosphere were being sampled. However, the NO_y levels did not continue to increase as would have been expected; instead they dropped precipitously. This behavior indicated to the investigators that the NO_y had been removed from the air in the polar vortex. This observed denitrification was one strong confirmation for the models that pointed to heterogeneous chemistry as the cause of the polar stratospheric ozone depletion that has been intensively investigated in the last few years.

Currently the vertical profiles of the full suite of nitrogen species in the stratosphere can be measured remotely from satellites, and many total column measurements can be obtained from the ground (*see* the preceding discussion). However, these results represent such large spatial averages that much of the power of simultaneous measurements is lost. In situ speciation of the nitrogen family is required. To obtain the required spatial information demands the adaptation of the measurements to an aircraft platform. The nitrogen species that are currently of most interest in the stratosphere and are currently not measured by such in situ, fast-response techniques are NO_2, N_2O_5, HNO_3, $ClONO_2$, and HO_2NO_2.

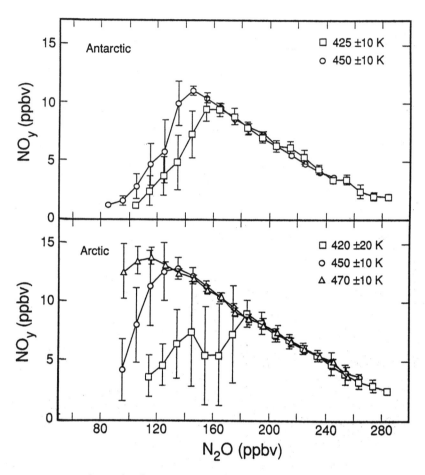

Figure 11. Relationship between NO$_y$ and N$_2$O in the polar stratosphere. The curves are labeled with the potential temperatures; the lower potential temperatures correspond to lower altitudes and, in these cases, lower absolute temperatures. (Reproduced with permission from reference 85. Copyright 1990 Macmillan Magazines Ltd.)

Chemical and Physical Characterization of Nitrogen-Containing Aerosol. Nitrogen-containing aerosols are important in both the troposphere and the stratosphere. Tropospheric aerosols contain ammonium nitrate and probably other nitrogen species (81). In the stratosphere the aerosols that form polar stratospheric clouds (PSCs) have been found to contain nitric acid (19, 82–84). They are believed to play essential roles in the reactions in the polar regions that release chlorine from the reservoir species ClONO$_2$ to form the active Cl forms that destroy O$_3$. Furthermore, the gravitational sedimentation of the aerosol removes the oxidized nitrogen species from the stratosphere and prevents the reservoir species from reforming.

The characterization of the chemistry and physics of atmospheric aerosols is a science in its infancy, and virtually the entire field is an open challenge for analytical chemists. Even techniques for properly collecting samples of atmospheric aerosols need development (*see*, for example, reference 86). Several challenges specific to nitrogen species can be listed. First, very little is known concerning the organic nitrate component of aerosols. A study from an urban area recently appeared (87) that indicates a photochemical source for such species. Second, the NO_y converter measures some fraction of the oxidized nitrogen in aerosols, but the fraction is variable and not well characterized. This variability contributes to the difficulty in characterizing the balance between the NO_y measurement and the sum of the individually measured species (88). Third, heterogeneous reactions on aerosols may be involved in the transformations of gaseous species. There is evidence for the reaction of N_2O_5 on wet aerosols to yield nitric acid (89). It is of interest to learn if the aqueous-phase nitric acid evaporates or remains as particulate nitrate when the aerosol dries or evaporates.

Summary

The application of analytical chemistry to the measurement of atmospheric nitrogen species is still a young field. In situ methods to measure species that we are not currently able to measure reliably (HNO_3, gaseous and particulate organic nitrates, NO_3, N_2O_5, and HONO in the troposphere and NO_2, N_2O_5, HNO_3, $ClONO_2$, and HO_2NO_2 in the stratosphere) must be developed. Even for species that investigators believe they can measure, additional methods should be developed, particularly approaches with definite specificity, rapid time response, and good spatial resolution. Then these different methods must be subjected to rigorous intercomparisons to ensure that each is free of interferences and artifacts. The methods, both current and yet to be developed, must be adapted not only to surface field studies, but also to aircraft and balloon platforms. Well-designed field studies that simultaneously measure as many of these species as possible need to be carried out. In addition, numbers should not simply be collected; the results must be analyzed in an imaginative manner to gain as much information as possible concerning atmospheric photochemical processes.

Acknowledgments

The authors thank David Fahey and Fred Fehsenfeld for critically reading preliminary drafts of this manuscript and Michael Trainer and Jerry Harder for many helpful discussions and suggestions.

References

1. Isaksen, I. S. A. *Tropospheric Ozone Regional and Global Scale Interactions*; D. Reidel Publishing: Dordrecht, Holland, 1988; p 422.
2. Logan, J. A. *J. Geophys. Res.* **1989**, *94*, 8511–8532.
3. Calvert, J. G.; Lazrus, A.; Kok, G. L.; Heikes, B. G.; Walega, J. G.; Lind, J.; Cantrell, C. A. *Nature (London)* **1985**, *317*, 27–35.
4. *1990 Integrated Assessment Report*; The U.S. National Acid Precipitation Assessment Program: Washington, DC, 1991.
5. *Ozone Depletion, Greenhouse Gases, and Climate Change*; National Academy: Washington, DC, 1989.
6. Mitchell, J. F. B. *Rev. Geophys.* **1989**, *27*, 115–139.
7. Logan, J. A. *J. Geophys. Res.* **1983**, *88*, 10,785–10,807.
8. Trainer, M.; Buhr, M. P.; Curran, C. M.; Fehsenfeld, F. C.; Hsie, E. Y.; Liu, S. C.; Norton, R. B.; Parrish, D. D.; Williams, E. J. *J. Geophys. Res.* **1991**, *96*, 3045–3063.
9. Atkinson, R.; Aschmann, S. M.; Carter, W. P. L.; Winer, A. M.; Pitts, J. N., Jr. *J. Phys. Chem.* **1982**, *86*, 4563–4569.
10. Platt, U.; Perner, D.; Patz, H. W. *J. Geophys. Res.* **1979**, *84*, 6329–6335.
11. Hanst, P. L.; Wong, N. W.; Bargin, J. *Atmos. Environ.* **1982**, *16*, 969–981.
12. Schiff, H. I.; Karecki, D. R.; Harris, G. W.; Hastie, D. R.; Mackay, G. I. *J. Geophys. Res.* **1990**, *95*, 10,147–10,153.
13. Finlayson-Pitts, B. J.; Pitts, J. N., Jr. *Atmospheric Chemistry: Fundamentals and Experimental Techniques*; John Wiley: New York, 1986; pp 319–347.
14. Sandholm, S. T.; Bradshaw, J. D.; Dorris, K. S.; Rodgers, M. O.; Davis, D. D. *J. Geophys. Res.* **1990**, *95*, 10,155–10,161.
15. Schendel, J. S.; Stickel, R. E.; van Dijk, C. A.; Sandholm, S. T.; Davis, D. D.; Bradshaw, J. D. *Appl. Opt.* **1990**, *29*, 4924–4937.
16. Fredriksson, K. A.; Hertz, H. M. *Appl. Opt.* **1984**, *23*, 1403–1411.
17. Parrish, D. D.; Norton, R. B.; Bollinger, M. J.; Liu, S. C.; Murphy, P. C.; Albritton, D. L.; Fehsenfeld, F. C.; Huebert, B. J. *J. Geophys. Res.* **1986**, *91*, 5379–5393.
18. Parrish, D. D.; Hahn, C. H.; Fahey, D. W.; Williams, E. J.; Bollinger, M. J.; Hübler, G.; Buhr, M. P.; Murphy, P. C.; Trainer, M.; Hsie, E. Y.; Liu, S. C.; Fehsenfeld, F. C. *J. Geophys. Res.* **1990**, *95*, 1817–1836.
19. Fahey, D. W.; Kelly, K. K.; Ferry, G. V.; Poole, L. R.; Wilson, J. C.; Murphy, D. M.; Loewenstein, M.; Chan, K. R. *J. Geophys. Res.* **1989**, *94*, 11,299–11,315.
20. Hoell, J. M., Jr.; Albritton, D. L.; Gregory, G. L.; McNeal, R. J.; Beck, S. M.; Bendura, R. J.; Drewry, J. W. *J. Geophys. Res.* **1990**, *95*, 10,047–10,054.
21. Webster, C. R.; May, R. D. *J. Geophys. Res.* **1987**, *92*, 11,931–11,950.
22. Webster, C. R.; May, R. D.; Toumi, R.; Pyle, J. A. *J. Geophys. Res.* **1990**, *95*, 13,851–13,866.
23. Farmer, C. B.; Toon, G. C.; Shaper, P. W.; Blavier, J. F.; Lowes, L. L. *Nature (London)* **1987**, *329*, 126.
24. Sanders, R. W.; Solomon, S.; Carroll, M. A.; Schmeltekopf, A. L. *J. Geophys. Res.* **1989**, *94*, 11,381–11,391.
25. Solomon, S.; Miller, H. L.; Smith, J. P.; Sanders, R. W.; Mount, G. H.; Schmeltekopf, A. L.; Noxon, J. F. *J. Geophys. Res.* **1989**, *94*, 11,041–11,048.
26. Smith, J. P.; Solomon, S. *J. Geophys. Res.* **1990**, *95*, 13,819–13,827.
27. Coffey, M. T.; Mankin, W. G.; Goldman, A. *J. Geophys. Res.* **1989**, *94*, 16,597–16,613.
28. Wahner, A.; Jakoubek, R. O.; Mount, G. H.; Ravishankara, A. R.; Schmeltekopf, A. L. *J. Geophys. Res.* **1989**, *94*, 16,619–16,632.

29. Russell, J. M., III; Farmer, C. B.; Rinsland, C. P.; Zander, R.; Froidevaux, L.; Toon, G. C.; Gao, B.; Shaw, J.; Gunson, M. *J. Geophys. Res.* **1988**, *93*, 1718–1736.
30. Gunson, M. R.; Farmer, C. B.; Norton, R. H.; Zander, R.; Rinsland, C. P.; Shaw, J. H.; Gao, B.-C. *J. Geophys. Res.* **1990**, *95*, 13,867–13,882.
31. Rinsland, C. P.; Toon, G. C.; Farmer, C. B.; Norton, R. H.; Namkung, J. S. *J. Geophys. Res.* **1989**, *94*, 18,341–18,349.
32. *Report of the International Ozone Trends Panel: 1988*; World Meteorological Organization: Washington, DC, 1988; Report No. 18; Global Ozone Research and Monitoring Project.
33. Kondo, Y.; Aimedieu, P.; Matthews, W. A.; Fahey, D. W.; Murcray, D. G.; Hofmann, D. J.; Johnston, P. V.; Iwasaka, Y.; Iwata, A.; Sheldon, W. R. *Geophys. Res. Lett.* **1990**, *17*, 437–440.
34. Kunde, V. G.; Brasunas, J. C.; Maguire, W. C.; Herman, J. R.; Massie, S. T.; Abbas, M. M.; Herath, L. W.; Shaffer, W. A. *Geophys. Res. Lett.* **1988**, *15*, 1177–1180.
35. Blatherwick, R. D.; Murcray, D. G.; Murcray, F. H.; Murcray, F. J.; Goldman, A.; Vanasse, G. A.; Massie, S. T.; Cicerone, R. J. *J. Geophys. Res.* **1989**, *94*, 18,337–18,340.
36. Abbas, M. M.; Glenn, M. J.; Kunde, V. G.; Brasunas, J.; Conrath, B. J.; Maguire, W. C.; Herman, J. R. *J. Geophys. Res.* **1987**, *92*, 8343–8353.
37. Fehsenfeld, F. C.; Drummond, J. W.; Roychowdhury, U. K.; Galvin, P. J.; Williams, E. J.; Buhr, M. P.; Parrish, D. D.; Hübler, G.; Langford, A. O.; Calvert, J. G.; Ridley, B. A.; Grahek, F.; Heikes, B. G.; Kok, G. L.; Shetter, J. D.; Walega, J. G.; Elsworth, C. M.; Norton, R. B.; Fahey, D. W.; Murphy, P. C.; Hovermale, C.; Mohnen, V. A.; Demerjian, K. L.; Mackay, G. I.; Schiff, H. I. *J. Geophys. Res.* **1990**, *95*, 3579–3597.
38. Gregory, G. L.; Hoell, J. M., Jr.; Torres, A. L.; Carroll, M. A.; Ridley, B. A.; Rodgers, M. O.; Bradshaw, J.; Sandholm, S.; Davis, D. D. *J. Geophys. Res.* **1990**, *95*, 10,129–10,138.
39. Fehsenfeld, F. C.; Dickerson, R. R.; Hübler, G.; Luke, W. T.; Nunnermacker, L. J.; Williams, E. J.; Roberts, J. M.; Calvert, J. G.; Curran, C. M.; Delany, A. C.; Eubank, C. S.; Fahey, D. W.; Fried, A.; Gandrud, B. W.; Langford, A. O.; Murphy, P. C.; Norton, R. B.; Pickering, K. E.; Ridley, B. A. *J. Geophys. Res.* **1987**, *92*, 14,710–14,722.
40. Williams, E. J.; Sandholm, S. T.; Bradshaw, J. D.; Schendel, J. S.; Langford, A. O.; Quinn, P. K.; LeBel, P. J.; Vay, S. A.; Roberts, P. D.; Norton, R. B.; Watkins, B. A.; Buhr, M. P.; Parrish, D. D.; Calvert, J. G.; Fehsenfeld, F. C. *J. Geophys. Res.* **1992**, *97*, 11,591–11,611.
41. Gregory, G. L.; Hoell, J. M., Jr.; Carroll, M. A.; Ridley, B. A.; Davis, D. D.; Bradshaw, J.; Rodgers, M. O.; Sandholm, S. T.; Schiff, H. I.; Hastie, D. R.; Karecki, D. R.; Mackay, G. I.; Harris, G. W.; Torres, A. L.; Fried, A. *J. Geophys. Res.* **1990**, *95*, 10,103–10,127.
42. Gregory, G. L.; Hoell, J. M., Jr.; Ridley, B. A.; Singh, H. B.; Gandrud, B.; Salas, L. J.; Shetter, J. *J. Geophys. Res.* **1990**, *95*, 10,077–10,087.
43. Fox, D. L.; Stockburger, L.; Weathers, W.; Spicer, C. W.; Mackay, G. I.; Schiff, H. I.; Eatough, D. J.; Mortensen, F.; Hansen, L. D.; Shepson, P. B.; Kleindienst, T. E.; Edney, E. O. *Atmos. Environ.* **1988**, *22*, 575–585.
44. Hering, S. V.; Lawson, D. R.; Allegrini, I.; Febo, A.; Perrino, C.; Possanzini, M.; Sickles, J. E., II; Anlauf, K. G.; Wiebe, A.; Appel, B. R.; John, W.; Ondo, J.; Wall, S.; Braman, R. S.; Sutton, R.; Cass, G. R.; Solomon, P. A.; Eatough, D. J.; Eatough, N. L.; Ellis, E. C.; Grosjean, D.; Hicks, B. B.; Womack, J. D.; Horrocks, J.; Knapp, K. T.; Ellestad, T. G.; Paur, R. J.; Mitchell, W. J.; Pleasant,

M.; Peake, E.; MacLean, A.; Pierson, W. R.; Brachaczek, W.; Schiff, H. I.; Mackay, G. I.; Spicer, C. W.; Stedman, D. H.; Winer, A. M.; Biermann, H. W.; Tuazon, E. C. *Atmos. Environ.* 1988, *22*, 1519–1539.

45. Talbot, R. W.; Vijgen, A. S.; Harriss, R. C. *J. Geophys. Res.* 1990, *95*, 7553–7561.

46. Gregory, G. L.; Hoell, J. M., Jr.; Huebert, B. J.; Van Bramer, S. E.; LeBel, P. J.; Vay, S. A.; Marinaro, R. M.; Schiff, H. I.; Hastie, D. R.; Mackay, G. I.; Karecki, D. R. *J. Geophys. Res.* 1990, *95*, 10,089–10,102.

47. Papenbrock, T. H.; Stuhl, F. *J. Atmos. Chem.* 1990, *10*, 451–469.

48. Schlager, H.; Arnold, F. *Geophys. Res. Lett.* 1990, *17*, 433–436.

49. Darnall, K. R.; Carter, W. P. L.; Winer, A. M.; Lloyd, A. C.; Pitts, J. N., Jr. *J. Phys. Chem.* 1976, *80*, 1948–1950.

50. Atkinson, R.; Lloyd, A. C. *J. Phys. Chem. Ref. Data* 1984, *13*, 315–444.

51. Liu, S. C. *EOS, Trans., Am. Geophys. Union* 1984, *65*, 833.

52. Shepson, P. B.; Edney, E. O.; Kleindienst, T. E.; Pittman, J. H.; Namie, G. R.; Cupitt, L. T. *Environ. Sci. Technol.* 1985, *19*, 849–854.

53. Stockwell, W. R. *Atmos. Environ.* 1986, *20*, 1615–1632.

54. Calvert, J. G.; Madronich, S. *J. Geophys. Res.* 1987, *92*, 2211–2220.

55. Atherton, C. S.; Penner, J. E. *Tellus* 1988, *40B*, 380–392.

56. Dlugokencky, E. J.; Howard, C. J. *J. Phys. Chem.* 1989, *93*, 1091–1096.

57. Roberts, J. M. *Atmos. Environ.* 1990, *24A*, 243–287.

58. Singh, H. B.; Salas, L. J.; Ridley, B. A.; Shetter, J. D.; Donahue, N. M.; Fehsenfeld, F. C.; Fahey, D. W.; Parrish, D. D.; Williams, E. J.; Liu, S. C.; Hübler, G.; Murphy, P. C. *Nature (London)* 1985, *318*, 347–349.

59. Gandrud, B. W.; Shetter, J. D.; Ridley, B. A.; Parrish, D. D.; Williams, E. J.; Buhr, M. P.; Norton, R. B.; Fehsenfeld, F. C.; Westberg, H. H.; Farmer, J. C.; Lamb, B. K.; Allwine, E. J. *EOS, Trans., Am. Geophys. Union* 1986, *67*, 884.

60. Ridley, B. A.; Shetter, J. D.; Walega, J. G.; Madronich, S.; Elsworth, C. M.; Grahek, F. E.; Fehsenfeld, F. C.; Norton, R. B.; Parrish, D. D.; Hübler, G.; Buhr, M.; Williams, E. J.; Allwine, E. J.; Westberg, H. H. *J. Geophys. Res.* 1990, *95*, 13,949–13,961.

61. Appel, B. R. *J. Air Pollut. Control Assoc.* 1973, *23*, 1042–1044.

62. Atlas, E. L. *Nature (London)* 1988, *331*, 426–428.

63. Buhr, M. P.; Parrish, D. D.; Norton, R. B.; Fehsenfeld, F. C.; Sievers, R. E.; Roberts, J. M. *J. Geophys. Res.* 1990, *95*, 9809–9816.

64. Flocke, F.; Volz-Thomas, A.; Kley, D. *Atmos. Environ.* 1991, *25A*, 1951–1960.

65. Lloyd, A. C.; Atkinson, R.; Lurmann, F. W.; Nitta, B. *Atmos. Environ.* 1983, *17*, 1931–1950.

66. Bandow, H.; Okuda, M.; Akimoto, H. *J. Phys. Chem.* 1980, *84*, 3604–3608.

67. Schuetzle, D.; Cronn, D.; Crittenden, A. L.; Charlson, R. J. *Environ. Sci. Technol.* 1975, *9*, 838–845.

68. Henderson, B. W. *Aviat. Week Space Technol.* 1990, *36*, 36–38.

69. Platt, U.; LeBras, G.; Poulet, G.; Burrows, J. P.; Moortgat, G. *Nature (London)* 1990, *348*, 147–149.

70. Wayne, R. P.; Barnes, I.; Biggs, P.; Burrows, J. P.; Canosa-Mas, C. E.; Hjorth, J.; Le Bras, G.; Moortgat, G. K.; Perner, D.; Poulet, G.; Restelli, G.; Sidebottom, H. *Atmos. Environ.* 1991, *25A*, 1–203.

71. Perner, D.; Platt, U. *Geophys. Res. Lett.* 1979, *6*, 917–920.

72. Pitts, J. N., Jr.; Biermann, H. W.; Winer, A. M.; Tuazon, E. C. *Atmos. Environ.* 1984, *18*, 847–854.

73. Rondon, A.; Sanhueza, E. *Tellus* 1989, *41B*, 474–477.

74. Notholt, J.; Hjorth, J.; Raes, F. *Atmos. Environ.* 1992, *26A*, 211–217.

75. Pitts, J. N., Jr.; Sanhueza, E.; Atkinson, R.; Carter, W. P. L.; Winer, A. M.; Harris, G. W.; Plum, C. N. *Int. J. Chem. Kinet.* **1984**, *16*, 919–939.
76. Vecera, Z.; Dasgupta, P. K. *Environ. Sci. Technol.* **1991**, *25*, 255–260.
77. Kitto, A.-M. N.; Harrison, R. M. *Atmos. Environ.* **1992**, *26A*, 235–241.
78. Appel, B. R.; Winer, A. M.; Tokiwa, Y.; Biermann, H. W. *Atmos. Environ.* **1990**, *24A*, 611–616.
79. Rodgers, M. O.; Davis, D. D. *Environ. Sci. Technol.* **1989**, *23*, 1106–1112.
80. Businger, J. A.; Delany, A. C. *J. Atmos. Chem.* **1990**, *10*, 399–410.
81. Seinfeld, J. H. *Atmospheric Chemistry and Physics of Air Pollution*; John Wiley: New York, 1986; p 738.
82. Toon, O. B.; Hamill, P.; Turco, R. P.; Pinto, J. *Geophys. Res. Lett.* **1986**, *13*, 1284–1287.
83. Pueschel, R. F.; Snetsinger, K. G.; Goodman, J. K.; Toon, O. B.; Ferry, G. V.; Oberbeck, V. R.; Livingston, J. M.; Verma, S.; Fong, W.; Starr, W. L.; Chan, K. R. *J. Geophys. Res.* **1989**, *94*, 11,271–11,284.
84. Gandrud, B. W.; Sperry, P. D.; Sanford, L.; Kelly, K. K.; Ferry, G. V.; Chan, K. R. *J. Geophys. Res.* **1989**, *94*, 11,285–11,297.
85. Fahey, D. W.; Solomon, S.; Kawa, S. R.; Loewenstein, M.; Podolske, J. R.; Strahan, S. E.; Chan, K. R. *Nature (London)* **1990**, *345*, 698–702.
86. Huebert, B. J.; Lee, G.; Warren, W. L. *J. Geophys. Res.* **1990**, *95*, 16,369–16,381.
87. Mylonas, D. T.; Allen, D. T.; Ehrman, S. H.; Pratsinis, S. E. *Atmos. Environ.* **1991**, *25A*, 2855–2861.
88. Atlas, E. L.; Ridley, B. A.; Hübler, G.; Walega, J. G.; Carroll, M. A.; Montzka, D. D.; Huebert, B. J.; Norton, R. B.; Grahek, F. E.; Schauffler, S. *J. Geophys. Res.* **1992**, *97*, 10,449–10,462.
89. Leaitch, W. R.; Bottenheim, J. W.; Strapp, J. W. *J. Geophys. Res.* **1988**, *93*, 12,569–12,584.

RECEIVED for review March 20, 1991. ACCEPTED revised manuscript June 22, 1991.

Analytical Methods Used To Identify Nonmethane Organic Compounds in Ambient Atmospheres

Hal Westberg[1] and Pat Zimmerman[2]

[1]Laboratory for Atmospheric Research, Washington State University, Pullman, WA 99164
[2]Atmospheric Chemistry Division, National Center for Atmospheric Research, Boulder, CO 80307

Nonmethane hydrocarbons are important participants in the atmospheric chemical reactions that cause photochemical smog, acid deposition, and greenhouse gases. In order to understand their involvement in these processes, researchers are using sophisticated air-quality simulation models. In most cases, these models require individual species information. Consequently, considerable effort is currently being directed toward the development and validation of analytical methods for measuring the concentration of nonmethane organic species in various ambient environments. Analytical methods currently used to define ambient concentrations of hydrocarbons and their oxygenated derivatives (carbonyls and organic acids) are summarized. This review emphasizes strengths and weaknesses of the various analytical techniques and shows where new or improved methods are needed.

THE METHODOLOGIES USED to determine the concentration of nonmethane organic compounds (NMOCs) in ambient atmospheres are reviewed in this chapter. The symposium from which this chapter evolved was designed to provide analytical chemists who are not directly involved in atmospheric measurements with a brief summary of measurement technologies currently used for determining nonmethane organic compounds, with the hope that interest in developing improved measurement methods could be

0065–2393/93/0232–0275$06.00/0
© 1993 American Chemical Society

generated. The volatile organic compounds of interest include true hydrocarbons (i.e., they contain C and H only) as well as various families of oxygenated hydrocarbons. Over the past 10 to 15 years, the true hydrocarbons have received the most attention and consequently are best characterized. Of the oxygenated hydrocarbons, those containing the alcohol functional group are least understood. Carbonyls and organic acids have attracted considerable interest in recent years; the result is a growing data base of concentration information in various ambient atmospheres.

Volatile organic compounds are important atmospheric constituents from both a chemical and biological standpoint. Biologically, they serve as (1) a carbon source for terrestrial microorganisms, (2) plant hormones (e.g., ethylene), (3) pheromones, and (4) a contributing factor in controlling plant selection and grazing pressure through vegetation palatability. Hydrocarbons play an important role in tropospheric chemistry. Figure 1 summarizes some

URBAN ATMOSPHERE

PHOTOCHEMICAL SMOG

$$NMOC + NO_X + SUNLIGHT \longrightarrow SMOG$$

REGIONAL ATMOSPHERE

PHOTOCHEMICAL SMOG

ACID DEPOSITION

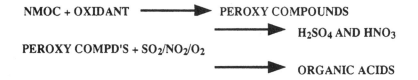

$$NMOC + OXIDANT \longrightarrow PEROXY\ COMPOUNDS$$
$$\longrightarrow H_2SO_4\ AND\ HNO_3$$
$$PEROXY\ COMPD'S + SO_2/NO_2/O_2$$
$$\longrightarrow ORGANIC\ ACIDS$$

GLOBAL ATMOSPHERE

GLOBAL WARMING

$$NMOC + OXIDANT \longrightarrow CO + CO_2$$

$$NMOC + NO_X + SUNLIGHT \longrightarrow OZONE$$

Figure 1. The role of hydrocarbons in atmospheric chemistry.

of the significant atmospheric functions of nonmethane organic species. Hydrocarbons are one of the primary ingredients in the chemical process that produces smog on the urban and regional scale. Organic acids contribute to the lowering of pH in acidic deposition processes. Hydrocarbons influence atmospheric acidity because products of their oxidation, such as peroxy radicals, facilitate the oxidation of sulfur and nitrogen oxides to sulfuric and nitric acid. On a global scale, hydrocarbon oxidation leads to products such as CO_2 and O_3, which absorb outgoing IR radiation and thus can contribute to climate warming. Carbon monoxide, which is a product of hydrocarbon oxidation, is not a primary greenhouse gas; however, it can affect climate change indirectly through its reaction with atmospheric hydroxyl radical. Increases in CO will reduce OH levels, which in turn will lead to an increase in atmospheric methane concentrations, because OH is the major sink for methane. Methane is one of the more important greenhouse gases in the troposphere.

In order to better understand nonmethane hydrocarbon involvement in the processes outlined in Figure 1, researchers are using sophisticated air-quality simulation models. These models usually require individual species information. Consequently, considerable effort is currently being directed toward the development and validation of analytical methods for measuring the concentrations of vapor-phase organic compounds in various environments.

Determination of Atmospheric Concentrations

All of the methods used to determine atmospheric hydrocarbon concentrations include the three distinct steps of

- collection,
- speciation, and
- detection.

The three phases are of roughly equal importance. If sample integrity is not maintained during collection, the result will not reflect true ambient conditions. Separation of the air matrix into individual components is clearly a requirement for meaningful analysis, and without sensitive and precise detectors, quantitation is not possible.

Collection. Procedures commonly used to collect vapor-phase organic compounds include whole-air, cryogenic, adsorption, absorption, and derivatization methods. Whole-air sampling involves the capture of an air parcel in a container. Stainless steel canisters or plastic bags constructed from an inert material, such as Teflon or Tedlar, are most commonly used. Each of these container types has advantages and disadvantages. Rigid metal canisters are easier to clean, less prone to leakage, and better for shipping samples from field sites to an analytical laboratory. However, canisters are expensive. In addition, because of their rigid structure, cans are not as useful as bags for the collection of time-integrated samples. In several recent field studies, a combination of bags and canisters was used. For example, Teflon bags, because of their minimal weight, can be attached to a tethered balloon line and filled at various altitudes in order to define vertical hydrocarbon profiles. Contents of the bags are then transferred to stainless steel canisters for storage and shipment to the laboratory. If storage times are short, all of the true hydrocarbons present in ambient air can be recovered at high efficiency from canisters and bags. This is not true, however, for the more polar oxygenated hydrocarbons. Organic acids and alcohols are recovered at very low efficiencies; carbonyls are intermediate between these compounds and the hydrocarbons. Canister losses appear to be due to wall losses and not chemical degradation. The usual procedure for filling canisters involves flushing the container for a few minutes and then closing the downstream valve and pressurizing to about 2 atm (200 kPa). Passage of the airstream through a pump removes most of the ozone, and what remains is quickly destroyed on contact with the container walls. Humidity in the canister reduces NMOC adsorptive losses (1). Apparently water occupies active wall sites more efficiently than NMOCs. Laboratory studies designed to quantify storage lifetimes of NMOCs in canisters must include the addition of water in order to simulate real atmosphere conditions. The inclusion of humidified dilution air is essential when NMOC standards are prepared in stainless steel canisters.

Cryogenic collection utilizes a glass, Teflon, or stainless steel trap that is cooled to subambient temperatures. A liquid argon or liquid oxygen cooled trap (−185 °C) will quantitatively retain all of the nonmethane organic compounds but permit passage of nitrogen and most of the oxygen in ambient air. This separation is important because a big plug of nitrogen or oxygen flushed onto a capillary column will disrupt the initial portion of the chromatogram. This is a good collection method for hydrocarbons, including many of the oxygenates. However, ozone and water vapor in ambient air can cause problems. Ozone will be concentrated in the cryogenic trap, and when the loop is warmed to transfer its contents to the gas chromatograph, ozonolysis reactions will occur. Thus, if olefins are to be measured, the ozone must be removed from the airstream before the sample enters the trap. A short precolumn of potassium carbonate or sodium sulfite will scavenge the ozone without affecting most of the organic compounds. Ice will restrict the amount of air that can be passed through the cryogenic trap.

The most common sorbent used to collect organic compounds in air is Tenax. It has the desired property of not retaining significant amounts of water, and the organic compounds that are adsorbed can be eluted by heating. The main advantage of the adsorbent method is that large volumes of ambient air can be processed; these large samples lead to very low detection levels for many organic compounds. Care must be exercised, however, when interpreting data acquired with Tenax or other adsorbents. Collection and desorption efficiencies must be established for each organic species. For example, the breakthrough volume is a few hundred cubic centimeters or less for highly volatile hydrocarbons (fewer than six carbons) when Tenax is used. Other sorbents have been used that will retain the lower molecular weight organic compounds, but with many of these, quantitative recovery of the higher molecular weight species is not possible. A single adsorbent that provides an acceptable medium for collecting hydrocarbons over the entire C_2–C_{12} range is not yet available. Solid adsorbents will produce some bleed peaks during the heat desorption process. Therefore, "blank" analyses must be performed on a regular basis, and these must be integrated into the final data interpretation. The adsorbent methodology is good for individual species if quality assurance studies show that interferences are absent.

Derivatization is commonly used to collect polar, oxygenated hydrocarbons. Most carbonyls can be extracted from an air sample by passage through a medium containing 2,4-dinitrophenylhydrazine (DNPH). The usual methodology involves coating a chromatographic material such as silica gel, Florisil, or C_{18} with an acidic solution of the DNPH. When ambient air containing carbonyls is passed through the derivatizing medium, hydrazones are formed. The hydrazones can be eluted with an appropriate organic solvent and quantified by conventional analytical methods. Organic acids can be collected as salts upon passage of an airstream through a filter impregnated with a base. Denuder tubes coated with a basic material such as sodium carbonate are an effective collection medium for organic acids as well.

Use of the derivatization methods requires that collection and recovery efficiencies be established for each species. In addition, extreme care must be exercised in order to exclude contamination during storage before and after exposure to ambient air. A series of blanks must be included that represents the various sources of contamination—solvents, transportation to collection site, and storage prior to elution for analysis. In addition to positive interferences due to contamination by the species of interest, negative interferences are also possible. For example, if ozone is present at concentrations in excess of 50 parts per billion by volume (ppbv), the ozone will cause a negative interference with formaldehyde when the latter is collected by passing air through a cartridge containing silica that is impregnated with DNPH (2). Studies have shown that formic acid can be formed on filters impregnated with base (3). This formation presumably occurs through the oxidation of formaldehyde, which is then trapped as formic acid. This reaction doesn't appear to be a problem with denuder tubes.

Absorption methods are currently being used to collect polar air contaminants that are soluble in water. Low-molecular-weight organic acids and peroxides are examples of species that can be removed from ambient air that contacts a water surface. Formic and acetic acid can be collected by injecting ambient air into a chamber containing a water vapor mist. With this methodology, the acids are reported to be scrubbed from the airstream with 100% efficiency (4). Another method in which air is allowed to diffuse past a condensation plate that is cooled below the dew point has been utilized for the collection of organic acids in ambient atmospheres (5). The gaseous acids are absorbed in the water layer that condenses on the cooled plate. Lazrus et al. collected airborne peroxides by passing air through a water trap (6). Various trap designs have been used for collecting the peroxides. The preferred procedure and that currently used involves scrubbing the peroxides from an airstream that is slowly pumped through a coil containing liquid water.

Collection technology for vapor-phase organic compounds can be summarized as follows:

- True hydrocarbons can be collected and stored in passivated stainless steel canisters without significant losses.

- Cryogenic collection of true hydrocarbons as well as some oxygenated hydrocarbons works well if precautions are taken to remove ozone.

- Adsorption of organic compounds on materials such as Tenax is best done on an individual species basis in which the collection and recovery efficiencies are established and blank analyses show no interferences from the adsorbent material.

- Derivatization and absorption are being used for collection of polar and reactive organic species that do not behave well when collected by the whole-air, cryogenic, or adsorption techniques.

Aside from whole-air collection in stainless steel canisters, additional analytical studies are needed with the other collection methodologies. For example, the cryogenic collection procedure potentially offers an excellent way of automating the collection of hydrocarbons. However, methods for removing ozone and water must be developed that are applicable for continuous operation. A single solid adsorbent material that will quantitatively retain and elute organic species would be highly desirable.

Speciation.   The method used to resolve a complex air matrix into individual species is dependent on the collection procedure that was used. Gaseous samples are separated into the individual components with gas chromatography, whereas samples in liquid media (derivatized and absorbed) are usually resolved on a liquid or ion chromatograph.

The favored method will always incorporate a gas chromatographic separation when possible because of the much better resolution achievable on gas chromatograph (GC) columns as compared to the liquid systems. Low-molecular-weight oxygenated compounds that don't behave well (low response or high polarity) in gas chromatographic systems are best resolved with liquid chromatographic conditions.

Gas Chromatography.   Fused silica capillary columns are used to separate gaseous mixtures. A preconcentrated (cryogenic or adsorbent) sample is injected onto the head of a capillary column that has been cooled to subambient temperature (~-50 °C). This setup provides a mechanism for focusing most of the hydrocarbons in a thin band at the head of the column. The GC oven is then programmed to increase the temperature to approximately 100 °C. A typical run would go from -50 to 100 °C at 4 °C/min, with a hold at the final temperature for an additional 15 min. Under these conditions, hydrocarbons containing up to 10 carbons can be eluted in 45 min. Figure 2 (a through c) shows examples of the separation that can be achieved with this procedure. The urban sample chromatogram, shown in Figure 2a, contains over 100 peaks. The C_2 hydrocarbons (ethane, ethylene, and acetylene) are not resolved, and resolution is marginal for the C_3 hydrocarbons. Dominant hydrocarbons in the C_4 to C_8 molecular weight range are well resolved. Separation is acceptable for anthropogenic hydrocarbons in the C_9 and C_{10} range, but quantifying trace amounts of monoterpenes that might be present in urban atmospheres would be difficult.

Complexity of the chromatogram is reduced in rural samples; about 25 major species are usually present. The use of fused silica capillary columns allows separation of locally emitted biogenic hydrocarbons from the background mix of long-lived anthropogenic species. As shown in Figure 2b, isoprene and the pinenes can be readily quantified. Samples representative of the free troposphere (Figure 2c) contain very few peaks because hydrocarbons in the C_4 and above molecular weight range are present at concentrations near or below the detection limit of most GC systems.

Hydrocarbons containing two and three carbons are generally separated on packed columns. Chemically bonded materials such as *n*-octane or phenyl isocyanate on Porasil have proven to be good separator systems for these highly volatile nonmethane hydrocarbons. However, these systems require a separate analysis from that employed for the C_4–C_{12} hydrocarbons (7). Recent developments include the use of capillary-type columns [e.g., Al_2O_3 porous layer open tubular (PLOT)] for separation of the lower molecular weight hydrocarbons (8).

Liquid–Ion Chromatography.   Oxygenated hydrocarbons, such as carbonyls, that have been converted to derivatives and that absorb light in the UV range can be quantified by liquid chromatographic procedures. Reversed-phase chromatographic techniques with gradient elution are normally used. When compared to gas chromatography separations, the liquid chro-

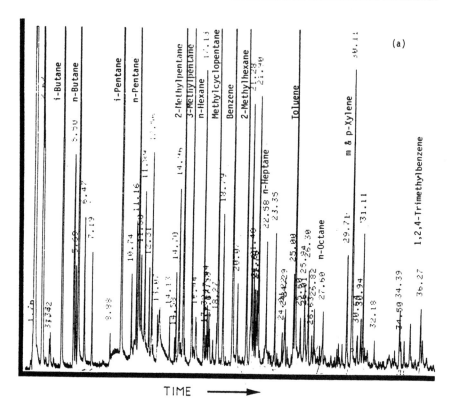

TIME ⟶

Figure 2. Chromatograms typical of a, an urban environment. GC conditions: 30-m DB–1 fused silica capillary column; oven temperature –50 to 100 °C at 4 °C/min. Continued on next page.

matography (LC) separations are poor. However, LC column eluent systems have been developed that provide adequate resolution of the low-molecular-weight carbonyl derivatives (9). Figure 3 illustrates the type of separation achievable for several of the more volatile carbonyl species present in urban atmospheres.

Organic acids have received considerable interest recently because of their potential for reducing the pH in precipitation (10). An ion chromatograph (IC) with a conductivity detector is best suited for determining organic acid concentrations. As indicated in the "Collection" section, the acids are removed from ambient air either through derivatization to an alkali metal salt (by a filter–denuder) or by absorption in water. In either case, a water mixture is injected onto the IC column, where separations as depicted in Figure 4 can be expected. Ion chromatography, like liquid chromatography, has much greater potential for interference than gas chromatography because of much poorer column resolution.

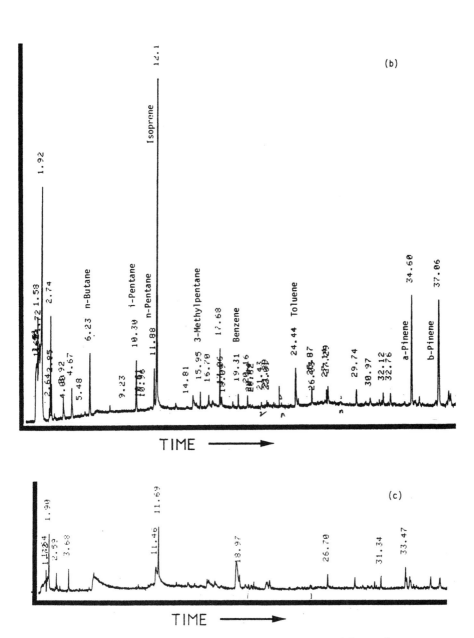

Figure 2. Continued. *Chromatograms typical of b, a rural forested environment, and c, a clean continental environment. GC conditions: 30-m DB–1 fused silica capillary column; oven temperature –50 to 100 °C at 4 °C/min.*

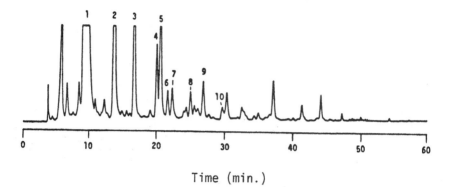

Time (min.)

Figure 3. LC chromatogram of carbonyls collected on a DNPH-impregnated silica gel cartridge. Peak identities: 1, DNPH; 2, formaldehyde; 3, acetaldehyde; 4, acrolein; 5, acetone; 6, propionaldehyde; 7, x-acrolein; 8, crotonaldehyde; 9, butyraldehyde; and 10, benzaldehyde. (Reproduced with permission from reference 8. Copyright 1992.)

Detection.   Nearly all of the vapor-phase organic compounds will respond when added to a flame ionization detector. Consequently, this detector is most commonly used. Other special-purpose detectors include photoionization, mass spectrometry, atomic emission, ion mobility, mercury oxide reduction, and chemiluminescence detectors.

Flame Ionization.   This has proven to be the best detection system for most organic compounds. Its nearly equal carbon response for the true hydrocarbons greatly facilitates calibration. For example, a single hydrocarbon (or a mixture of a few hydrocarbons) can be used to determine the response-versus-concentration curve for calibration of a GC system. It is then a simple matter to determine the concentrations of all the hydrocarbons in a complex mixture, such as that represented by the chromatogram in Figure 2a. In addition to being easy to calibrate, flame ionization detectors (FIDs) are robust systems that can be transported to and operated at remote field sites. The wide linear response range and low detection limit make the FID an ideal detector for quantifying hydrocarbons in urban atmospheres. The practical detection limit for a typical flame system is approximately 10 parts per trillion by volume (for an air volume of 1 L); however, precision deteriorates rapidly as concentrations fall below 100 pptrv. Thus, the FID is marginally acceptable for the analysis of free troposphere samples in which most of the hydrocarbons are present at low parts-per-trillion levels. An enhancement of FID sensitivity by a factor of 10 to 100 would be very beneficial for the determination of hydrocarbon concentrations in clean atmospheres.

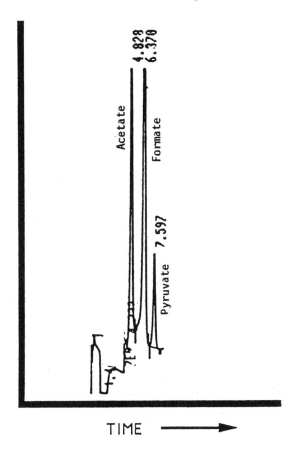

TIME ———————➤

Figure 4. IC chromatogram showing resolution of organic acids. A polysty-rene–divinylbenzene (HPIC–AS4A) separator column was used; the eluent solution was 0.0013 M sodium tetraboratedecahydrate at a flow rate of 2 mL/min.

Mass Spectrometry.   The use of a quadrupole mass spectrometer as a GC detector for nonmethane hydrocarbon analysis has come of age in recent years. Development of capillary columns with low carrier gas flows has greatly facilitated the interfacing of the GC and mass spectrometer (MS). The entire capillary column effluent can be dumped directly into the MS ion source to maximize system sensitivity. GC–MS detection limits are compound-specific but in most cases are similar to those of the flame ionization detector. Quantitation with a mass spectrometer as detector requires individual species calibration curves. However, the NMOC response pattern as represented by a GC–MS total ion chromatogram is usually very similar to the equivalent FID chromatogram. Consequently, the MS detector can be used to establish hydrocarbon identities and an FID system for quantitation.

The coupling of two mass spectrometer systems has received attention in recent years. This system can be operated in an atmospheric pressure mode by passing the air matrix directly into the ionization source (11). This method minimizes sample contamination and degradation problems. Detection limits are compound-dependent and can vary over more than an order of magnitude for different families of hydrocarbons. For example, an aromatic hydrocarbon such as toluene cannot be detected at levels below 5 ppbv, whereas most aldehydes are detectable at levels as low as 50 pptrv. The tandem MS–MS system has the potential to be a useful detection system for organic compounds that do not store well in collection containers.

Atomic Emission. Coupling an atomic emission detector (AED) with a gas chromatograph has the potential for selective and sensitive determination of nonmethane hydrocarbons. Organic compounds exiting the GC column are excited in a microwave plasma that yields light emissions at wavelengths characteristic of the elements present. By measuring the intensity of light emitted, chromatograms selective for various elements (e.g., C, H, O, and so forth) are obtained. A GC–AED system is now commercially available, but no reports have appeared documenting its use in the determination of trace organic compounds in the atmosphere. Literature provided by the manufacturer (Hewlett-Packard) indicates that detection limits for carbon and hydrogen are lower than those of the flame ionization detector. The system appears to have an acceptable linear dynamic range ($\sim 10^4$) as well. The ability to verify the presence of C, H, and O in a single analysis should prove to be very beneficial for the identification of trace organic compounds in the atmosphere. Quantitation will require individual species calibration.

Special-Purpose Detectors. Ion mobility, HgO reduction, and chemiluminescence are used as special-purpose detection systems that can be utilized for NMOC analysis. Hill and co-workers (12) have described the use of a capillary column gas chromatograph coupled to an ion mobility detector for trace organic analysis. Effluent from the GC column enters the detector, where the organic molecules are ionized by a proton-transfer mechanism involving $(H_2O)_n H^+$. The ionized organic compound then passes through a drift tube at a particular velocity that depends on factors such as collision frequency with drift-gas molecules, temperature, charge on the ion, and so forth. The ion mobility detector is reported to have picogram sensitivity and can provide very selective detection.

The mercuric oxide reduction detector was originally described for monitoring carbon monoxide in clean atmospheres (13). However, because the principle of detection relies only on the transformation of HgO to Hg vapor, any species that will effect this reduction can be detected. Organic molecules that contain unsaturated bonds will respond. O'Hara and Singh used the gas reduction cell to measure acetaldehyde and acetone concentrations (*14*). They report a wide linear range (10^3) and sensitivity 20 to 40 times that of a flame ionization detector. Zimmerman and Greenberg have used a gas chromatograph equipped with a gas reduction cell for analysis of biogenic hydrocarbons in forest environments (*15*). Isoprene and monoterpenes can be detected at low parts-per-billion levels with a 1-cm^3 injection. This sensitivity is advantageous for rural studies because it eliminates the need for cryogenic preconcentration. The contents of a 1-cm^3 sample loop can be transferred directly to a megabore-type capillary column. An additional positive feature of this detector is that it doesn't require flame support gases (hydrogen and oxygen).

The reaction between olefins and ozone produces light that can be measured and related to the concentration of the reactants. One of the preferred methods for measuring ambient ozone concentrations utilizes the chemiluminescence generated in the ozone–ethylene reaction for detection. Recently, Hills and Zimmerman (*16*) described the use of this detection principle for determining hydrocarbon concentrations. They utilized the chemiluminescence created when ozone reacts with isoprene for development of a continuous, fast-response isoprene analyzer. This real-time isoprene system is reported to be linear over three orders of magnitude and to have a detection limit of about 1 ppbv. Because the system doesn't include a preseparation of hydrocarbons, interferences from other olefins (ethylene, propylene, and so forth) could occur. Thus far the chemiluminescent detector has been used to monitor isoprene emissions under conditions in which the concentrations of olefins that could interfere are negligible compared to those of the biogenic hydrocarbon.

Future Directions

State-of-the-art monitoring systems for nonmethane hydrocarbons are currently available and very adequate for defining the qualitative and quantitative NMOC composition in urban environments. Future development efforts for urban monitoring need to be directed toward automation and improved speciation of oxygenated hydrocarbons. Recent monitoring efforts in Atlanta have utilized an automated system that collects ambient hydrocarbons on an adsorbent trap and then causes a thermal desorption and automatically transfers the sample to a capillary GC column. Information concerning the success of this analytical procedure should be available in the near future. Scientists in the National Oceanic and Aviation Administration (NOAA) Aeronomy Laboratory are perfecting an automated system for determination of C_3–C_{10} hydrocarbons in clean, rural environments (P. Goldan, unpublished data). The aim is to develop a system that will provide hourly speciated hydrocarbon concentrations on a continuous basis. A capillary column–FID gas chromatographic system is being utilized along with cryogenic collection–preconcentration methods. The real challenge comes in designing a collection system that will retain the organic compounds but continuously remove water, ozone, and any other interfering species from an ambient air sample.

The procedures used to determine ambient carbonyl concentrations involve a collection step with silica or C_{18} cartridges impregnated with 2,4-dinitrophenylhydrazine. Contamination is inevitable with this system, and blanks must be used to compensate for the degree of contamination. Selection of the appropriate blank values to subtract is a difficult and uncertain process. Consequently, development of a gas chromatographic system that will resolve and respond to the low-molecular-weight aldehydes and ketones is needed. The mercuric oxide and atomic emission detectors should provide adequate response for the carbonyls.

Simplified analytical procedures for determination of gas-phase organic acids would be very beneficial. Currently, the acids are collected by using impregnated filters, denuder tubes, or water absorption techniques and then an ion chromatographic analysis. Normally, the collection and analysis steps are decoupled in time (i.e., samples collected at a field site are returned to a home laboratory for IC analysis). Once again, blank samples must be utilized to compensate for contamination during transport and storage prior to analysis.

Development of fast-response techniques for measurement of NMOC fluxes is badly needed. Detectors with specificity for a compound and the speed to be networked with fast eddy correlation micrometeorological techniques would be very useful. Present development activities in this area are aimed at coupling existing fast micrometeorological sensors with slow analytical methods for the NMOCs. Figure 5 illustrates a conditional sampling system designed to provide hydrocarbon flux information. The inlet for hydrocarbon sampling is colocated with the sensor unit of a sonic anemometer. A computer-controlled solenoid network is designed to channel hydrocarbon sample line flow into one of three collection containers, depending on the direction of air movement. When eddies are moving upward as determined by the anemometer, ambient air flows into the up container while neutral air motion fills the stagnant collector; during downward motion, air is channeled to the remaining container. The flux is proportional to the difference in concentrations in the up and down collectors times the vertical wind speed fluctuations (σ_w). The NMOC concentrations are measured with a conventional GC–FID system.

Methods are also needed for establishing accuracy of NMOC analysis. At present, each research group making NMOC measurements must prepare its own calibration standards. Measurement accuracy is then judged by intercomparing the results obtained when two or more laboratories analyze the same samples. When the results of urban samples have been intercom-

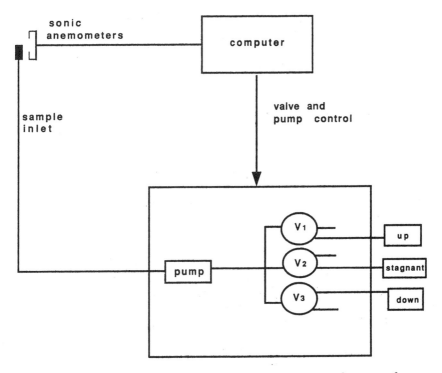

Figure 5. Block diagram of a conditional sampling system for quantifying NMOC fluxes.

pared, agreement between laboratories has been generally quite good. The relative standard deviation for all of the dominant hydrocarbons was 20% or less for samples collected in several northeastern United States cities and analyzed by three independent laboratories (*17*). Hydrocarbon concentrations in the urban intercomparison studies generally varied between 10 and 100 ppb C. Laboratory intercomparability for C_2–C_5 hydrocarbons in samples collected in the remote atmosphere is not very good. The percent relative standard deviation for ethane (~2 ppbv) from 10 laboratories was about 20% but increased to greater than 100% for many hydrocarbons present at 500 pptrv or less (*18*).

References

1. Pate, B.; Jayanty, R. K. M.; Peterson, M. R.; Evans, G. F. *J. Air Waste Management Assoc.* **1992**, *42*, 460–462.
2. Arnts, R. R.; Tejada, S. B. *Environ. Sci. Technol.* **1989**, *23*, 1428.
3. Keene, W. C.; Talbot, R. W.; Andreae, M. O.; Beecher, K.; Berresheim, H.; Castro, M.; Farmer, J. C.; Galloway, J. N.; Hoffman, M. R.; Li, S. M.; Maben, J. R.; Munger, J. W.; Norton, R. B.; Pszenny, A. A. P.; Puxbaum, H.; Westberg, H.; Winiwarter, W. *J. Geophys. Res.* **1989**, *94*, 6457–6460.

4. Cofer, W. R., III; Collins, V. G.; Talbot, R. W. *Environ. Sci. Technol.* **1985,** *19,* 557–560.
5. Dawson, G. A.; Farmer, J. C. *J. Geophys. Res.* **1988,** *93,* 5200–5206.
6. Lazrus, A. L.; Kok, G. L.; Lind, J. A.; Gitlin, S. N.; Heikes, B. G.; Shetter, R. E. *Anal. Chem.* **1986,** *58,* 594–597.
7. Greenberg, J. P.; Zimmerman, P. R. *J. Geophys. Res.* **1984,** *89(D),* 4767–4778.
8. Rudolph, J.; Khedim, A.; Bonsang, B. *J. Geophys. Res.* **1992,** *97(D),* 6181–6186.
9. Tejada, S. B. *Int. J. Environ. Chem.* **1986,** *26,* 167–185.
10. Talbot, R. W.; Beecher, K. M.; Harriss, R. C.; Cofer, W. R., III *J. Geophys Res.* **1988,** *93(D),* 1638–1652.
11. Dumdei, B. E.; Kenny, D. V.; Shepson, P. B.; Kleindienst, T. E.; Nero, C. M.; Cupitt, L. T.; Claxton, L. D. *Environ. Sci. Technol.* **1988,** *22,* 1493–1498.
12. St. Louis, R. H.; Siems, W. F.; Hill, H. H., Jr. *LC–GC* **1987,** *6,* 810–814.
13. Robbins, R. C.; Borg, K. M.; Robinson E. *J. Air Pollut. Control Assoc.* **1968,** *18,* 106–110.
14. O'Hara, D.; Singh, H. B. *Atmos. Environ.* **1988,** *22,* 2613–2615.
15. Zimmerman, P. R.; Greenberg, J. P. *EOS, Trans., Am. Geophys. Union* **1988,** *69,* 1056.
16. Hills, A. J.; Zimmerman, P. R. *Anal. Chem.* **1990,** *62,* 1055–1060.
17. Westberg, H.; Lonneman, W.; Holdren, M. *Identification and Analysis of Organic Pollutants in Air;* Keith, L. H., Ed.; Butterworth Publishers: Boston, 1984; pp 323–337.
18. Carsey, T. P.; Bachmann, K.; Blake, D. R.; Blake, N. J.; Bonsang, B.; Dalluge, B.; Greenberg, J.; Harvey, G. R.; Kanakidou, M.; Laird, C. K.; Lightman, P.; Penkett, S.; Rasmussen, R. A.; Rowland, S.; Rudolph, J.; Westberg, H.; Zimmerman, P. R. *J. Atmos. Chem.* **1990,** submitted.

RECEIVED for review March 20, 1991. ACCEPTED revised manuscript September 16, 1992.

Measurement Methods for Peroxy Radicals in the Atmosphere

Chris A. Cantrell, Richard E. Shetter, Anthony H. McDaniel, and Jack G. Calvert

Atmospheric Kinetics and Photochemistry Group, Atmospheric Chemistry Division, National Center for Atmospheric Research, P.O. Box 3000, Boulder, CO 80307–3000

The measurement of peroxy radicals (RO_2) in the atmosphere is an important and challenging problem. Determining the concentrations of HO_2 and RO_2 has been accomplished in the atmosphere and in the laboratory with systems that may be broadly grouped into two categories: chemical and spectroscopic. Several chemical conversion techniques and the use of spectroscopic methods in various wavelength regions are described. These approaches are critically evaluated for their potential use as atmospheric monitoring tools, primarily in the troposphere, although stratospheric applications are also mentioned.

THE CHEMISTRY OF THE TROPOSPHERE is an intertwining of cycles involving gas-phase, condensed-phase, and multiple-phase reactions (*1–8*). In order to understand the distributions (spatial and temporal) of a chemical species, the important factors (i.e., sources, sinks, and chemical reactions) that govern its behavior must be understood. The roles played by free radicals in the earth's atmosphere are many and varied. One radical family of particular interest is the odd hydrogen family ($HO + HO_2$). The organic peroxy and oxy radicals (RO_2 and RO; $R = CH_3$, C_2H_5, etc.) are chemically similar to the odd hydrogen radicals and are important intermediates in the oxidation of organic compounds in the atmosphere (*9*).

Peroxy radicals are formed in the troposphere through the interaction of sunlight with certain molecules or as products of other radical reactions.

0065–2393/93/0232–0291$09.00/0

Important peroxy radical sources include the reactions of the hydroxyl radical with various compounds, for example, carbon monoxide:

$$HO + CO \longrightarrow H + CO_2 \tag{1}$$

alkanes:

$$HO + RH \longrightarrow H_2O + R \tag{2}$$

$$R + O_2 \xrightarrow{M} RO_2 \tag{3}$$

or alkenes:

$$HO + RCH{=}CHR \xrightarrow{M} RCH(OH)CHR \tag{4}$$

$$RCH(OH)CHR + O_2 \longrightarrow RCH(OH)C(O_2)HR \tag{5}$$

These equations demonstrate the link that is expected between HO and RO_2 in the troposphere. Hydroxyl radicals are formed in processes initiated by photolysis of various precursors, for example, the ultraviolet photolysis of ozone (O_3) or nitrous acid (HONO).

$$O_3 + h\nu \longrightarrow O\,(^1D) + O_2 \tag{6}$$

$$O\,(^1D) + H_2O \longrightarrow 2\,HO \tag{7}$$

$$HONO + h\nu \longrightarrow HO + NO \tag{8}$$

Peroxy radicals are also formed in the troposphere through the photolysis of aldehydes (10, 11) and through nitrate radical (NO_3) reactions (12–14). The hydrogen atom and formyl radical that are formed then react with molecular oxygen (O_2) (reactions 11 and 12) under tropospheric conditions.

$$CH_2O + h\nu \longrightarrow H + HCO \tag{9a}$$

$$H + HCO \longrightarrow H_2 + CO \tag{9b}$$

$$CH_2O + NO_3 \longrightarrow HCO + HNO_3 \tag{10}$$

$$H + O_2 \xrightarrow{M} HO_2 \tag{11}$$

$$HCO + O_2 \longrightarrow HO_2 + CO \tag{12}$$

The reactions of alkenes with ozone are also very important sources of tropospheric peroxy radicals (*15, 16*):

$$H_2C=CH_2 + O_3 \longrightarrow H_2COO^* + CH_2O \tag{13}$$

$$CH_2OO^* \longrightarrow HCO_2H^* \tag{14}$$

$$HCO_2H^* \longrightarrow 2H + CO_2 \tag{15a}$$

$$HCO_2H^* \longrightarrow H_2 + CO_2 \tag{15b}$$

$$HCO_2H^* \longrightarrow H_2O + CO \tag{15c}$$

$$HCO_2H^* \xrightarrow{M} HCO_2H \tag{15d}$$

The H atoms formed in reaction 15a can react with O_2 (reaction 11) to form HO_2. The stabilized Criegee intermediate (CH_2OO) can participate in further reactions, some of which will result in the formation of peroxy radicals. Larger alkenes react with ozone to produce organic peroxy radicals.

Peroxy radicals play many roles in the troposphere. A reaction of crucial importance is the oxidation of nitric oxide (NO) by peroxy radicals.

$$RO_2 + NO \longrightarrow RO + NO_2 \tag{16a}$$

A second pathway in this reaction results in organic nitrate formation (*9, 17, 18*). The size and structure of the organic group controls the yield of reaction 16b relative to 16a.

$$RO_2 + NO \xrightarrow{M} RONO_2 \tag{16b}$$

These alkyl nitrate compounds have been measured in the troposphere and constituted about 1.5% of the total odd nitrogen budget at a rural eastern U.S. site (*19*). Reaction 16a competes with the oxidation of NO by ozone in the troposphere.

$$O_3 + NO \longrightarrow O_2 + NO_2 \tag{17}$$

The O_3 that is formed in the troposphere is controlled approximately by the rate of its generation by NO_2 photodecomposition (reactions 18 and 19) and by the rate of its removal by reaction with NO; the concentration is roughly $[O_3] = j_{NO_2}[NO_2]/(k_{17}[NO])$ (*20*). This balance is influenced by the presence of peroxy radicals because of the reaction shown in equation 16a.

$$NO_2 + h\nu \longrightarrow NO + O \tag{18}$$

$$O + O_2 \xrightarrow{M} O_3 \tag{19}$$

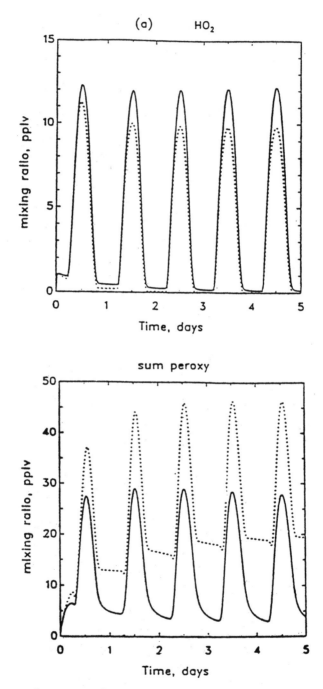

Figure 1. *Hydroperoxy and organic peroxy radical concentrations as simulated for the marine boundary layer with (solid line) and without (dotted line) peroxy radical permutation reactions. (Reproduced with permission from reference 22. Copyright 1990 American Geophysical Union.)*

Equations 17 through 19 govern the so-called NO–NO$_2$–O$_3$ photostationary state system in the atmosphere. Measurements of key components in this system have been used to infer peroxy radical concentrations.

Peroxy radicals are the only gas-phase source of peroxide compounds in the troposphere.

$$HO_2 + HO_2 \longrightarrow H_2O_2 + O_2 \tag{20}$$

$$RO_2 + HO_2 \longrightarrow ROOH + O_2 \tag{21}$$

Peroxy radicals are intermediates in the atmospheric oxidation of virtually all organic compounds. HO$_2$ is soluble in aqueous aerosols (*21*) and can participate in a number of oxidation reactions in the aerosols. The overall importance of the aqueous-phase processes compared to the gas-phase chemistry is uncertain.

The results of computer simulations can be used to estimate the degree of sensitivity required for measurement of the peroxy radicals in the troposphere. Madronich and Calvert (*22*) gave results of 5-day simulations for free tropospheric ("clean") and Amazon boundary layer ("moderately polluted") conditions (Figures 1 and 2, respectively). The solid and dotted lines show the simulations with and without reactions among the peroxy radicals

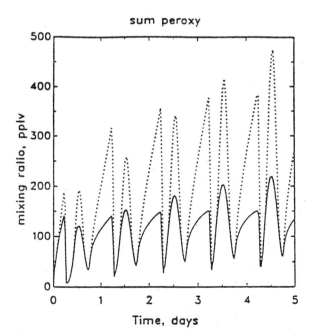

Figure 2. Organic peroxy radical concentrations as simulated for the moderately polluted Amazon boundary layer with (solid line) and without (dotted line) peroxy radical permutation reactions. (Reproduced with permission from reference 22. Copyright 1990 American Geophysical Union.)

themselves, respectively. For this discussion, we concentrate on the solid lines. The total midday maximum peroxy radical concentrations vary from about 7×10^8 molecules/cm^3 (28 parts per trillion (ppt) by volume mixing ratio) for the clean conditions to around 5×10^9 molecules/cm^3 (200 ppt) for the more polluted case. The HO$_2$ concentration is about one-third and one-fifth of the total peroxy radical concentration for the two cases, respectively. Models of stratospheric concentrations of HO$_2$ (*see* Figure 8, details discussed later) that use kinetics from the evaluations of Demore et al. (23) indicate concentrations of 3×10^6, 1×10^7, and 3×10^6 molecules/cm^3 for altitudes of 20, 40, and 60 km, respectively. Thus, very sensitive techniques will be required to measure the diurnal cycles and spatial variations of the peroxy radicals in the troposphere or in the stratosphere.

Measurement Methods

The discussion that follows is divided into two sections. The Spectroscopic Methods section includes those measurement techniques that involve the interaction of a photon with a peroxy radical. The Chemical Conversion Methods section describes the measurement of another molecule or radical to which a peroxy radical has been converted.

Spectroscopic Methods.   HO$_2$ and the other peroxy radicals have characteristic absorptions due to various molecular processes. In principle, these spectroscopic features could be used to determine atmospheric concentrations of peroxy radicals. The discussion of spectroscopic techniques in the measurement of peroxy radicals is divided into descriptions of specific spectral regions. General issues related to the use of spectroscopy for quantitative analysis are presented next.

The basis of absorption spectroscopy is straightforward. Radiation of the desired wavelength is passed through a cell (or atmospheric air mass) containing the gas of interest. The amount of light absorbed is determined by comparison with a reference measurement taken with the cell empty or with the gas of interest removed (more difficult for atmospheric sampling!). From the Beer–Lambert law relating the cell path length (l), the absorption cross section (σ) at wavelength λ, and the intensity of light for the reference (I_0) and the sample (I), the concentration (C) can be determined:

$$C = \frac{\ln (I_0/I)}{\sigma(\lambda) l} \tag{22}$$

This procedure is repeated for all the wavelengths of interest. For this approach to be effective, one must either be able to find a region of the spectrum that is free of interference due to absorption by other atmospheric molecules or one must be able to compensate for the interfering absorption by some method (spectral subtraction or spectral fitting, for example). The problem of spectral overlap with other molecules often limits the use of absorption spectroscopy in atmospheric measurements. For atmospheric measurements where measurement of I_0 can be difficult or impossible, differential absorption can be used. Here the difference in absorption between the peak and valley in a spectral feature is used, and this approach eliminates the requirement for an accurate reference spectrum. The degree of spectral overlap can also be minimized by narrowing the spectral features. This step is often accomplished through reducing the total pressure by pumping the sample to be measured into a measurement cell. At lower total pressures, not only are the line widths smaller, but the peak cross sections are also larger; thus, this approach yields a double benefit.

Thermal emission spectroscopy can be used in middle- and far-infrared spectral regions to make stratospheric measurements, and it has been applied to a number of important molecules with balloon-borne and satellite-based detection systems. In this approach, the molecules of interest are promoted to excited states through collisions with other molecules. The return to the ground state is accompanied by the release of a photon with energy equal to the difference between the quantum states of the molecule. Therefore, the emission spectrum is characteristic of a given molecule. Calculation of the concentration can be complicated because the emission may have originated from a number of stratospheric altitudes, and this situation may necessitate the use of computer-based inversion techniques (24–27) to retrieve a concentration profile.

Visible and ultraviolet fluorescence spectroscopy can also be used in certain instances. In this case the molecule is promoted to an excited state with a photon of energy that matches a transition. After a time, the molecule returns to the ground state, sometimes by emission of a photon that has been red-shifted (is of lower energy) from the wavelength of the exciting photon. This shift is due to the vibrational relaxation that occurs as the molecule loses energy in collisions, usually placing the molecule in the lowest vibrational level of an excited electronic state. Fluorescence spectroscopy can be very sensitive, but there can be problems with quenching and with differentiating photons scattered by air molecules and by aerosols from those photons that are due to fluorescence. Indeed, the net quantum yield for fluorescence can be a strong function of total pressure, and this observation has prompted research of fluorescence spectroscopy in cells at reduced pressure.

Ultraviolet Spectral Region. HO_2 and the other peroxy radicals have an absorption in the middle ultraviolet spectral region due to a transition of the nonbonding electron to a σ-antibonding orbital ($n-\sigma^*$). Unfortunately, these absorptions show few structural features, as shown in Figure 3. This lack of structure probably arises from the dissociative nature of the absorption

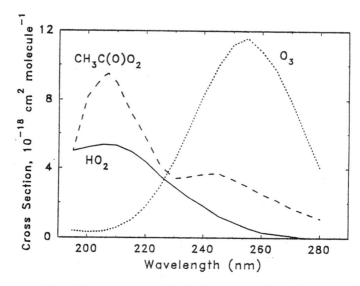

Figure 3. Ultraviolet cross sections for hydroperoxy and acetylperoxy radicals of Moortgat et al. (51) compared with cross sections for ozone of Molina and Molina (134).

process. From an atmospheric standpoint absorption due to ozone dominates the ultraviolet region from 200 to 300 nm. The problem is compounded by the fact that the concentration of ozone is typically 50 to 2000 times greater than the peroxy radical concentration, so even at the minimum of the ozone spectrum (210 nm) the absorption due to ozone would be 7 to 300 times greater than that due to HO_2. Other ultraviolet-absorbing compounds (e.g., various hydrocarbons) with unstructured spectra in this region make the use of absorption spectroscopy for the measurement of atmospheric peroxy radicals impractical. Ultraviolet absorption spectroscopy has been used successfully in laboratory studies (11, 28–57), where the reaction mixture can be controlled (i.e., in the absence of ozone) and I_0 can easily be measured.

Ultraviolet absorption spectroscopy has been applied to the measurement of some tropospheric molecules, including CH_2O, HO, O_3, NO_2, and HONO (58–60), in the spectral region from about 300 to 400 nm. These measurements are most reliable if the resolution of the instrument is less than or equal to the absorber line widths for the given experimental conditions. Figure 4 shows a section of the highly structured formaldehyde spectrum (61). This structure is crucial in order to unambiguously identify and quantify a tropospheric molecule. The potential problems of spectral overlap with interfering species can be ignored for the purpose of calculating a detection limit for HO_2 in an absorption experiment conducted under ideal conditions. In this case, for a 200-m optical path, a minimum measurable absorbance of 10^{-3}, and a wavelength of 210 nm, about 1×10^{10} molecules/ cm^3 of HO_2 could be detected. This number is approximately 5 times the

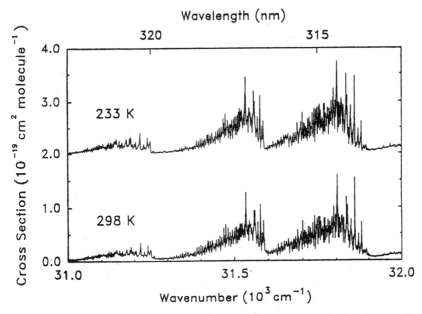

Figure 4. Ultraviolet cross sections for formaldehyde from the study of Cantrell et al. (61).

total peroxy radical concentration expected for a moderately polluted troposphere (*see* Figure 2). Researchers making the atmospheric measurements discussed for other important molecules have often used optical paths of 5 km or more and used instrumentation capable of measuring absorbances in the 10^{-4} range. Using a 5-km path length results in a calculated detection limit of 4×10^{8} molecules/cm^3 of HO$_2$, sufficiently low to make tropospheric measurements under moderately polluted conditions. In the use of ultraviolet absorption spectroscopy for the measurement of tropospheric HO$_2$, the problem of spectral interference (overlap) is of more concern than the relatively small strength of the absorbance. This spectral interference problem precludes the use of ultraviolet absorption for tropospheric HO$_2$ and RO$_2$ measurement.

Ultraviolet fluorescence spectroscopy as described earlier cannot be applied to peroxy radicals because the absorption leads almost exclusively to dissociation. Photodissociation has been exploited in the laboratory to measure HO$_2$ and CH$_3$O$_2$, because the fragments (HO, CH$_3$O, etc.) formed in the photodissociation can be found in electronically excited states, which can then emit measurable radiation. This so-called "photofragmentation" technique has been successfully applied to the measurement of a number of molecules in the atmosphere, including nitric acid (HNO$_3$), for which a detection limit of about 2.5×10^{9} molecules/cm^3 (0.1 ppb) has been reported (62). The use of photofragmentation to measure atmospheric peroxy radicals has not been reported.

Near-Infrared Spectral Region.   Absorptions due to the peroxy radicals do not occur in the visible spectral region. Peroxy radicals do have a weak near-infrared absorption due to a forbidden transition to a low-lying electronic state, which in HO_2 is designated $^2A' \leftarrow {}^2A''$. These absorptions have been observed by Hunziker and Wendt (63) for HO_2 and DO_2, as well as for CH_3O_2 and $CH_3CH_2O_2$ (64) (*see* Figure 5). The band-peak cross sections seem to be very weak ($\sigma \approx 10^{-20}$ cm^2/molecule), although the exact values are uncertain. The weakness of this absorption may preclude its use for atmospheric monitoring, although the recent technological improvements in near-infrared diode lasers may make absorption or fluorescence with near-infrared transitions viable options. The various organic peroxy radicals (e.g., CH_3O_2, $C_2H_5O_2$, etc.) may have slightly different near-infrared absorption features, and therefore speciation of various atmospheric peroxy radicals may also be possible.

If near-infrared diode lasers have low-noise characteristics similar to those of mid-infrared diode lasers, and thus minimum absorbances of 10^{-5} or less are possible, then an approximate detection limit can be calculated for an absorption experiment. For a 200-m optical path, the calculated detection limit is 5×10^{10} molecules/cm^3, which is well above levels of HO_2 expected to be found in the atmosphere. An absorption experiment in this spectral region apparently would require extremely long optical path lengths, and, indeed, a calculation with a 5-km path yields a calculated detection limit of 2×10^9 molecules/cm^3, still rather high for tropospheric measurements. Other issues associated with the use of diode lasers in absorption spectroscopy are discussed in the next section.

Middle-Infrared Spectral Region.   The middle-infrared spectral region shows several features due to peroxy radicals, known at least since 1963 (65–74). The ν_3 band of HO_2 is shown in Figure 6. Infrared absorption spectroscopy has been used successfully in laboratory measurements of HO_2 kinetics (68, 75–78). Diode-laser-based infrared absorption spectroscopy has also been applied to measurements of a number of trace gases in the atmosphere, including CH_4, NO, NO_2, N_2O, CH_2O, HNO_3, HCl, H_2O_2, O_3, H_2O, and CO_2 (79–83). This diode laser absorption technique has advantages over other infrared-based techniques because of the low noise levels associated with the laser diode as a light source, allowing absorptions of 10^{-5} or lower to be measured, and also because of the line narrowing (and concomitant peak cross section increase) associated with either pumping the air sample into a cell to a total pressure of a few torrs (millipascals) or making measurements in the stratosphere directly. This line narrowing leads to enhanced selectivity (and increased sensitivity) because of reduced overlap with possible interfering absorptions.

The infrared spectrum of HO_2 is highly structured, and this feature potentially allows the absorption due to HO_2 to be differentiated from other atmospheric species. Some infrared absorption data are also available for

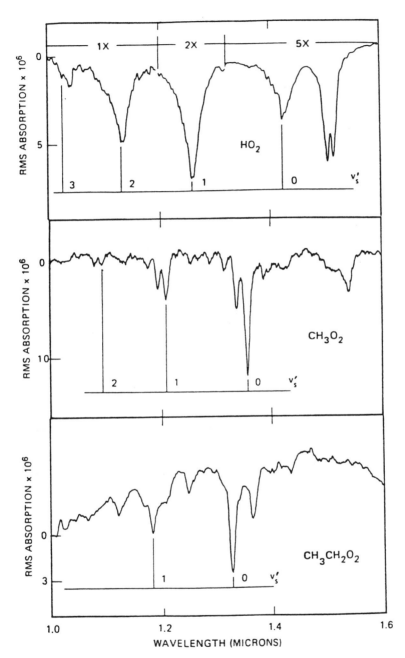

Figure 5. Near-infrared absorption spectra of HO₂ (top), CH₃O₂ (middle), and CH₃CH₂O₂ (bottom) from the study of Hunziker and Wendt (64). (Reproduced with permission from reference 64. Copyright 1976 American Institute of Physics.)

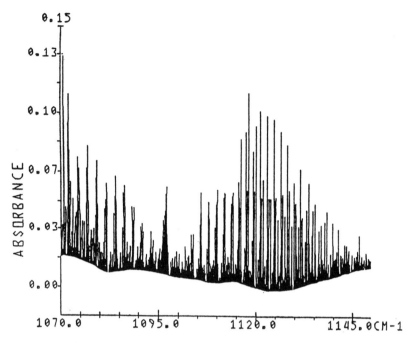

Figure 6. Infrared absorption spectrum of ν_3 band of HO_2 obtained by Burk-holder (74).

organic peroxy radicals (84). Zahniser et al. (73) reported line strengths for HO_2 for nearly coincident line pairs in the ν_2 band of HO_2 at 1371.927 and 1411.180 cm^{-1} and suggest they could possibly be used for laboratory and field measurements of HO_2. The line pairs have combined line intensities of 1.2 × 10^{-20} cm^2 molecule^{-1} cm^{-1}. A peak cross section of 5 × 10^{-18} cm^2/molecule is estimated for a Doppler-width line of this strength (the peak cross section could be smaller depending on the total pressure; this represents the best case). This estimate results in a calculated detection limit for HO_2 of 2.5 × 10^9 molecules/cm^3 (ambient concentration) in a diode laser absorption experiment with a minimum detectable absorbance of 10^{-5}, a cell pressure of 30 torr (3390 Pa), and a 200-m optical path length. This concentration is approximately what is expected for a midday maximum value in a moderately polluted atmosphere. Clearly, longer optical paths or lower absorbance limits or both would be required to measure atmospheric HO_2 by diode-laser-based infrared absorption spectroscopy with reasonable signal-to-noise ratios. The detection limit for a 5-km optical path is 1 × 10^8 molecules/cm^3.

Far-Infrared and Millimeter-Wave Spectral Regions. At least two measurements of HO_2 have been reported from the measurement of emission from thermally populated rotational levels in HO_2. Traub et al. (85) used a balloon-borne far-infrared spectrometer to measure HO_2 from an altitude of 19 to 49 km. A balloon or satellite platform is required for this approach, which is applied to stratospheric measurements because the troposphere is essentially opaque in the far-infrared region. They evaluated R-branch rotational lines at a resolution of 0.04 cm^{-1} from 142 to 147 cm^{-1} and compared each line to a synthetic spectrum generated from a layered model atmosphere (Figure 7). The amount of HO_2 in the model was adjusted until the least-squares difference between the model and the measurement was minimized. Estimated detection limits for this technique are about 1×10^7 molecules/cm^3 (6 ppt) near 20 km and about 5×10^5 molecules/cm^3 (30 ppt) near 50 km. The derived day and night HO_2 profiles were compared with results of a photochemical model based on kinetics from the evaluations of Demore et al. (23) (Figure 8). The daytime modeled profile agrees with measurements up to an altitude of 40 km, and is 30% below the measurements above 40 km. The nighttime measurements are much less than those from the daytime, as expected from theory, and are in general agreement with the model.

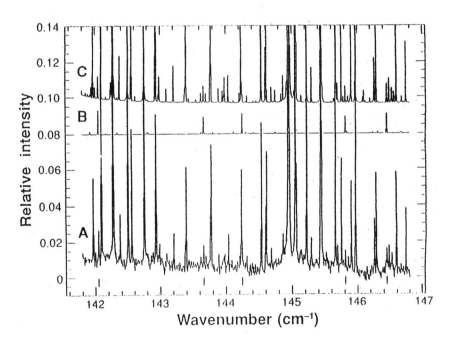

Figure 7. A, Atmospheric emission spectrum of region near that used by Traub et al. (85) to quantify stratospheric HO_2 concentrations. B, A laboratory HO_2 spectrum. C, The best-fit calculated spectrum. (Reproduced with permission from reference 85. Copyright 1990 American Association for the Advancement of Science.)

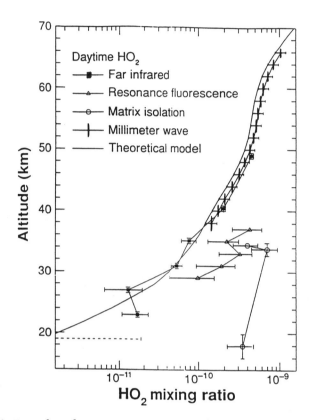

Figure 8. Far-infrared emission measurements of HO₂ of Traub et al. (85) (■), millimeter-wave emission measurements of de Zafra et al. (86) (+), resonance fluorescence measurements of Anderson et al. (107) (△), and matrix isolation– EPR measurements of Helten et al. (96) (○). (Reproduced with permission from reference 85. Copyright 1990 American Association for the Advancement of Science.)

de Zafra et al. *(86)* measured HO_2 by using ground-based millimeter-wave spectroscopy and three rotational emission lines near 265.8 GHz (~8.9 cm⁻¹). They compared their results to a photochemical model based on kinetics from the evaluations of Demore et al. *(23)* and found good agreement above 35 km. Traub et al. *(85)* compared their far-infrared measurements to the millimeter-wave measurements and to two in situ results, to be discussed in following sections. They found the millimeter-wave measurements to be about 18% larger than those of the theoretical profile from 37 to 67 km. The estimated accuracy of this technique is about ±25% throughout the range from 37 to 67 km. This accuracy corresponds to an uncertainty of about 4×10^6 molecules/cm³ at 37 km to 6×10^5 molecules/cm³ at 67 km. This method of HO_2 measurement is sensitive to the column abundance and the broad shape of the HO_2 distribution in the upper stratosphere and lower mesosphere, but not to the details of the distribution. As seen in Figure 8, the millimeter-wave results match the shape of the theoretical curve well.

No data have been presented to date on the utility of far-infrared or millimeter-wave spectroscopy for other peroxy radicals or on its possible use in the troposphere when combined with a low-pressure absorption cell.

Laser Magnetic Resonance and Electron Paramagnetic Resonance. This category in "spectroscopic" techniques of peroxy radical measurement is somewhat different from the others, in that the sample must be placed in a magnetic field in order for the absorptions to occur. Laser magnetic resonance of HO_2 is based on the absorption of far-infrared laser radiation (87). Typically, a gaseous sample is pumped through a section of the cavity of a continuous-wave gas laser operating at far-infrared wavelengths. A spectrum is obtained by scanning the magnetic field strength to bring the accidental near-resonance of the molecular transitions into resonance with a fixed wavelength laser line through the Zeeman effect. The amount of absorption is determined by monitoring the laser power with an infrared detector. This technique has been limited to laboratory studies of the spectroscopy and kinetics of HO_2 and other radicals (88–92) and may not be applicable to ambient measurements because of the mass and power required to generate the magnetic field. The detection limit reported for the Howard and Evenson study (88) was 2×10^9 molecules/cm^3, although they point out that this value is dependent on the experimental conditions, such as the gas pressure and the magnetic modulation amplitude.

The problem of bringing a large magnet into the field for ambient measurements has been overcome in electron paramagnetic resonance (EPR, also called electron spin resonance, ESR) by Mihelcic, Helten, and coworkers (93–99). They combined EPR with a matrix isolation technique to allow the sampling and radical quantification to occur in separate steps. The matrix isolation is also required in this case because EPR is not sensitive enough to measure peroxy radicals directly in the atmosphere. EPR spectroscopy has also been used in laboratory studies of peroxy radical reactions (100, 101).

EPR is based on the splitting of magnetic energy levels caused by the action of a magnetic field on an unpaired electron in a molecule. Typically the magnetic field strength is on the order of 3500 G, and sweep coils allow the field to be varied over a small range so the radiation can be made resonant with the transition. The radiation is in the microwave region at about 9.5 GHz (0.3 cm^{-1}). Other techniques such as magnetic field modulation with lock-in detection improve the signal-to-noise ratio of these measurements. Because of the interaction of the electron spin with nearby nuclear spins, hyperfine splitting patterns in the spectra allow differentiation between the various radicals present. This property has been exploited in the most recent of the atmospheric measurements of Mihelcic et al. (99).

The matrix isolation procedure relies on the condensation of H_2O and CO_2 on a liquid-nitrogen-cooled cold finger to form a stable matrix for radicals and other atmospheric species (Figure 9). Typically about 20 L of air are required to achieve the desired sensitivity for ambient measurements. A matrix of deuterated water (D_2O) narrowed the EPR line widths and improved the signal-to-noise ratio, and thus this matrix has been used for measurements since October 1982. Recent improvements in the use of this matrix isolation–EPR technique have been in the analysis of the spectra.

The analysis of the spectrum is a multiple-step process. The first step is the removal of the contribution to the spectrum due to the sampling apparatus. The next step is subtraction of the relatively large contribution due to NO_2. The remaining absorptions are due to HO_2, other organic peroxy radicals, and possibly other unknown radicals. A numerical fitting procedure

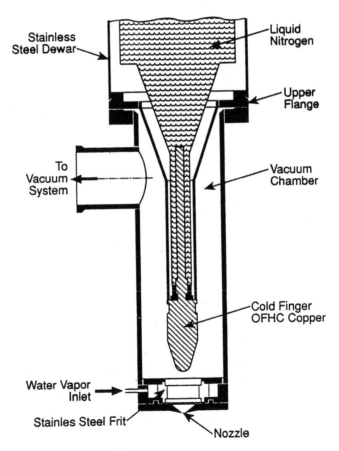

Figure 9. Cold finger sampling apparatus used in matrix isolation–EPR measurements of Mihelcic et al. (97). (Reproduced with permission from reference 97. Copyright 1985 Kluwer Academic.)

that has been developed successfully quantifies the amounts of the HO_2, $CH_3C(O)O_2$, and the sum of the other RO_2 radicals (98). This sequence is shown graphically in Figure 10. The early calibrations of this method were performed by using the thermal decomposition of peroxyacetyl nitrate (PAN), which generates peroxyacetyl radicals, although calibration of other peroxy radicals (HO_2, CH_3O_2, etc.) has since been performed. The PAN decomposition reaction yields one peroxy radical and one NO_2 molecule that is also measured, and thus results, in principle, in an absolute calibration.

The matrix isolation–EPR technique (MIESR) has proven to be a useful method of peroxy radical measurement in the atmosphere. The matrix isolation results shown in Figure 8 are from measurements reported in 1984 (96), and significant improvements have since been made in this technique, although new concentration profiles such as those in Figure 8 have not yet been reported. The drawbacks to this method are a moderate detection limit (a few parts per trillion by volume mixing ratio, approximately 10^8 molecules/ cm^3), which limits its usefulness in remote measurement situations; the requirement of removing the absorption due to NO_2, which limits measurements to situations with relatively low NO_2 levels; and the fairly long integration times required (of the order of 0.5 to 2 h).

The technique of spin-trapping radicals has been applied to the measurement of atmospheric hydroxyl by Watanabe et al. (102), although there are no reports of its use for peroxy radicals. The principle involves the reaction of the radical of interest with an organic nitrone immobilized on a filter paper or other substrate. The sample is returned to the laboratory, and the nitrone–radical product is dissolved in a suitable solvent and measured with EPR. The disadvantages of the spin-trapping technique are difficulty in finding suitable organic nitrone compounds and the fact that most of these molecules are photochemically unstable.

Mass Spectrometry. Mass spectrometric detection was used in early laboratory studies of HO_2 (103, 104) and has also been used in more recent investigations (105). The HO_2 peak at the mass-to-charge ratio $m/e = 33$ is useful in laboratory identification and quantification, but in the atmosphere, most likely multiple species will interfere because of, for example, fragmentation of hydrocarbons, hydrogen peroxide, or oxygen isotopes.

Chemical Conversion Methods. *Laser-Induced and Resonance Fluorescence of HO.* Considerable effort has been applied to the measurement of HO in the stratosphere and troposphere. Ultraviolet fluorescence techniques based on lasers or resonance lamps have received a great deal of attention and study. Because HO concentrations are typically factors of one-tenth to one-hundredth those of HO_2 in the atmosphere, the difficulties associated with making HO measurements by using fluorescence [low signal-to-noise ratio, laser-generated HO, background fluorescence, etc.; *see* the

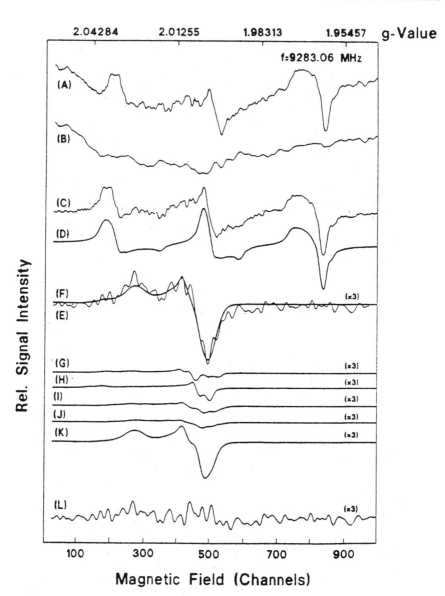

Figure 10. Illustration of spectral analysis from the matrix isolation–EPR measurements of Mihelcic et al. (99). The curves correspond to A, an ESR spectrum of a sample collected on July 5, 1986, from 11:39 to 12:09 (CET); B, spectrum of sample holder; C, difference of A and B; D, NO_2 reference spectrum; E, Difference of C and D magnified by a factor of 3; F, sum of HO_2 (13 ppt), $CH_3C(O)O_2$ (16 ppt), CH_3O_2 (15 ppt), $C_2H_5O_2$ (11 ppt), and $C_4H_9O_2$ (60 ppt) as retrieved by the fit (magnified by a factor of 3); G through K, amounts of HO_2, $CH_3C(O)O_2$, CH_3O_2, $C_2H_5O_2$, and $C_4H_9O_2$, all magnified by a factor of 3, as retrieved by the multiple fit; and L, residuals after subtraction of F from E. (Reproduced with permission from reference 99. Copyright 1990 Kluwer Academic.)

discussion of Smith and Crosley (*106*)] become much less important when a method of converting HO_2 to HO quantitatively is found. The reaction of HO_2 with NO is one means of accomplishing this conversion.

$$HO_2 + NO \longrightarrow HO + NO_2 \tag{23}$$

Anderson et al. (*107*) reported a series of HO_2 measurements in the stratosphere between 29 and 37 km; they induced reaction 23 and recorded resonance fluorescence detection of HO at 309 nm. The exciting radiation for this measurement was generated by microwave discharge of a He–H_2O mixture. The measured mixing ratios were compared to the photochemical model of Logan et al. (*1*). The mean HO_2 levels were systematically higher than those of the model, although the range of values measured overlapped the model results. A comparison with computer simulations is also shown in Figure 8 [model results reported by Traub et al. (*85*)], again indicating measured values higher than those of the model. One aspect of this technique, which was apparently not recognized in the original paper, is that methylperoxy radicals are also converted to HO by the following sequence along with reaction 24, although reaction 25 may be sufficiently slow under stratospheric conditions (low temperature and low O_2 concentration) that its contribution is negligible.

$$CH_3O_2 + NO \longrightarrow CH_3O + NO_2 \tag{24}$$

$$CH_3O + O_2 \longrightarrow HO_2 + CH_2O \tag{25}$$

The use of laser-induced fluorescence (LIF) for tropospheric HO and HO_2 measurements was reported by Hard and co-workers (*108–110*), who developed a fluorescence technique based on pumping the air sample into a low-pressure cell (FAGE) and exciting it with a copper vapor laser-pumped dye laser with a high repetition rate. Their HO_2 measurements were not made in conjunction with enough other supporting measurements to allow an accurate test of photochemical models from the results.

Photofragment-Induced Emission. Lee and co-workers at San Diego State University (*111–113*) used the photodissociation of HO_2 at 147 nm to produce excited HO radicals. The emission of radiation from HO has been used to quantify the HO_2 concentration in laboratory kinetics experiments:

$$HO_2 + h\nu \longrightarrow HO\,(A^2\Sigma^+) + O \tag{26}$$

$$HO\,(A^2\Sigma^+) \longrightarrow HO\,(X^2\Pi) + h\nu\ (306\text{–}320\ nm) \tag{27}$$

A similar approach was described by Hartmann et al. (*114*) for methylperoxy radicals in laboratory photodissociation experiments. The excimer laser photolysis at 248 nm produces HO $(A^2\Sigma^+)$, which emits radiation as shown in reaction 27. These photofragmentation techniques may not be applicable to atmospheric measurements because of the large number of possible interferents at the photolyzing wavelengths, although such interferents have not yet been demonstrated.

Near-Infrared Chemiluminescence. The technique of chemiluminescence has been most successfully applied in the atmosphere to the measurement of NO and other oxides of nitrogen through various conversion procedures. Glaschick-Schimpf et al. (*115*) and Holstein et al. (*116*) reported results of laboratory kinetic studies in which the HO_2 concentration was determined through the chemiluminescence reaction system shown in equations 28 and 29. It involves the same molecular transitions in the near-infrared as discussed previously (for HO_2: emission from the $^2A'$ state).

$$HO_2 + O_2\,(^1\Delta) \longrightarrow HO_2\,(A^2A') + O_2 \tag{28}$$

$$HO_2\,(A^2A') \longrightarrow HO_2\,(A^2X'') + h\nu\ (1.43\ \mu m) \tag{29}$$

Thus, in a constant concentration of $O_2\,(^1\Delta)$ that is much greater that the concentration of HO_2, the radiation intensity at 1.43 μm is proportional to the HO_2 concentration. This information has not been applied to atmospheric measurements or to other peroxy radicals to date.

Chemical Amplification. The measurement of a small electrical signal is often accomplished by amplification to a larger, more easily measured one. This technique of amplification can also be applied to chemical systems. For peroxy radicals, Cantrell and Stedman (*117*) proposed, as a "possible" technique, the chemical conversion of peroxy radicals to NO_2 with amplification (i.e., more than one NO_2 per peroxy radical). This method has also been used for laboratory studies of HO_2 reactions on aqueous aerosols (*21*). The following chemical scheme was proposed as the basis of the instrument:

$$HO_2 + NO \longrightarrow HO + NO_2 \tag{23}$$

$$HO + CO \longrightarrow H + CO_2 \tag{1}$$

This chain reaction converts NO and CO to NO_2 and CO_2 at a rate proportional to the sum of the HO_2 and HO concentrations. The technique also measures certain organic peroxy radicals, because they are converted to HO_2 after a few steps. Peroxy radicals are converted to HO_2 according to the following reactions for primary peroxy radicals:

$$RCH_2O_2 + NO \longrightarrow RCH_2O + NO_2 \tag{30}$$

$$RCH_2O + O_2 \longrightarrow RCHO + HO_2 \tag{31}$$

and for secondary peroxy radicals:

$$RR'CHO_2 + NO \longrightarrow RR'CHO + NO_2 \qquad (32)$$

$$RR'CHO + O_2 \longrightarrow RR'CO + HO_2 \qquad (33)$$

Tertiary peroxy radicals have no alpha hydrogen that can be abstracted by O_2 in the second step. However, a unimolecular decomposition of the alkoxy radical results in the formation of a peroxy radical, which can be measured.

$$RR'R''CO_2 + NO \longrightarrow RR'R''CO + NO_2 \qquad (34)$$

$$RR'R''CO \longrightarrow R + R'R''CO \qquad (35)$$

$$R + O_2 \xrightarrow{M} RO_2 \qquad (3)$$

Aryl peroxy radicals may not be as effectively converted because the analogous unimolecular decomposition reaction may not occur. Acyl peroxy radicals are converted to HO_2 in a similar fashion, shown for acetylperoxy radicals:

$$CH_3C(O)O_2 + NO \longrightarrow CH_3C(O)O + NO_2 \qquad (36)$$

$$CH_3C(O)O \longrightarrow CH_3 + CO_2 \qquad (37)$$

$$CH_3 + O_2 \xrightarrow{M} CH_3O_2 \qquad (38)$$

The methyl peroxy radical that is formed converts to HO_2 in the sequence for primary peroxy radicals shown in equations 30 and 31. The chain reaction is subject to termination steps that include the following:

$$OH + NO \xrightarrow{M} HONO \qquad (39)$$

$$HO_2 + wall \longrightarrow nonradical\ products \qquad (40)$$

$$HO_2 + HO_2 \longrightarrow H_2O_2 + O_2 \qquad (20)$$

$$HO_2 + NO_2 \xrightarrow{M} HO_2NO_2 \qquad (41)$$

For most conditions employed in the current instrument, the chain is terminated by reactions 39 and 40, although reaction 41 can be important for certain circumstances. For optimum conditions of about 3 ppm by volume NO, 10% v/v CO, and 5 s of reaction time, chain lengths (amplification factors) of 500 to 1000 are possible (*118–122*).

The measurement of NO_2 (typically ppb by volume levels) in the presence of large NO and CO concentrations is the greatest analytical challenge in the use of this technique. Early attempts involved the use of optoacoustic spectroscopy in which a chopped 488-nm argon ion laser beam induces a pressure pulse proportional in strength to the NO_2 concentration (*see* references 123–125). However, this approach was not sensitive enough for tropospheric peroxy radical measurements. The measurements reported in the literature are all based on the measurement of NO_2 with luminol chemiluminescence (*126, 127*). Although the presence of NO and CO does affect the strength of the chemiluminescent signal, NO_2 calibration can be performed in their presence. The luminol instrument uses an aqueous, pH 12.5, 0.01 M Na_2SO_3, 1.0×10^{-4} M luminol solution that is flowed over a disk of filter paper, and the emission from the excited 3-aminophthalate molecules produced in the reaction is viewed through a glass window by an end-on photomultiplier tube. Luminol can be oxidized by molecules other than NO_2, including O_3, PAN (peroxyacetyl nitrate) (*127, 128*), and H_2O_2 (*129*). These interferences are not a problem in the measurement of peroxy radicals because of the modulation approach, which is described in the following discussion.

Other molecules in the atmosphere oxidize NO to NO_2 aside from peroxy radicals, ozone being the most abundant. Therefore, a large background is present from the NO_2 produced because of the NO–ozone reaction as well as from ambient NO_2. This background must be periodically measured to ensure accurate determination of the peroxy radical signal. Typically, one-third to one-half of the time N_2 is substituted for CO, and all those species that do not participate in the chain reaction are measured. Thus, the peroxy radical measurement is the difference between this background and the signal with CO present.

A schematic diagram of the chemical amplifier system is shown in Figure 11. Typical raw signals are shown in Figure 12. Ambient NO_2 modulation data from the 1988 Scotia Range field study (a rural site near Scotia, Pennsylvania) (*130*) are shown in Figure 13. Absolute radical calibration is an area of research that will help bring the chemical amplifier technique to fruition. The reports in the literature have used the second-order decay of HO_2 (reaction 16) in the laboratory to determine the peroxy radical sensitivity (*119–122*). A full-field calibration procedure currently under investigation utilizes a titration procedure to determine the radical concentration in a synthetic radical source. Detection limits for a 1-min averaging time with a stable background signal are of the order of 1×10^8 molecules/cm^3 with the current system.

Missing Oxidant from Photostationary State Measurements.

As was discussed in the introduction, the photolysis of NO_2, the reaction of NO with O_3, and the combination of O atoms with O_2 form the photostationary state system. If differential rate equations are written and solved assuming

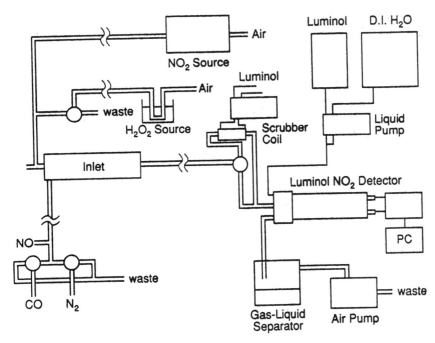

Figure 11. Schematic diagram of chemical amplifier system of Cantrell et al.
(130).

a steady-state concentration for either NO or NO_2, a relation is derived that relates the concentrations of NO, NO_2, O_3, and two kinetic parameters, namely j_{NO_2} and k_{17}:

$$\frac{j_{NO_2}[NO_2]}{k_{17}[NO]} = [O_3] \tag{42}$$

When atmospheric measurements are performed for [NO], $[NO_2]$, $[O_3]$, j_{NO_2}, and temperature (to determine k_{17}), the ratio on the left of equation 42 is typically larger than the ozone concentration. In other words, the $[NO_2]$/ [NO] ratio is larger than would be expected on the basis of the other values in the equation. This finding can be interpreted in terms of reaction 16a, which is the oxidation of NO to NO_2 by a peroxy radical. If reaction 16a is included in the photostationary state system, the five measurements mentioned yield an estimate of the total peroxy radical concentration (weighted inversely by the rate coefficient for the reaction with NO).

$$[RO_2] = \frac{k_{17}}{k_{16a}} \left(\frac{j_{NO_2}[NO_2]}{k_{17}[NO]} - [O_3] \right) \tag{43}$$

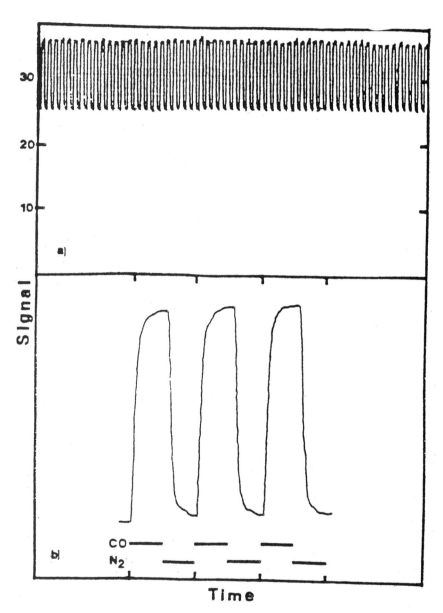

Figure 12. Sample modulation data of chemical amplifier of Cantrell et al. (130), showing stability over a 1-h time period (top) and the time dependence on a shorter time scale (3 min).

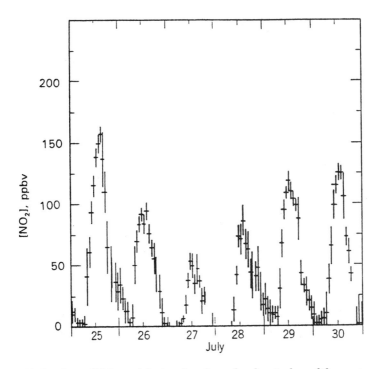

Figure 13. Six days of NO_2 modulation data from the chemical amplifier system from the Scotia Range, Pennsylvania, study, summer 1988 (130).

The term in large brackets yields the so-called "missing oxidant" in units of ozone concentration, and the rate constant ratio (k_{17}/k_{16a}) converts to units of peroxy radical concentration. Near room temperature, k_{17}/k_{16a} is about 1/500 for HO_2 and CH_3O_2, a fact indicating the much greater efficiency of peroxy radicals in oxidizing NO, per unit concentration, as compared to O_3. Because the calculation may involve a relatively small difference between two larger values (equation 43), highly precise and accurate concentration and photolytic rate data are required. Several photostationary state measurements have been reported (131–133). As an example, missing oxidant concentrations calculated by using data from the Scotia Range field study of 1988 (130) are shown in Figure 14. They can be compared with RO_2 measurements made by using the chemical amplifier (discussed in a preceding section) during the same time period (Figure 15). Within the scatter of the measurements, the agreement is good. This result demonstrates that the value of peroxy radical measurements is enhanced by comparison with other pertinent species, although clearly the missing oxidant method for the determination of peroxy radical concentrations is rather indirect and only usable in daylight hours. The detection limit for this method depends on a variety

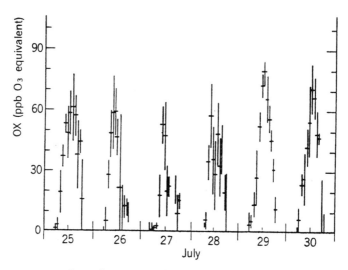

Figure 14. Six days of missing oxidant calculations from measurements at the Scotia Range, Pennsylvania, study, summer 1988 (130).

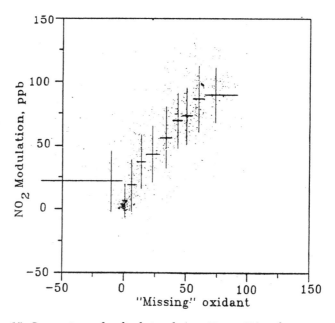

Figure 15. Comparison of radical signals (see Figure 13) and missing oxidant values from the Scotia Range, Pennsylvania, study, summer 1988 (130).

Table I. Summary of Peroxy Radical Measurement Methods

Technique	Averaging Time	Detection Limit[a]	L or A[b]	S or T[c]	Comments
Spectroscopic methods					
UV absorption	minutes	4×10^6	L	—[d]	O_3 interference
Near IR absorption	minutes	2×10^9	L	—	cross section uncertain
Middle IR absorption	minutes	1×10^6	L	—	diode laser possible
Far IR absorption	—	—	—	—	not tested
Millimeter-wave absorption	—	—	—	—	not tested
UV fluorescence	—	—	—	—	not possible—photodissociation
Near IR fluorescence	—	—	—	—	not tested
Middle IR emission	—	—	—	—	not tested for peroxy radicals
Far IR emission	0.6–0.8 h	0.05×10^7–1×10^7	A	S	proven for upper atmosphere
Millimeter-wave emission	—	0.6×10^6–4×10^6	A	S	proven for upper atmosphere
Laser magnetic resonance	minutes	2×10^9	L	—	large magnet
Matrix isolation–ESR	0.5–2.0 h	1×10^6	LA	ST	NO_2 interference, HO_2 and RO_2
Spin trapping–ESR	—	—	A	—	used for HO only
Mass spectroscopy	—	—	L	—	possible interference problems
Chemical conversion methods					
HO resonance fluoresence	—	8×10^7	A	S	higher values than spectral
HO laser-induced fluorescence	0.1–1.0 h	1×10^6	A	T	promising
Photofragment spectroscopy	—	2.5×10^9	LA	—	not tested in atmosphere for RO_2
Near IR chemiluminescence	—	—	L	—	sensitivity questions
Chemical amplification	1–30 min	1×10^6	LA	T	total RO_2, calibration problems
Missing oxidant calculation	1–30 min	2×10^9	A	T	daytime only, total RO_2

[a]This number describes minimum detectable concentration in molecules per cubic centimeter; conditions are discussed in the text.

[b]Code describes whether the technique has been used in the laboratory (L), the atmosphere (A), or both (LA).

[c]Code describes whether atmospheric measurements were performed in the stratosphere (S), troposphere (T), or both (ST).

[d]A dash indicates that insufficient information is known about a system to provide an entry in this column or that known problems preclude its use.

of factors, including the stability of the photolysis rate and species concentrations, but an estimated detection limit of 2×10^8 molecules/cm^3 is reasonable.

Summary

Several potential peroxy radical measurement techniques exist in the realm of atmospheric chemistry studies, although most have been used only in the laboratory. The techniques are summarized in Table I. Possibly, some laboratory methods could be applied to atmospheric measurements. The database for ambient peroxy radical concentrations in the troposphere and stratosphere is meager. Much of the available stratospheric data yield concentrations of HO_2 higher than those calculated with computer models. The reasons for this systematic difference are not known. In the troposphere, more measurements are called for in conjunction with other related species such as ozone, NO_x, NO_y, j_{NO_2}, and j_{O_3}. It will also be appropriate to develop multiple methods, and, when they have reached maturity, to perform intercomparison studies.

Acknowledgments

Thanks to Lenny Newman for the invitation to participate in the Measurement Challenges in Atmospheric Chemistry symposium at the American Chemical Society meeting in Boston, Massachusetts, April 1990. Thanks also to Fred Fehsenfeld, Dave Parrish, and Marty Buhr of the National Oceanic and Aviation Administration Aeronomy Lab for renewing our interest in the chemical amplifier technique in the late 1980s. Finally, a big thank you to Geoffrey Tyndall, John Lind, John Orlando, Selena Slyter, and Steve Massie for assistance in the preparation of this manuscript.

References

1. Logan, J. A.; Prather, M. J.; Wofsy, S. C.; McElroy, M. B. *J. Geophys. Res.* **1979**, *86*, 7210–7254.
2. Fishman, J.; Carney, T. A. *J. Atmos. Chem.* **1984**, *1*, 351–376.
3. Ko, M. K. W.; Dak Sze, N. *J. Geophys. Res.* **1984**, *89*, 11619–11632.
4. Calvert, J. G.; Lazrus, A.; Kok, G. L.; Heikes, B. G.; Walega, J. G.; Lind, J.; Cantrell, C. A. *Nature (London)* **1985**, *317*, 27–35.
5. Lurmann, F. W.; Lloyd, A. C.; Atkinson, R. *J. Geophys. Res.* **1986**, *91*, 10905–10936.
6. Seigneur, C.; Wegrecki, A. M. *Atmos. Environ.* **1990**, *24A*(5), 989–1006.
7. Liu, S. C.; Trainer, M. *J. Atmos. Chem.* **1988**, *6*, 221–233.
8. Thompson, A. M.; Huntley, M. A.; Stewart, R. W. *J. Geophys. Res.* **1990**, *95*, 9829–9844.
9. Calvert, J. G.; Madronich, S. *J. Geophys. Res.* **1987**, *92*, 2211–2220.

10. Calvert, J. G. *Proceedings of the NATO Advanced Study Institute on Atmospheric Ozone: Its Variation and Human Influences*; U.S. Department of Transportation: Federal Aviation Administration: Washington, DC, 1979; pp 153–190; Report No. FAA–EE–80–20.
11. Moortgat, G. K.; Cox, R. A.; Schuster, G.; Burrows, J. P.; Tyndall, G. S. *J. Chem. Soc., Faraday Trans. 2* **1989**, *85*, 809–829.
12. Atkinson, R.; Plum, C. N.; Carter, W. P. L.; Winer, A. M.; Pitts, J. N., Jr. *J. Phys. Chem.* **1984**, *88*, 1210–1215.
13. Cantrell, C. A.; Stockwell, W. R.; Anderson, L. G.; Busarow, K. L.; Perner, D.; Schmeltekopf, A.; Calvert, J. G.; Johnston, H. S. *J. Phys. Chem.* **1985**, *89*, 139–146.
14. Cantrell, C. A.; Davidson, J. A.; Busarow, K. L.; Calvert, J. G. *J. Geophys. Res.* **1986**, *91*, 5347–5353.
15. Su, F.; Calvert, J. G.; Shaw, J. H. *J. Phys. Chem.* **1980**, *84*, 239–246.
16. Kan, C. S.; Su, F.; Calvert, J. G.; Shaw, J. H. *J. Phys. Chem.* **1981**, *85*, 2359–2363.
17. Atkinson, R.; Aschmann, S. M.; Carter, W. P. L.; Winer, A. M.; Pitts, J. N., Jr. *J. Phys. Chem.* **1982**, *86*, 4563–4569.
18. Roberts, J. M. *Atmos. Environ.* **1990**, *24A*(2), 243–287.
19. Buhr, M. P.; Parrish, D. D.; Norton, R. B.; Fehsenfeld, F. C.; Sievers, R. E.; Roberts, J. M. *J. Geophys. Res.* **1990**, *95*, 9809–9816.
20. Calvert, J. G.; Stockwell, W. R. *Can. J. Chem.* **1983**, *61*, 983–992.
21. Mozurkewich, M.; McMurry, P. A.; Gupta, A.; Calvert, J. G. *J. Geophys. Res.* **1987**, *92*, 4163–4170.
22. Madronich, S.; Calvert, J. G. *J. Geophys. Res.* **1990**, *95*, 5697–5715.
23. Demore, W. B.; Molina, M. J.; Sander, S. P.; Golden, D. M.; Hampson, R. F.; Kurylo, M. J.; Howard, C. J.; Ravishankara, A. R. *Chemical Kinetics and Photochemical Data for Use in Stratospheric Modeling*; Jet Propulsion Laboratory: Pasadena, CA, 1987; JPL Publication 87–41.
24. Gille, J. C.; House, F. B. *J. Atmos. Sci.* **1971**, *28*, 1427–1438.
25. Abbas, M. M.; Glenn, M. J.; Kunde, V. G.; Brasunas, V.; Conrath, B. J.; Maguire, W. C.; Herman, J. R. *J. Geophys. Res.* **1987**, *92*, 8343–8353.
26. Massie, S. T. et al. *J. Geophys. Res.* **1987**, *92*, 14806–14814.
27. Brasunas, J. C.; Kunde, V. G.; Herath, L. W. *Appl. Opt.* **1988**, *27*, 4964–4976.
28. Hochanadel, C. J.; Ghormley, J. A.; Ogren, P. J. *J. Chem. Phys.* **1972**, *56*, 4426–4433.
29. Parkes, D. A.; Paul, D. M.; Quinn, C. P.; Robson, R. C. *Chem. Phys. Lett.* **1973**, *23*, 425–429.
30. Parkes, D. A. *Int. J. Chem. Kinet.* **1977**, *9*, 451–469.
31. Hochanadel, C. J.; Ghormley, J. A.; Boyle, J. W.; Ogren, P. J. *J. Phys. Chem.* **1977**, *81*, 3–7.
32. Anastasi, C.; Smith, I. W. M.; Parkes, D. A. *J. Chem. Soc., Faraday Trans. 1* **1978**, *74*, 1693–1701.
33. Kan, C. S.; McQuigg, R. D.; Whitbeck, M. R.; Calvert, J. G. *Int. J. Chem. Kinet.* **1979**, *11*, 921–933.
34. Cox, R. A.; Tyndall, G. S. *Chem. Phys. Lett.* **1979**, *65*, 357–360.
35. Sanhueza, E.; Simonaitis, R.; Heicklen, J. *Int. J. Chem. Kinet.* **1979**, *11*, 907–914.
36. Cox, R. A.; Tyndall, G. S. *J. Chem. Soc., Faraday Trans. 2* **1980**, *76*, 153–163.
37. Adachi, H.; Basco, N.; James, D. G. L. *Int. J. Chem. Kinet.* **1980**, *12*, 949–977.

38. Addison, M. C.; Burrows, J. P.; Cox, R. A.; Patrick, R. *Chem. Phys. Lett.* **1980**, *73*, 283–287.
39. Sander, S. P.; Watson, R. T. *J. Phys. Chem.* **1980**, *84*, 1664–1674.
40. Sander, S. P.; Watson, R. T. *J. Phys. Chem.* **1981**, *85*, 2960–2964.
41. Basco, N.; Parmar, S. S. *Int. J. Chem. Kinet.* **1985**, *17*, 891–900.
42. Pilling, M. J.; Smith, M. J. C. *J. Phys. Chem.* **1985**, *89*, 4713–4720.
43. Kurylo, M. J.; Ouellette, P. A.; Laufer, A. H. *J. Phys. Chem.* **1986**, *90*, 437–440.
44. McAdam, K.; Veyret, B.; Lesclaux, R. *Chem. Phys. Lett.* **1987**, *133*, 39–44.
45. Kurylo, M. J.; Wallington, T. J.; Ouellette, P. A. *J. Photochem.* **1987**, *39*, 201–213.
46. Kurylo, M. J.; Dagaut, P.; Wallington, T. J.; Neuman, D. M. *Chem. Phys. Lett.* **1987**, *139*, 513–518.
47. Kurylo, M. J.; Wallington, T. J. *Chem. Phys. Lett.* **1987**, *138*, 543–547.
48. Jenkin, M. E.; Cox, R. A.; Hayman, G. D.; Whyte, L. J. *J. Chem. Soc., Faraday Trans. 2* **1988**, *84*, 913–930.
49. Wallington, T. J.; Dagaut, P.; Kurylo, M. J. *J. Photochem. Photobiol., A* **1988**, *42*, 173–185.
50. Dagaut, P.; Wallington, T. J.; Kurylo, M. J. *J. Photochem. Photobiol., A* **1989**, *48*, 187–198.
51. Moortgat, G. K.; Veyret, B.; Lesclaux, R. *Chem. Phys. Lett.* **1989**, *160*, 443–447.
52. Moortgat, G. K.; Veyret, B.; Lesclaux, R. *J. Phys. Chem.* **1989**, *93*, 2362–2368.
53. Dagaut, P.; Kurylo, M. J. *J. Photochem. Photobiol., A* **1990**, *51*, 133–140.
54. Lightfoot, P. D.; Roussel, P.; Veyret, B.; Lesclaux, R. *J. Chem. Soc., Faraday Trans.* **1990**, *86*, 2927–2936.
55. Lightfoot, P. D.; Lesclaux, R.; Veyret, B. *J. Phys. Chem.* **1990**, *94*, 700–707.
56. Lightfoot, P. D.; Veyret, B.; Lesclaux, R. *J. Phys. Chem.* **1990**, *94*, 708–714.
57. Simon, F.-G.; Schneider, W.; Moortgat, G. K. *Int. J. Chem. Kinet.* **1990**, *22*, 791–813.
58. Platt, U.; Perner, D.; Pätz, H. W. *J. Geophys. Res.* **1979**, *84*, 6329–6335.
59. Platt, U.; Perner, D. *J. Geophys. Res.* **1980**, *85*, 7453–7458.
60. Hübler, G.; Perner, D.; Platt, U.; Tönnissen, A.; Ehhalt, D. *J. Geophys. Res.* **1984**, *89*, 1309–1319.
61. Cantrell, C. A.; Davidson, J. A.; McDaniel, A. H.; Shetter, R. E.; Calvert, J. G. *J. Phys. Chem.* **1990**, *94*, 3902–3908.
62. Papenbrock, Th.; Stuhl, F. *J. Atmos. Chem.* **1990**, *10*, 451–469.
63. Hunziker, H. E.; Wendt, H. R. *J. Chem. Phys.* **1974**, *60*, 4622–4625.
64. Hunziker, H. E.; Wendt, H. R. *J. Chem. Phys.* **1976**, *64*, 3488–3490.
65. Milligan, D. E.; Jacox, M. E. *J. Chem. Phys.* **1963**, *38*, 2627–2631.
66. Ogilvie, J. F. *Spectrochim. Acta* **1967**, *23A*, 737–750.
67. Jacox, M. E.; Milligan, D. E. *J. Mol. Spectrosc.* **1972**, *42*, 495–513.
68. Paukert, T. T.; Johnston, H. S. *J. Chem. Phys.* **1972**, *56*, 2824–2838.
69. Nagai, K.; Endo, Y.; Hirota, E. *J. Mol. Spectrosc.* **1981**, *89*, 520–527.
70. Yamada, C.; Endo, Y.; Hirota, E. *J. Chem. Phys.* **1983**, *78*, 4379–4384.
71. Buchanan, J. W.; Thrush, B. A.; Tyndall, G. S. *Chem. Phys. Lett.* **1983**, *105*, 167–168.
72. Zahniser, M. S.; Stanton, A. C. *J. Chem. Phys.* **1984**, *80*, 4951–4960.
73. Zahniser, M. S.; McCurdy, K. E.; Stanton, A. C. *J. Phys. Chem.* **1989**, *93*, 1065–1070.
74. Burkholder, J., NOAA Aeronomy Laboratory, Boulder, CO, unpublished results.
75. Thrush, B. A.; Tyndall, G. S. *J. Chem. Soc., Faraday Trans. 2* **1982**, *78*, 1469–1475.

76. Thrush, B. A.; Tyndall, G. S. *Chem. Phys. Lett.* **1982**, *92*, 232–235.
77. Worsnop, D. R.; Zahniser, M. S.; Kolb, C. E.; Sharfman, L. R.; Gardner, J. A.; Davidovits, P. *EOS Trans., Am. Geophys. Union* **1987**, *68*, 270.
78. Martin, N. A.; Thrush, B. A. *Chem. Phys. Lett.* **1988**, *153*, 200–202.
79. Webster, C. R.; May, R. D. *J. Geophys. Res.* **1987**, *92*, 11931–11950.
80. May, R. D.; Webster, C. R. *J. Geophys. Res.* **1989**, *94*, 16343–16350.
81. Webster, C. R.; May, R. D.; Toumi, R.; Pyle, J. A. *J. Geophys. Res.* **1990**, *95*, 13851–13866.
82. Mackay, G. I.; Schiff, H. I.; Wiebe, A.; Anlauf, K. *Atmos. Environ.* **1988**, *22*, 1555–1564.
83. Harris, G. W.; Mackay, G. I.; Iguchi, T.; Mayne, L. K.; Schiff, H. I. *J. Atmos. Chem.* **1989**, *8*, 119–137.
84. Ase, P.; Bock, W.; Snelson, A. *J. Phys. Chem.* **1986**, *90*, 2099–2109.
85. Traub, W. A.; Johnston, D. G.; Chance, K. V. *Science (Washington, D.C.)* **1990**, *247*, 446–449.
86. de Zafra, R. L.; Parrish, A.; Solomon, P. M.; Barrett, J. W. *J. Geophys. Res.* **1984**, *89*, 1321–1326.
87. Radford, H. E.; Evenson, K. M.; Howard, C. J. *J. Chem. Phys.* **1974**, *60*, 3178–3183.
88. Howard, C. J.; Evenson, K. M. *Geophys. Res. Lett.* **1977**, *4*, 437–440.
89. Thrush, B. A.; Wilkinson, J. P. T. *Chem. Phys. Lett.* **1981**, *81*, 1–3.
90. Brune, W. H.; Schwab, J. J.; Anderson, J. G. *J. Phys. Chem.* **1983**, *87*, 4503–4514.
91. Rozenshtein, V. B.; Gershenzon, Yu. M.; Il'in, S. D.; Kishkovitch, O. P. *Chem. Phys. Lett.* **1984**, *112*, 473–478.
92. Sinha, A.; Lovejoy, E. R.; Howard, C. J. *J. Chem. Phys.* **1987**, *87*, 2122–2128.
93. Mihelcic, D.; Ehhalt, D. H.; Klomfass, J.; Kulessa, G. F.; Schmidt, U.; Trainer, M. *Ber. Bunsen-Ges. Phys. Chem.* **1978**, *82*, 16–19.
94. Mihelcic, D.; Ehhalt, D. H.; Kulessa, G. F.; Klomfass, J.; Trainer, M.; Schmidt, U.; Röhrs, H. *Pure Appl. Geophys.* **1978**, *116*, 530–536.
95. Mihelcic, D.; Helten, M.; Fark, H.; Müsgen, P.; Pätz, H. W.; Trainer, M.; Kempa, D.; Ehhalt, D. *2nd Symposium, Composition of the Nonurban Troposphere*; Williamsburg, VA; American Meteorological Society: Boston, 1982; pp 327–329.
96. Helten, M.; Pätz, W.; Trainer, M.; Fark, H.; Klein, E.; Ehhalt, D. H. *J. Atmos. Chem.* **1984**, *2*, 191–202.
97. Mihelcic, D.; Müsgen, P.; Ehhalt, D. *J. Atmos. Chem.* **1985**, *3*, 341–361.
98. Volz, A.; Mihelcic, D.; Müsgen, P.; Pätz, H. W.; Pilwat, G.; Geiss, H.; Kley, D. In *Tropospheric Ozone*; Isaksen, I. S. A., Ed.; D. Reidel: Dordrecht, 1988; pp 293–301.
99. Mihelcic, D.; Volz-Thomas, A.; Pätz, H. W.; Kley, D.; Mihelcic, M. *J. Atmos. Chem.* **1990**, *11*, 271–297.
100. Carlier, M.; Sochet, L.-R. *J. Chem. Soc., Faraday Trans. 1* **1983**, *79*, 815–821.
101. Laverdet, G.; Le Bras, G.; Mellouki, A.; Poulet, G. *Chem. Phys. Lett.* **1990**, *172*, 430–434.
102. Watanabe, T.; Yoshida, M.; Fujiwara, S.; Kazuhisa, A.; Onoe, A.; Hirota, M.; Igarashi, S. *Anal. Chem.* **1982**, *54*, 2470–2474.
103. Foner, S. N.; Hudson, R. L. *J. Chem. Phys.* **1953**, *21*, 1608–1609.
104. Ingold, K. U.; Bryce, W. A. *J. Chem. Phys.* **1956**, *21*, 360–364.
105. Sander, S. P. *J. Phys. Chem.* **1984**, *88*, 6018–6021.
106. Smith, G. P.; Crosley, D. R. *J. Geophys. Res.* **1990**, *95*, 16427–16442.
107. Anderson, J. G.; Grassl, H. J.; Shetter, R. E.; Margitan, J. J. *Geophys. Res. Lett.* **1981**, *8*, 289–292.

108. Hard, T. M.; O'Brien, R. J.; Chan, C. Y.; Mehrabzadeh, A. A. *Environ. Sci. Technol.* **1984**, *18*, 768–777.
109. Hard, T. M.; Chan, C. Y.; Mehrabzadeh, A. A.; O'Brien, R. J. *Nature (London)* **1986**, *322*, 617–620.
110. Chan, C. Y.; Hard, T. M.; Mehrabzadeh, A. A.; George, L. A.; O'Brien, R. J. *J. Geophys. Res.* **1990**, *95*, 18569–18576.
111. Suto, M.; Lee, L. C. *J. Chem. Phys.* **1984**, *80*, 195–200.
112. Manzanares, E. R.; Suto, M.; Lee, L. C. *J. Chem. Phys.* **1986**, *85*, 5027–5034.
113. Wang, X.; Suto, M.; Lee, L. C. *J. Chem. Phys.* **1988**, *88*, 896–899.
114. Hartmann, D.; Karthäuser, J.; Zellner, R. *J. Phys. Chem.* **1990**, *94*, 2963–2966.
115. Glaschick-Schimpf, I.; Leiss, A.; Monkhouse, P. B.; Schurath, U.; Becker, K. H.; Fink, E. H. *Chem. Phys. Lett.* **1979**, *67*, 318–323.
116. Holstein, K. J.; Fink, E. H.; Wildt, J.; Winter, R.; Zabel, F. *J. Phys. Chem.* **1983**, *87*, 3943–3948.
117. Cantrell, C. A.; Stedman, D. H. *Geophys. Res. Lett.* **1982**, *9*, 846–849.
118. Stedman, D. H.; Cantrell, C. A. *2nd Symposium, Composition of the Nonurban Troposphere*; Williamsburg, VA; American Meteorological Society: Boston, 1982; pp 68–71.
119. Cantrell, C. A.; Stedman, D. H.; Wendel, G. J. *Anal. Chem.* **1984**, *56*, 1496–1502.
120. Buhr, M. Master's Thesis, University of Denver, 1986.
121. Stedman, D. H.; Walega, J. G.; Cantrell, C. A.; Burrows, J. P.; Tyndall, G. *Chemistry of Multiphase Atmospheric Systems*; NATO ASI Series G6, 1986; pp 351–366.
122. Ghim, B. T. Master's Thesis, University of Denver, 1988.
123. Fried, A.; Hodgeson, J. *Anal. Chem.* **1982**, *54*, 278–282.
124. Rooth, R. A.; Verhage, A. J. L.; Wouters, L. W. *Appl. Opt.* **1990**, *29*, 3643–3653.
125. Harren, F. J. M.; Reuss, J.; Woltering, E. J.; Bicanic, D. D. *Appl. Spectrosc.* **1990**, *44*, 1360–1368.
126. Maeda, Y. K.; Aoki, K.; Munemori, M. *Anal. Chem.* **1980**, *52*, 307–311.
127. Wendel, G. J.; Stedman, D. H.; Cantrell, C. A.; Damrauer, L. *Anal. Chem.* **1983**, *55*, 937–940.
128. Fehsenfeld, F. C. et al. *J. Geophys. Res.* **1990**, *95*, 3579–3597.
129. Kok, G. L.; Holler, T. P.; Lopez, M. B.; Nachtrieb, H. A.; Yuan, M. *Environ. Sci. Technol.* **1978**, *12*, 1072–1076.
130. Cantrell, C. A.; Shetter, R. E.; McDaniel, A. H.; Davidson, J. A.; Calvert, J. G.; Parrish, D. D.; Buhr, M. B.; Fehsenfeld, F. C.; Trainer, M. *EOS Trans., Am. Geophys. Union* **1988**, *69*, 1056.
131. Stedman, D. H.; Jackson, J. O. *Int. J. Chem. Kinet.* **1975**, *1*, 493–501.
132. Ritter, J. A.; Stedman, D. H.; Kelly, T. J. In *Nitrogenous Air Pollutants*; Grosjean, D., Ed.; Ann Arbor Scientific: Ann Arbor, MI, 1979; pp 325–343.
133. Parrish, D. D.; Trainer, M.; Williams, E. J.; Fahey, D. W.; Hübler, G.; Eubank, C. S.; Liu, S. C.; Murphy, P. C.; Albritton, D. L.; Fehsenfeld, F. C. *J. Geophys. Res.* **1986**, *91*, 5361–5370.
134. Molina, L. T.; Molina, M. J. *J. Geophys. Res.* **1986**, *91*, 14501.

RECEIVED for review January 29, 1991. ACCEPTED revised manuscript June 22, 1992.

12

Tropospheric Hydroxyl Radical

A Challenging Analyte

Robert J. O'Brien and Thomas M. Hard

Chemistry Department and Environmental Sciences Program, Portland State University, Portland, OR 97207

The methods for the direct measurement of tropospheric hydroxyl radical, HO, are reviewed, and the technical hurdles that remain to be surmounted are discussed in the light of theoretical and experimental results. Sensitivities, advantages, and disadvantages of several HO methods are compared, and a way to compare many of the existing HO methods experimentally is presented.

T ROPOSPHERIC HYDROXYL RADICAL IS FORMED in sunlight and reacts rapidly with a variety of trace atmospheric constituents, usually converting them from water-insoluble to water-soluble forms. (Hydroxyl radical, here termed HO, is widely called OH in the atmospheric science community. We prefer the IUPAC name HO because it is consistent with other hydrogen oxides, H_2O, H_2O_2, and HO_2, and because it better distinguishes HO from OH^-, a species often confused with HO by nonatmospheric scientists.) Hydroxyl greatly assists wet deposition or precipitation-scavenging of natural and anthropogenic atmospheric species. Although daytime HO concentrations are much higher than nighttime levels (1), even the daytime tropospheric HO concentrations are low (e.g., a sea level number density between 10^5 and 10^7 molecules per cubic centimeter or a mole fraction of 4×10^{-15} to 4×10^{-13}). This low concentration is partly a result of the high chemical reactivity of HO. Depending on the atmospheric nitric oxide (NO) concentration, HO is often regenerated in chain sequences such as this:

$$CO + HO\cdot \longrightarrow CO_2 + H\cdot \qquad (1)$$

$$H\cdot + O_2 \longrightarrow HO_2\cdot \qquad (2)$$

0065–2393/93/0232–0323$13.25/0

$$HO_2\bullet + NO\bullet \longrightarrow NO_2\bullet + HO\bullet \qquad (3)$$

$$NO_2\bullet + h\nu \longrightarrow NO\bullet + O \qquad (4)$$

$$O + O_2 \longrightarrow O_3 \qquad (5)$$

If the NO concentration is appreciable, the efficiency of trace gas oxidation (free radical chain length) is high for this species. In remote air masses, very low NO concentrations limit recycling of HO, and the HO oxidation rate approaches the prime HO generation rate, for instance by these reactions:

$$O_3 + h\nu \longrightarrow O_2\,(a^1\Delta_g) + O\,(a^1D); \lambda < 320 \text{ nm} \qquad (6)$$

$$O\,(^1D) + H_2O \longrightarrow 2HO\bullet \qquad (7)$$

In the global troposphere, the role of HO has been to maintain the low and relatively constant trace gas composition that apparently has persisted for at least 10,000 years (2). More recently, however, measured increases in tropospheric methane (3–9) have led to concern about humanity's possible influence on the natural abundance of HO (10).

In remote regions of the troposphere, nonmethane hydrocarbons and oxides of nitrogen are quite low in concentration in large measure because of HO-catalyzed removal of these species during the process of long-range transport from more polluted regions. In polluted atmospheric regions ranging from continental to urban, these same natural cleansing processes generate concentrations of intermediate products that collectively have been called photochemical smog. These products include ozone, peroxyacetyl nitrate (PAN), H_2O_2, and other oxidants; aldehydes and ketones; sulfuric, nitric, and organic acids; and free radicals in addition to HO (11–14). Thus, hydroxyl radical plays a key role in chemical models for either the remote or the polluted troposphere, and accurate assessment of the reactions that control its concentration is a prerequisite for credible modeling of atmospheric processes.

Although HO was already known as an agent active in combustion, the 1960s brought the realization of HO's tropospheric reactivity. Leighton (14) speculated about the "nature and indeed the reality of [the] apparent excess rate of olefin [alkene] consumption" in photochemical smog studies. This excess rate was soon identified with the presence of hydroxyl radical in addition to the recognized reactive species ozone and oxygen atoms. In fact, a current estimation (15) of the relative rates of removal of several classes of hydrocarbons by O_3, HO, and NO_3 (Table I) indicates that (except for certain alkenes) HO dominates the daytime removal of most hydrocarbons. (Although very reactive, O atoms are rapidly scavenged by O_2 to generate O_3.)

Interest in tropospheric HO concentrations has been high since at least the late 1960s, when Weinstock (16) suggested that tropospheric CO might

Table I. Trace Gas Rate Constants and Lifetimes for Reaction with Ozone, Hydroxyl Radical, and Nitrate Radical

Species	Rate Constants (molecules cm^{-3} s^{-1})			Tropospheric Lifetimes (days)		
	HO	O_3	NO_3	HO	O_3	NO_3
Methane	8.5×10^{-15}	7.0×10^{-24}	4.0×10^{-19}	1362	1.7×10^6	1.2×10^5
CO	2.8×10^{-13}	4.0×10^{-25}	4.0×10^{-16}	42	3.0×10^7	116
CH_2O	9.8×10^{-12}	2.1×10^{-24}	6.8×10^{-16}	1	5.6×10^6	68
Methanol	9.3×10^{-13}	very small	2.1×10^{-16}	12	NA	220
Ethane	2.7×10^{-13}	1.0×10^{-20}	8.0×10^{-18}	43	1200	5787
Propane	1.1×10^{-12}	1.0×10^{-20}	very small	11	1200	NA
Isoprene	1.0×10^{-10}	1.4×10^{-20}	8.2×10^{-13}	0.12	843	0.06
α-Pinene	5.0×10^{-11}	8.0×10^{-17}	6.2×10^{-12}	0.23	0.15	0.01
n-Butane	2.7×10^{-12}	1.0×10^{-20}	5.5×10^{-17}	4	1200	843
Ethene	2.5×10^{-12}	1.9×10^{-18}	2.0×10^{-16}	5	6.2	231
Propene	2.5×10^{-11}	1.1×10^{-17}	9.5×10^{-15}	0.46	1	5
Acetylene	7.0×10^{-13}	7.8×10^{-21}	5.1×10^{-17}	17	1513	908
Benzene	1.2×10^{-12}	7.0×10^{-23}	3.2×10^{-17}	10	1.7×10^5	1448
Toluene	6.4×10^{-12}	1.5×10^{-22}	6.9×10^{-17}	2	7.9×10^4	671
$(CH_3)_2S$	9.8×10^{-12}	8.0×10^{-19}	9.7×10^{-13}	1	15	0.05
Ammonia	1.6×10^{-13}	very small	6.0×10^{-16}	72	NA	77
NO_2	1.1×10^{-11}	3.2×10^{-17}	1.2×10^{-12}	1	0.37	0.04

NOTE: Lifetimes are based upon $[O_3]$ = 40 ppbv, $[HO^\bullet]$ = 1.0×10^6 molecules per cubic centimeter (daytime), and $[NO_3^\bullet]$ = 10 pptv (nighttime). NA, not available.

be removed by reaction with HO. Reactions 1 through 5 provide a succinct illustration of HO catalysis. The result of such a cycle is to split molecular oxygen; HO is conserved. The net of reactions 1–5 is

$$CO + 2O_2 + h\nu \longrightarrow CO_2 + O_3 \cdot \qquad (8)$$

In remote tropospheric air, where NO concentrations can be quite low (17), the HO + CO oxidation mechanism can follow other pathways, leading to net ozone destruction rather than formation (18, 19). Reactions 1 through 5 typify the more complex catalytic reactivity of HO with hydrocarbons, which produce a complex array of oxidation products while generating ozone photochemically (11–13).

Interest in actually measuring the concentration of tropospheric HO followed Weinstock's (16) 1969 discussion of the role of HO in controlling tropospheric CO and Levy's (20–22) description of the production of HO throughout the troposphere in reactions 6 and 7. This mechanism is believed to dominate HO production in the least polluted air and remains a significant source in polluted air where a number of other sources also exist.

The importance of HO as a theoretically interesting small free radical and as an important catalyst in combustion and in atmospheric chemistry has led to measurements of rate constants for its reaction with a wide array of other gases. These measurements have achieved, over the last several decades, a high degree of sophistication and accuracy, and they are regularly reviewed and evaluated (13, 23).

The role of HO as the "sole reactant" for many tropospheric trace gases has allowed the measurement, under simulated atmospheric conditions, of the relative reaction rate constants of HO with a host of hydrocarbons, other organic compounds (24), and other trace gases, including NO_2 (25). These rate constants are in good agreement with those measured by the more direct techniques, and they corroborate the concept of HO control of homogeneous daytime tropospheric chemistry. This evidence for HO control of trace gas lifetimes can form the basis for indirect measurement of ambient HO concentrations and can serve as a means of calibrating HO measurement devices (discussed later in the chapter). Condensed-phase atmospheric oxidations (in aerosols, fogs, and clouds) that convert SO_2 (aq) to SO_4^{2-} (aq), although not directly influenced by HO, may be indirectly coupled to it, because gaseous HO may determine the concentration of oxidants such as ozone or hydrogen peroxide, which have significant aqueous solubility. Although HO reacts rapidly at many surfaces and may be thereby removed, it may also be absorbed or emitted from atmospheric aerosols and droplets. Zellner et al. (26) have measured aqueous HO generation rates from solutions of nitrate, nitrite, and hydrogen peroxide. On the basis of their measurements they suggested that in situ HO generation rates within aerosols may be comparable to HO deposition rates. Thus, even the direction of the HO flux at aerosol surfaces may be in doubt. The high heterogeneous and homogeneous reactivity of HO has important ramifications for HO measurements in sampled ambient air, for relative-rate measurement of HO rate constants under simulated atmospheric conditions, for indirect measurement of ambient HO concentrations, for chemical removal of HO from sampled air during HO background signal measurements, and for the calibration of ambient measurement instruments.

HO Chemical Lifetime: Implications for Measurement

Aside from those species that react with atmospheric oxygen, nitrogen, or water vapor (e.g., H, O (^3P), O (^1D), or organic radicals), hydroxyl radical is apparently the atmosphere's most reactive species, and its short chemical lifetime has various implications for measurement strategies. In rough terms, a single HO radical survives for about 1 s in the cleanest regions of the troposphere but only about 1 ms in the most polluted regions. These lifetimes are inversely proportional to the concentrations and HO reactivities of those species with which HO reacts. In remote marine air [nonmethane hydrocarbon concentrations less than 25 parts per billion by volume (ppbv) (27), CO is the dominant reactant with HO and methane is second, whereas in polluted air an immense array of anthropogenic and biogenic hydrocarbons dominate HO removal. The lifetime of an individual HO radical should be distinguished from the decay of HO concentration when the photolytic production terms (e.g., reactions 6 and 7) are reduced—for instance, when a cloud passes in front of the sun—or stopped—as at sunset or when air enters a dark sampling system. The chemical lifetime of an individual HO molecule is given by $(\Sigma k_i[T_i])^{-1}$, where $[T_i]$ indicates the concentration of an atmospheric trace gas that reacts with HO and k_i is the rate constant. In contrast to this short lifetime, the concentration response time, characterizing the decline of HO concentration when darkness is imposed, depends upon the reactivity of the individual HO entity, the regeneration of HO by chain processes, and the collapse of labile HO reservoirs, which in general store a much larger concentration of HO than is present in the free radical state. Similar considerations apply to the increase in HO concentration when irradiance is increased.

In contrast with the typical 1-s lifetime of HO in the cleanest tropospheric air, a significant fraction of the reacting HO can be regenerated via reactions 2 and 3, so the observed decay of HO concentration is prolonged. The importance of the regeneration process depends on the relative concentrations of the various HO radical sinks. One predominant radical sink is NO_2, which stoichiometrically converts HO to the stable nonradical species HNO_3 as follows:

$$HO\bullet + NO_2\bullet \longrightarrow HNO_3 \qquad (9)$$

Important HO reservoirs are HO_2, HNO_4, PAN, and others. In clean air, the concentration of HO_2 is about 100 times that of HO (1), and HO_2 continues to generate HO in darkness by reaction 3 and other steps. Labile nonradical reservoirs dissociate to produce HO in several steps. For instance,

$$HNO_4 \longrightarrow HO_2 \cdot + NO_2 \cdot \qquad\qquad (10)$$

and this reaction can then be followed by reaction 3 to produce HO.

Short HO radical lifetimes in the dark prevent the collection of ambient air samples for later analysis, except perhaps for the case of filter collection techniques that use a spin-trapping reagent (28). Furthermore, rapid reactivity of HO with surfaces mandates careful attention to the sampling train used to move ambient air into a more amenable analysis environment.

Two goals can be envisioned when ambient HO measurements are contemplated, both related to the reciprocal control that HO maintains upon other trace gases and that they, in turn, maintain upon the HO concentration. This dichotomy arises because HO comes into rapid equilibrium with its chemical surroundings and at the same time alters this chemical environment on a much longer time scale ranging from hours to years. Thus HO equilibrates with CO and CH_4 in the cleanest tropospheric air in a matter of seconds but continues to remove CO and CH_4, which have chemical lifetimes of about 2 months and 7 years, respectively. In polluted air HO bears the same relationship with a multitude of hydrocarbons, NO, and NO_2, which have chemical lifetimes that range from less than 1 hour for the most reactive with HO to the 7-year lifetime of methane, the least reactive hydrocarbon. The simplest goal in HO determinations is to obtain HO concentrations that are broadly representative of the air mass being sampled. Even in the remote troposphere, removed from local emission sources, [HO] may vary significantly if there are fluctuations in its controlling chemical species (CO, O_3, H_2O, NO_2, NO, and so forth) because of vertical convection. If the more abundant trace gases have relatively constant concentrations, for instance in the mixed layer, HO would be expected to have a stable concentration over large regions of the atmosphere. In polluted regions, the proximity of sources and surface sinks results in much greater variability in these controlling chemical entities; this variability is expected to result in larger variability in HO concentrations, so more care must be used in defining a "representative" HO concentration. Nevertheless, it is useful to have measured values of [HO] representative of the "midlatitude free troposphere", the "remote continental mixed layer", the layer "below the inversion height in downtown Los Angeles", or elsewhere.

A second goal in HO measurements is more specifically oriented toward comparing the predictions of chemical models with measured HO concentrations. Now, the concentrations of all relevant chemical species, light fluxes, and so forth must be concurrently measured. The rapid equilibration of HO in photostationary equilibrium with its surroundings facilitates such comparisons. Thus, if the proximity of the ground or another surface has altered the atmospheric albedo or the total chemical composition of an air sample so that it is not representative of any particular air mass, meaningful model comparisons can still be made as long as ancillary measurements are made of the relevant chemical entities with which HO has equilibrated. An atmospheric chemical model may then be tested for its ability to correctly predict HO concentrations from an input of the relevant atmospheric conditions (light fluxes and dominant trace gas species concentrations).

A key element in measuring such a low-concentration, short-lived, chemical species is the averaging time required to obtain acceptable signal-to-noise ratios. The ultimate desire is to achieve sampling times shorter than the HO concentration response time itself. Such sensitivity would allow following short-term fluctuations in HO brought about by naturally occurring or deliberately introduced perturbations. Except for the complexity of atmospheric composition, these experiments would be the open-atmosphere equivalent of relaxation kinetics widely employed in the laboratory by chemical kinetics scientists. In the laboratory, a simple chemical composition allows unambiguous correlation of reaction rates with rate constants. In the open atmosphere, the more difficult goal is to relate measured HO fluctuations with the entire suite of trace gases with which HO may react or from which it could be formed. Sufficient sensitivity has yet to be achieved in ambient HO measurements because the continuous measurements have required long averaging times and the point measurements have not be made with sufficiently high frequency.

Tropospheric HO: A Challenging Analyte

Although the focus of this chapter is tropospheric HO measurements, it is worthwhile to mention techniques that have proven useful in the laboratory or in other regions of the atmosphere. As a small molecule in the gas phase, HO has a much-studied and well-understood discrete absorption spectrum in the near UV (29), shown in Figure 1, that lends itself to a variety of absorption and fluorescence techniques. The total atmospheric HO column density has been measured (30–32) from absorption of solar UV radiation, observed with a high-resolution scanning Fabry–Perot spectrometer. Long-path measurements of stratospheric HO from its thermal emission spectra in the far infrared have been reported (33–35). Long absorption paths in the atmospheric boundary layer have been used for HO detection from its UV absorption (36–42).

Fluorescence measurements of HO have been a common feature of laboratory kinetics studies of the reaction-rate coefficients of HO with various molecules and of studies of this free radical in combustion systems (24). In fact, although direct tropospheric fluorescence HO measurements were first

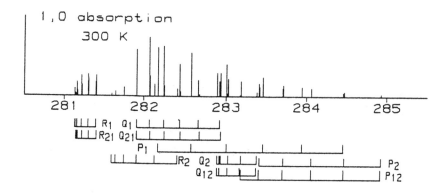

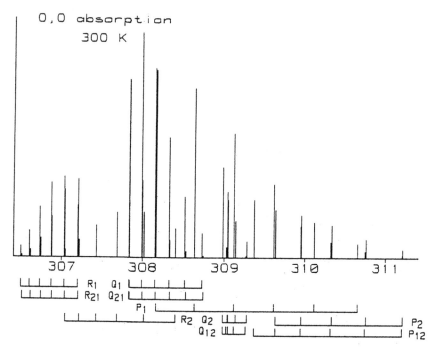

Figure 1. Portions of the UV excitation spectra of HO ($N'' = 1$ to 6) at 300 K; line positions are from Dieke and Crosswhite (29), and Einstein B coefficients are from Chidsey and Crosley (47). (Top) $1 \leftarrow 0$ band near 282 nm; (bottom) $0 \leftarrow 0$ band near 308 nm.

proposed by Baardsen and Terhune (43), HO had already been determined in the mesosphere by fluorescence that was excited by solar UV radiation (44). As with the upper atmosphere concentration measurements, most laboratory studies of HO kinetics that use fluorescence have been carried out at low pressure, where the higher fluorescence yield for a given absolute concentration allows better detection sensitivity.

Atmospheric chemical models provide essential information about expected concentrations of HO as a target analyte. Typical predictions are shown in Figure 2, which gives a calculated vertical profile (*45*) and a diurnal concentration cycle (*1*). For an atmospheric trace gas, both the absolute and relative concentrations are of interest. Absorption techniques provide a measurement of the number of absorbing molecules within the propagation path of the probe light beam (i.e., of absolute concentration). Fluorescence measurements of excited states produced by an exciting light source, on the other hand, are linear in the analyte mole fraction when the total air pressure is above the analyte's quenching half pressure (discussed later in the chapter). The dichotomy between absorption and fluorescence is illustrated by the following simplified absorption mechanism for HO near 308 nm via the transition $A^2\Sigma$ (v' = 0) ← $X^2\Pi$ (v" = 0):

$$\text{Absorption} \qquad \text{HO} + h\nu \xrightarrow{\sigma I_0} \text{HO*} \qquad (11)$$

$$\text{Fluorescence} \qquad \text{HO*} \xrightarrow{k_f} \text{HO} + h\nu \qquad (12)$$

$$\text{Quenching} \qquad \text{HO*} + \text{M} \xrightarrow{k_q} \text{HO} + \text{M} \qquad (13)$$

Here M denotes any other molecule with which HO may collide. If saturation is absent, the UV absorption rate is given by $\sigma I_0(n/n_0)[\text{HO}]$, where σ is the absorption cross section, I_0 is the incident UV flux, and n/n_0 represents the fractional population of HO in the absorbing state. For HO, this state is described by appropriate lambda coupling between rotational and spin states (*29, 46–48*).

In absorption, the path length L determines the magnitude of light attenuation from I_0 to I, and [HO] is given by the Beer–Lambert law as

$$[\text{HO}] = \frac{\ln(I/I_0)}{\sigma L(n/n_0)} \qquad (14)$$

It is essential to correctly evaluate the absorption cross-section σ relative to the laser line profile, the spectral resolution of the light collection optics, and the natural HO line width as influenced by Doppler, Voigt, or collisional broadening. The principles governing absorption measurements of HO over a distance through the atmosphere are discussed by Hübler et al. (*38*).

In order to obtain sufficient attenuation by ambient HO, absorption path lengths of 1 to 20 km have been used, and three groups of researchers have made such measurements. All of the early measurements and many of the current ones have used a single retroreflector, which allows the laser beam to retrace its base path. These measurements determine a path-length-averaged HO concentration. This averaging may be a disadvantage when the result is to be compared with model-calculated HO concentrations. Although early attempts to measure HO with a folded-path absorption cell were not

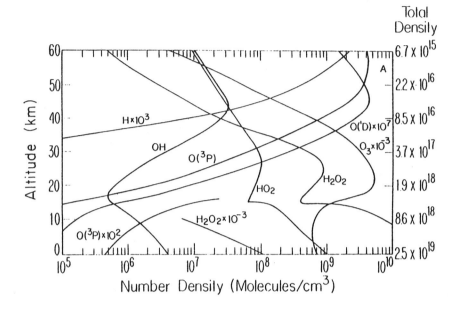

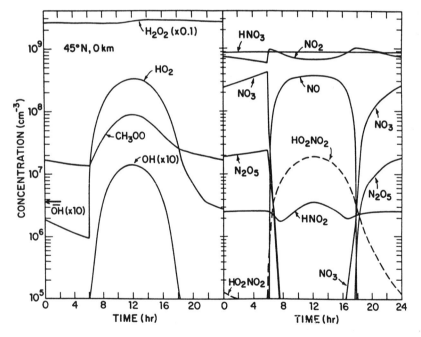

Figure 2. Atmospheric distributions of HO. A, vertical distribution. (Reproduced with permission from reference 45. Copyright 1978.) B, diurnal cycle at 45°N and 0-km altitude. (Reproduced with permission from reference 1. Copyright 1981.)

very successful (*49*, *50*), more recent measurements (*51–54*) in such a cell have successfully used a base path length of 6 m and a total path of 1.2 km with a rapidly scanned, narrow-band, frequency-doubled dye laser. Absorption has the advantage of not requiring instrumental calibration so long as the proper laser or HO absorption line profiles are known or measured. Absorption measurements are also less prone to some types of interferences (discussed later in the chapter). The primary difficulty lies in deconvoluting the HO spectral feature from other atmospheric UV absorbers (e.g., SO_2 or CH_2O), most of which are present at optical densities comparable to, or greater than, that of HO. This deconvolution is difficult even for the known absorbers with known line shapes—because the low tropospheric concentration of HO results in optical densities of $\sim 10^{-4}$ over a several-kilometer path length—when the measurements are taken in the presence of atmospheric turbulence and haze. In addition to known absorption features of other atmospheric constituents in this region, the presence of unidentified absorbers at any HO spectral feature makes the problem even more difficult and necessitates the use of at least two HO lines as confirmation that HO is responsible for light absorption (discussed later in the chapter).

Although generally believed to be more sensitive than absorption measurements, fluorescence measurements are complicated by molecular processes subsequent to reaction 11. Still, coupling of a fluorescence signal of known spectral and temporal characteristics with the known absorption profile can allow a greater degree of specificity for HO than possible in absorption measurements. However, the number of operative parameters makes an a priori calculation of HO from measured signals much more difficult than with absorption measurements. Fluorescence (reaction 12) and quenching (reaction 13) combine to produce an effective fluorescence yield given by

$$Y_f\,(0 \leftarrow 0)\,=\,k_f/(k_f\,+\,\Sigma k_{qi}[M]_i) \tag{15}$$

where the index i indicates the sum must be taken over all atmospheric species that quench the excited-state fluorescence. In air, these are principally N_2, O_2, and H_2O, the latter present at varying mixing ratios. In addition, geometric and other parameters associated with an absolute intensity measurement must be accurately measured, or the system must be calibrated in a procedure that removes some or all of these parameters from individual determination. Even with full calibration at known HO concentrations, significant variation of atmospheric pressure and water vapor content mandates a knowledge of the quenching rate coefficients k_{qi} for the most accurate interpretation of fluorescence signals. Equation 15 is appropriate for the $0 \leftarrow 0$ excitation and fluorescence now employed by most research groups for tropospheric measurements.

The early tropospheric HO measurements from fluorescence were made from red-shifted fluorescence generated by exciting the $1 \leftarrow 0$ transition near 282 nm and observing $0 \leftarrow 0$ fluorescence near 309 nm. Additional molecular processes associated with the competition between electronic quenching (k_{ql}) of the $v' = 1$ excited state and vibrational relaxation (k_v) to $v' = 0$ should be characterized for the most accurate measurements. The fluorescence yield for varying total pressure and water vapor concentration can be calculated with known HO quenching or vibrational relaxation parameters by using the two- and three-level fluorescence yield expressions given by equations 15 and 16.

$$Y_f(1 \leftarrow 0) = \frac{\Sigma k_{vi}}{(k_{f1} + \Sigma k_{qli} + \Sigma k_{vi})} \frac{k_{f0}}{(k_{f0} + \Sigma k_{q0i})} \tag{16}$$

In addition to being determined from the spectroscopic techniques of absorption and fluorescence, HO concentrations have been inferred from the measured rate of reaction of a species with which HO is (presumably) the sole reactant. This measurement was first carried out in smog chambers in the evaluation of relative HO rate constants (55–57). These measurements have been adequately quantified and corroborated [see the reviews by Atkinson et al. (13, 24)]. Such measurements have served as methods of inferring ambient HO in a variety of approaches, and we believe they can serve as a very useful method of corroborating or calibrating more direct HO techniques. This method has been used in the open atmosphere as a technique for measuring local ambient HO concentrations (58, 59). Related methods were used to obtain global tropospheric HO concentrations by evaluating the budgets of halocarbons (60–63), hydrocarbons (62, 64), SF$_6$ (62), or ^{14}CO (65). If changes in a more abundant species concentration are solely due to HO reaction, then chemical kinetics provides a simple relationship for the HO concentration:

$$[\text{HO}] = \frac{d\ln[\text{T}]}{k \, dt} \tag{17}$$

where [T] is the concentration of the HO tracer. [The status of hydrocarbon and HO rate constants k is frequently reviewed (13, 23, 24)]. Equation 17 may be solved in a variety of ways, depending upon the appropriate experimental circumstances. In the simple situation in which a chemical reactant T has no local sources and transport is not significant, [HO] may be obtained from the slope of a semilogarithmic plot of [T] versus time.

The method of deriving effective ambient HO concentrations from measured hydrocarbon distributions or from changes in such distributions has been applied both to the global troposphere and to regional air masses as mentioned earlier in the chapter. The calculation from hydrocarbon distributions requires knowledge of source terms for an HO chemical tracer (methylchloroform has been a popular tracer) and produces an "average" concentration, weighted over the appropriate pressure, temperature, and concentration regimes of the tropospheric regions involved. Such measurements have been used to estimate transport (of tracers) between the two hemispheres (*60–62*) and on finer grid scales as well (*63*).

For chemical tracer HO measurements in more limited regions, the goal has been to avoid considerations of transport and sources by looking at relative distributions of hydrocarbons of varying HO reactivity or at changes in concentration with time (e.g., reference 58). Such measurements have had limited success, but the reasons for their failure have been treated in detail (*66*) and need not be due to competition by reactions of the reacting tracer with species other than HO but have been attributed to previously unrecognized coupling between chemistry and transport. Highly reactive tracers yield excess concentration gradients that accelerate diffusion and result in underestimation of [HO]. McKeen et al. (*66*) found only two cases that can avoid chemistry–transport coupling errors. The least restrictive case is pulsed release of inert and reactive tracers that have no extraneous sources or sinks (other than HO reaction for the reactive tracer) between release and detection points. This approach has been proposed by Prinn (*67*), but apparently not yet performed for tropospheric HO. The other reliable case is continuous tracer release with downwind detection, with the restrictions that the reactive tracer concentration must have reached its steady-state concentration at the detection point and that the product of the transport time and first-order chemical loss rate of the reactive species must be less than 1.

The uniqueness of HO-tracer reactivity has led several groups to infer local ambient HO concentrations by measuring the production of a characteristic HO-tracer reaction product, as discussed later in the chapter. The development of HO determination techniques has been supported in the United States by the National Science Foundation and the National Aeronautics and Space Administration (NASA). Jointly, these agencies have periodically sponsored workshops to discuss the current status of technique development. These conferences have resulted in assessment documents on a regular basis (*68–70*).

Tropospheric HO Measurement Approaches

Ambient HO measurements have employed both chemical tracer and direct spectroscopic approaches, the latter including both fluorescence and absorbance. The history of ambient HO measurements has been one of uneven success, as might have been expected from the attempt to develop an analytical system for an important but difficult analyte. Nevertheless, real progress has been made, and all three approaches now provide viable alternatives for providing useful ambient HO data.

With the low concentration of tropospheric HO and the desire for measurements of fairly short time duration, all techniques have struggled with signal-to-noise ratio problems, and users of each method have gone to some lengths to improve this ratio. Additionally, interference problems, which

produce spurious signals, have affected some of the techniques. This overview considers the techniques in the order of their introduction to the
scientific literature: fluorescence (71–81), absorption (36–39), and chemical
tracer (82–88).

Fluorescence HO Measurements.   Fluorescence measurements of
HO have been attempted in a number of different configurations. The earliest
atmospheric HO measurement was by Anderson in 1971 (44), who used
rocket-borne passive detection of HO fluorescence excited by sunlight in
the mesosphere. The simplest approach, a balloon-borne HO resonance lamp
used to excite HO fluorescence, has been successful in the stratosphere
down to 30 km (89). At lower altitudes, these methods fail, principally because the fluorescence yield of HO decreases and the atmospheric scattering
background increases with decreasing altitude. In contrast, pulsed laser
excitation employs higher photon flux to produce short-term fluorescence
signals more easily discriminated from ambient light. Laser collimation reduces scattering by surfaces in the fluorescence detector's necessary field of
view. Thus stratospheric HO has been detected by Heaps et al. (90–92)
using a Nd:YAG-pumped tunable dye laser and by Stimpfle et al. (93, 94)
using a copper-vapor laser to pump the dye laser; in all cases the dye laser
output was frequency-doubled to 282 nm in the ultraviolet to excite the
$1 \leftarrow 0$ vibrational transition in HO.

HO measurement techniques in the troposphere are the principal focus
of this chapter. All reported fluorescence measurements of HO in the troposphere have used lasers. Most of the early tropospheric laser fluorescence
work used a frequency-doubled Nd:YAG laser to pump a Rh6G dye laser
that was also frequency-doubled and tuned to either the $P_1(1)$ line or the
$Q_1 1, 1' + R_2 3$ line group of the $A^2\Sigma$ $v' = 1 \leftarrow X^2$ $v'' = 0$ band of HO at 282
nm. Fluorescence is detected from the $A^2\Sigma$ $v' = 0 \rightarrow X^2$ $v'' = 0$ band at 309
nm. Although the three groups to report tropospheric HO data by this
method have often used similar lasers and have usually pumped the same
transitions, the air sampling configurations have been quite different.

LIDAR Measurements.   Wang, Davis, and co-workers of the Ford Motor Company research staff were the first to report tropospheric HO data
(71), taken outside their laboratory in Dearborn, Michigan, in a LIDAR
(laser radar) configuration. In the LIDAR mode, the pulsed 282-nm laser
beam is sent into the open atmosphere, and part of the red-shifted fluorescence is collected with a telescope located near the excitation source. Early
measurements used too high a laser flux, which produced HO radical (detected on the same or subsequent pulses) largely through the sequence of
reactions 6 and 7. This subject is treated in a separate section later in the
chapter. Subsequently, these workers expanded the excitation beam, reducing the production of spurious HO but degrading the signal-to-noise ratio

because of the increasing diffuseness of the signal relative to the solar scattering background at 308 nm. In airborne implementations of the LIDAR technique (72–74) the laser beam was sent through a quartz aircraft window toward a backstop painted onto the top surface of the wing to limit the solar scattering background. The backscattered fluorescence was collected through the window. Aircraft motion prevented the accumulation of photolytic HO from previous laser pulses, in contrast with ground-based LIDAR in which such accumulation can be a significant problem.

The initial attempt to measure HO in the unperturbed atmosphere, over a limited volume, is clearly the most desirable approach from an atmospheric chemistry standpoint, as such data would provide the best comparison with modeled concentrations (assuming other photochemical variables are also measured in the same limited volume). The chief difficulties with this approach lie in the attainable signal-to-noise ratios and in absolute calibration. Wang and co-workers (71–74) used a calculated calibration factor based upon a knowledge of the physical and chemical parameters in equation 16. They did, however, minimize the number of individual parameters by using the ratio of the signal to the Raman backscatter from atmospheric nitrogen, which takes into account the laser intensity and the optical and electronic parameters associated with signal collection and processing but introduces a new parameter (derived from the literature), the Raman cross section, which was appropriate for the 180° geometry employed. Davis et al. (73) showed how scattered sunlight, the principal background, was measured and subtracted. (This necessary subtraction degrades the signal-to-noise ratio, of course). Other potential sources of noise were passage of some 282-nm radiation by the 309-nm filtration device or broad-band fluorescence from other atmospheric constituents. Subsequent to their LIDAR measurements, Wang and co-workers converted their system for excitation in a sampled airstream at low pressure, as described later in the chapter.

Enclosed Atmospheric Pressure Fluorescence.   Georgia Tech researchers Davis, Bradshaw, Rodgers, and co-workers have been the major proponents of laser-induced fluorescence (LIF) measurements of HO in a flowing airstream at ambient pressure, or EAPF (76, 77). Depending upon the excitation mode, their system has been called single-photon laser-induced fluorescence (1-λ LIF) or two-wavelength LIF (2-λ LIF); EAPF is used here to distinguish it from other techniques that use single photons or two wavelengths. Designed for airborne operation—for instance, during the NASA CITE 1 experiments (95)—their system used large-diameter tubing to sample a high flow rate of ambient air available from the forward aircraft motion. Although the intake of ambient air into a sampling pipe has at least the potential for modification of ambient HO concentrations via contact with surfaces, Rodgers et al. (77) presented experimental data and calculations that indicate these effects were minimal. The chief advantages of the enclosed volume sampling system are the minimization of ambient light at the detection wavelength, the possibility of increasing collection efficiency by using up to 12 photomultiplier tubes along the excitation pathway, and the ability to generate a calibration HO signal via photolysis of elevated ozone concentrations in the flow tube. Starting in 1983, signal and background were determined nearly simultaneously with two pulsed lasers operating on and off the absorption wavelength (2-λ LIF) (77). The first of these lasers was tuned to the peak of the HO $Q_1 1, 1'$ absorption; the second laser was tuned to a nearby wavelength not absorbed by HO, and its pulses were timed to occur about 500 μs after those of the first. The latter precaution ensured that the signal and background were measured in the same air parcel during flight. This system's large-volume flow rate makes it difficult to sample from any container of finite volume. These efforts initially suffered from spurious HO produced and detected by the exciting pulse (96, 97), but the workers turned this to their advantage indirectly by using photolysis of high ozone concentrations to generate calibrating signals from HO produced by reactions 6 and 7. Uncertainties in the kinetic parameters associated with reactions 6 and 7 as well as the second-order dependence upon laser flux (discussed later in the chapter) could be sources of uncertainty in applying ozone calibration signals to ambient HO signals. These difficulties appear to have been overcome by Rodgers et al. (77), who showed agreement within 20% between ozone and hydrogen peroxide photolysis as HO instrument calibration sources. Their use of a reference HO chamber for normalization removes the second-order dependence on flux.

On the basis of their success with two-photon LIF detection of tropospheric nitric oxide (98–100), the Georgia Tech group proposed the two-step excitation of tropospheric HO fluorescence by successive photons at infrared and near-UV wavelengths. Two possible pumping schemes are illustrated in Figure 3 (100). The idea behind each is to generate HO fluorescence that is blue-shifted from the excitation wavelengths. This approach has obvious advantages in reducing the scattered background relative to lasers operating near 282 or 309 nm. Further, because neither of the pump lasers has sufficient photon energy to generate O (^1D) from ozone (reaction 7), the problem of spurious HO generation from ozone should be eliminated altogether. Bradshaw et al. (100) presented detailed calculations with a laboratory feasibility study of the sequential two-photon approach. Their experiments used the observed fluorescence from bottled laboratory air, and they noted that up to 5 or 10% fluctuation in background fluorescence would be inconsequential relative to the natural HO fluorescence. They concluded that the two-photon technique should be a viable approach for tropospheric HO determination, provided that an infrared laser of sufficient power at the correct wavelength can be found. Currently, no existing laser has the requisite properties, but developmental work may enable future measurements with this technique.

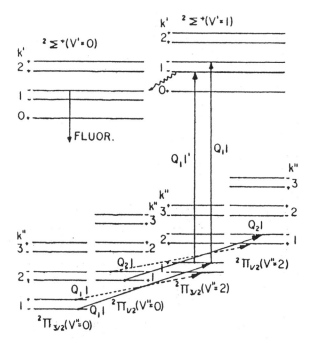

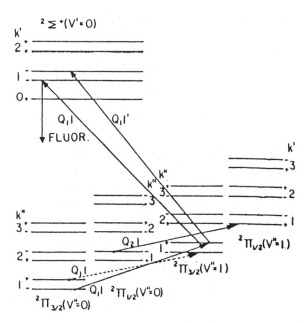

Figure 3. Transition diagrams for two methods of sequential two-photon laser excitation of HO. Top, 2, 0 scheme; bottom, 1, 0 scheme. (Reproduced with permission from reference 100. Copyright 1984.)

Low-Pressure Fluorescence.   The fluorescence yield expressions (reactions 15 and 16) suggest an alternative approach to ambient pressure sampling for HO (*101, 102*). The fluorescence signal is the product of the HO
excitation rate, the photon collection efficiency E_c, and the fluorescence
yield. For a two-level system the fluorescence signal is

$$S_f\,(0 \leftarrow 0) \,=\, \frac{\sigma I_o (n/n_o)[\mathrm{HO}]}{k_f \,+\, \Sigma k_{qi}[\mathrm{M}]_i} \tag{18a}$$

Any fluorescent species may be characterized by its quenching half
density, defined by $[M_{1/2}] \equiv k_f/k_q$, where k_q represents a weighted average
of the HO-quenching rate coefficients of the gases present. ($[M_{1/2}]$ is termed
a half pressure here: $P_{1/2} \,=\, RT[M_{1/2}]$), where R and T are the gas constant
and temperature. At the half pressure, half of the excited molecules are
quenched and half of them fluoresce. Substitution of $[M_{1/2}]$ into equation
18a gives a simplified expression for the fluorescence signal S_f:

$$S_f \,=\, \frac{E_c \sigma I_o (n/n_o) X_{\mathrm{HO}}}{1/[M] \,+\, 1/[M_{1/2}]} \tag{18b}$$

Equation 18b shows that whenever the sample's pressure is much higher
than the quenching half pressure ($[M] >> [M_{1/2}]$), then the fluorescence
signal is a function of the analyte mole fraction (X_{HO}) but is independent of
its absolute concentration. A similar expression can be obtained for three-
level fluorescence (*78*). Many atmospheric trace gases, including HO, have
quenching half pressures for electronic transitions that are two or more
decades below atmospheric. Thus, if the ambient sample is expanded to low
pressure in an enclosed flow system, the signal from the analyte will remain
constant until the total pressure approaches the vicinity of the half pressure,
whereas other interfering signals will be reduced with the pressure (Figure
4). The main advantages of low-pressure excitation (*78*) are reduced Rayleigh
and Mie scattering; a lengthening of the HO fluorescence waveform in time
(sufficient to make delayed gating and photon-counting practical and thus
further reduce the laser-excited background); and a 1 to 2 order of magnitude
reduction in background photons from false HO due to laser photolysis of
HO precursors. As with atmospheric pressure sampling, the enclosed volume
reduces the solar background at the detection wavelength. On the other
hand, low pressure slows collisional relaxation into the probed rotational
level and out of the excited state, so a greater propensity to saturate the
HO transition results (*78, 80, 103*).

Low-pressure determination of tropospheric HO (FAGE, fluorescence
assay with gas expansion) was first implemented by Hard, O'Brien, and co-
workers at Portland State University, who have improved the technique
through three generations (FAGE1 through FAGE3) (*78–81*). The FAGE

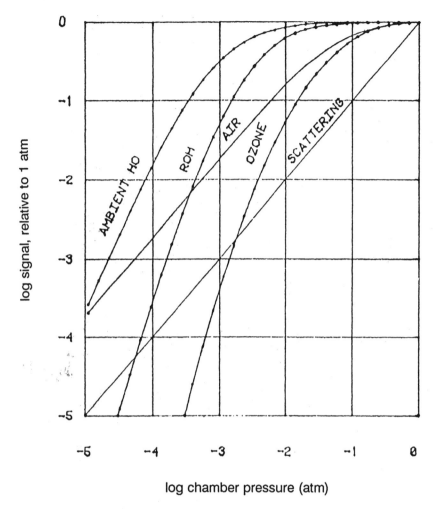

Figure 4. Signal and background sources versus detection pressure after air expansion in FAGE. Ambient HO: net HO signal after expansion, excitation, and fluorescence quenching; ROH: interference from photolysis of an HO-containing precursor; air: nonresonant fluorescence of non-HO species; ozone: interference due to ozone photolysis and subsequent HO production via reactions of O (^{1}D); and scattering: Rayleigh and Mie scattering by air sample. (Reproduced from reference 78. Copyright 1984 American Chemical Society.)

technique cannot increase fluorescence signals by low-pressure sampling, but it does significantly improve signal-to-noise ratios over atmospheric pressure detection by reducing the various backgrounds and interferences. Thus it allows shorter averaging times as illustrated in Figure 5. FAGE has also been applied to atmospheric NO_2 determination (*104*). Wang and co-workers (*75*) did experiments with low-pressure sampling; more recently, low-pres-

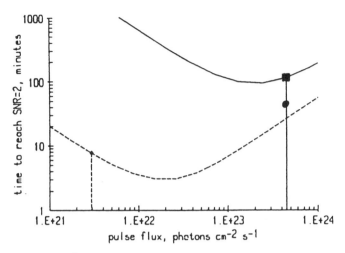

Figure 5. Required averaging times to achieve SNR = 2 at [HO] = 1 × 10⁶ cm³, for FAGE2 (solid line) and FAGE3 (dashed line) instruments versus laser power. For FAGE2: ■, *single pass;* ●, *multipass. (Reproduced with permission from reference 80. Copyright 1992.)*

sure sampling has been adopted by Hofzumahaus and co-workers at the Nuclear Research Institute in Jülich, Germany, and by Brune and co-workers at Pennsylvania State University.

In low-pressure fluorescence HO measurement, as with all other spectroscopic methods for atmospheric HO, background signals (B) are present and must be measured in the absence of the HO signal. Moreover, because the background varies with time, it is desirable to measure B at the same time as the gross signal $G = S + B$, or as closely in time as possible. When the periods of measuring G and B alternate in time, as in spectral modulation, Heaps (105) has shown that the optimum fraction of observation time devoted to measuring B reduces to 50% when $S \ll B$. (The latter result is limited to the case when Poisson noise is the only source of uncertainty.) The process of alternating measurements of gross signal and background is called modulation. Because laser-based methods make use of HO's narrow-band vibronic absorption transitions, it is natural to measure the background by tuning the laser to a nearby wavelength not absorbed by HO; this procedure results in spectral modulation. Another method, chemical modulation, makes use of HO's high chemical reactivity by injection of a chemical reagent into the sample flow to remove HO. Chemical modulation requires a substantial ambient air flow to clear the reagent prior to gross signal measurement; such a flow is available in low-pressure sampling. With care in the choice of reagent and reagent flow, chemical modulation yields smaller net interferences from laser photolysis of HO precursors than does spectral modulation (78, 80).

Low-pressure HO fluorescence measurements of atmospheric HO were also undertaken by Shirinzadeh et al. (*75*), who used spectral modulation to subtract the nonresonant background. They also subtracted values for ozone interference based on measurements with ozone-enriched ambient air obtained before and after the ambient HO measurements. Wang and co-workers (*106*) correctly pointed out the presence of substantial ozone-photolytic HO backgrounds in the work of Hard et al. (*78, 79*) and suggested a mechanism by which chemical modulation might not cancel these backgrounds. However, Wang and co-workers (*106*) overestimated the net positive interference of this mechanism by a factor of 10^3 (*107*) and obtained the wrong sign for the actual net interference. However, other more significant interference mechanisms (*80, 103, 107*) (discussed later in the chapter) justify the more general concerns of Wang and co-workers (*106*) and have led to changes in the method of laser excitation in FAGE (*81*).

Long-Path Absorption.   Long-path absorption (LPA) measurements of HO, like the fluorescence techniques, are straightforward in principle and suffer chiefly from low HO optical densities and the difficulty of propagating a laser beam through several kilometers of atmosphere in the presence of turbulence and haze. Such measurements were reported in a series of papers by Perner, Ehhalt, and co-workers at the Institute for Atmospheric Chemistry of the Nuclear Research Institute in Jülich, Germany (*36–41*). All measurements have monitored pressure-broadened rotational line groups near either $Q_1(2)$ or $Q_1(3)$ in the 307.9- to 308.2-nm region. An Ar^+-pumped, frequency-doubled dye laser is used to generate sufficient spectral width so that the desired absorption lines (Voigt width, 0.2 cm^{-1}) are captured along with the surrounding spectral region. Average UV power is 10–20 mW. The original experiment exposed a photographic plate, soon replaced by a rapid-scanning exit slit and in 1985 by a photodiode array. An experimental schematic of the German LPA system is shown in Figure 6 (*38*). The laser beam is expanded to about 0.25 m and reflected from a plane mirror at a distance of 1.5–5 km, depending upon the atmospheric optical depth. On some occasions no HO absorption is detected, while on others an HO spectrum is visible. Absorption features due to SO_2 and occasionally CH_2O are observed, but sufficient detail is often present that these features can be quantitatively deconvolved from the ambient spectrum. Bands not assignable to known species are assigned to the baseline with a resultant degradation of signal-to-noise ratio.

The LPA instrument of Mount (*42*) uses a XeCl laser to monitor absorption near the $Q_1(3)$ line group. The laser beam, expanded in a telescope to an initial area of 150 cm^2, is reflected from a retroreflector array for a total optical path of 20.6 km. The returned beam and a portion of the outgoing beam follow symmetric paths through an echelle spectrograph to a pair of photodiode array detectors, thus providing both I and I_0 spectra for the

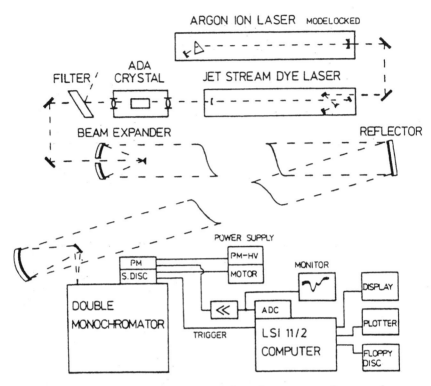

Figure 6. Diagram of LPA instrument. The reflector is at a distance of up to 5 km from the rest of the instrument. The slotted disk–photomultiplier (PM-S.DISC) has been replaced by a diode array. (Reproduced with permission from reference 38. Copyright 1984.)

Lambert–Beer equation (equation 14). With the provision of a reference HO absorption spectrum, and with care to avoid local instrumental artifacts that affect the two beams differently, this design allows the removal of extraneous atmospheric absorption features without requiring assignment to known absorbing species.

The Frankfurt LPA instrument (51–53) departs from both of these instruments in two principal ways: it achieves the necessary path length within a 6-m folded-path cell, and it rapidly scans a narrow-band frequency-doubled dye laser across the spectral region of interest (the $Q_1(2)$ line group) in a process sometimes called differential optical absorption spectrometry (DOAS). The scanning rate is sufficient to ensure that the observed air volume is chemically and physically stationary during each scan (the baseline standard deviation is less than 2×10^{-4} for a 0.2-ms scan). The laser output is actively feedback-stabilized to provide a flat spectral baseline, and a detection limit better than 10^{-5} in optical density has been claimed. A summary of published LPA configurations is given in Table II.

Table II. HO Determination by Long-Path Absorption

Group	Ref.	Laser Source	HO Lines	Background	Reference HO	Path Length	t_{at}	Detector	Resolution (cm^{-1})
Jülich	41	Frequency-doubled broadband tunable dye laser pumped by continuous wave Ar^+ ion laser	$Q_1 2$ $Q_{21} 2$ $R_{21} 2$ $Q_1 3$ $Q_{21} 3$ $P_1 1$	Indirect, via numerical fit to two or more HO lines	Stored spectrum (from experimental O_3–H_2O photolysis)	5.8 km	20 min at 3×10^6 cm^{-3}	photodiode array	0.9
NOAA	88	XeCl pulsed	$Q_1 3$ $Q_{21} 3$ $P_1 1$	I_o direct to parallel photodiode array	Propane flame	20.6 km	15 min at 1×10^6 cm^{-3}	photodiode array	0.06
Frankfurt	54	Fast-scanning feedback-stabilized frequency-doubled ring dye laser pumped by Ar^+ ion laser	$Q_1 2$ $Q_{21} 2$ $R_{21} 2$	Stabilized to less than 2×10^{-4} in region of interest	Absorption cell (HO from microwave dissociation of H_2O)	1.2 km in 6-m White cell	1 h at 3×10^5 cm^{-3}	PIN photodiode	0.026

NOTE: To date, none of the above groups has reported a calculation of steady-state spurious HO production relative to ambient HO production with a simple transport-free model, and none has implemented both real-time reference background measurement and real-time reference HO measurement. The averaging times listed above are those displayed for ambient HO in recent publications. All of these methods are undergoing continuous improvement.

Local Chemical Tracer Measurements. The high reactivity of HO has led several groups of investigators to make local HO concentration measurements by determining the concentration of characteristic tracer reaction products when the tracer is released under carefully controlled conditions. The high sensitivity associated with radionuclide counting techniques suggested to Campbell, Sheppard, and co-workers a method of radiocarbon determination of HO (82–86). Alternatively, Eisele (87) noticed an HO-induced diurnal pattern in measuring atmospheric ion concentrations and subsequently developed a method of chemically converting ambient HO concentrations to measurable ion currents in a quadrupole mass spectrometer.

Radiocarbon Characteristic Tracer Measurements. In the radiocarbon HO determination, ^{14}CO is used as the HO tracer and has been mixed with ambient air in several configurations. The oxidative reaction is

$$^{14}CO + HO \xrightarrow{\text{O}_2} {}^{14}CO_2 + HO_2 \tag{19}$$

$$k_{298} = 1.5 \times 10^{-13}(1 + P_{atm}) \text{ cm}^{-3}\text{s}^{-1}$$

(where P_{atm} is the pressure in atmospheres; 1 atm = 101.325 kPa), and the measurement is made by comparing the count rates of the tracer CO with those of the product CO_2, which must be chemically separated with high efficiency. Typical counting times are 20 min. For conditions where sampling and CO addition for a contact time t do not perturb the ambient photochemistry, equation 17 reduces to (84) $[HO] = [^{14}CO]/kt[^{14}CO_2]$. Numerous precautions must be employed in the successful implementation of this technique (83). Included in these are maintaining the purity of the tracer CO with respect to other ^{14}C-containing compounds (e.g., $^{14}CH_4$) or other radionuclides (e.g., ^{222}Rn) and the absence of surfaces that can catalytically oxidize CO to CO_2 during any stage of the process. (Stainless steel surfaces were found to be catalytically active.) It is also essential that surfaces in the sampling train do not unduly perturb the ambient HO concentration or that, if so perturbed, the HO returns to a steady-state concentration characteristic of the nearby open atmosphere.

This technique has been applied in three configurations. In the earliest, ambient air was drawn into a static reactor to which the tracer ^{14}CO was admixed. Samples were taken in from 10- to 100-s contact times, and $^{14}CO_2$ was separated cryogenically, purified, and later counted in the laboratory. Theoretical estimates of wall loss of HO were made, but no mention was made of potential offgassing of wall contaminants that could react with HO (108, 109) or of spurious HO generation within illuminated reaction vessels (110, 111), both potential problems known to air pollution chemists. For instance, air pollution chemists typically minimize wall effects by "conditioning" Teflon [poly(tetrafluoroethylene)] film bags with long irradiation periods, whereas Campbell et al. (*83*) frequently substituted fresh bags. Reported HO concentrations ranged from less than 2×10^5 (the stated detection limit) to 8.6×10^7 cm^{-3}. No uncertainties were quoted. Concentrations in the range of greater than 5×10^7 cm^{-3} seem quite high relative to atmospheric model calculations, and the potential exists for influences by the Teflon reactor. This method of ambient HO measurement is perhaps the most straightforward and least expensive of any proposed, particularly if nonradioactive chemical tracers could be used to eliminate the expense and complications of sample taking, storage, and later counting and to allow near real-time HO determination. However, before such a method can gain credence, it should be shown that the reactor walls do not influence enclosed HO concentrations. The short static reactor residence times used by Campbell et al. (*83*) (10–100 s) should have minimized wall problems. Nevertheless, Campbell and co-workers no longer use the static reactor; instead, they favor a flow reactor approach. The static reactor is similar in principle to the FAGE calibration procedure of Hard and co-workers (*79, 81*), who do not rely on the Teflon-film reactor to reproduce ambient concentrations but rather to generate a sample airstream containing an HO concentration that can be calculated from tracer consumption. Calibration of HO instruments by static and flow reactor sources of HO is discussed in the "HO Intercomparison" section.

The flow reactor of Felton et al. (*84*) samples ambient air with a minimum degree of perturbation (Figure 7). Ambient air is drawn by a sample pump into a quartz tube that faces the prevailing wind. The UV/vis-transparent

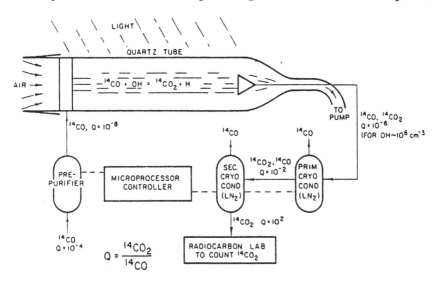

Figure 7. Diagram of CTM flow reactor. Values of the efficiency of separation Q are for ideal operating conditions. (Reproduced with permission from reference 83. Copyright 1986 Kluwer Academic Publishers.)

quartz tube should cause no perturbation of the radiation field; the external reflections are compensated for by internal reflections (112). The tracer CO is added from a number of nozzles, and flow modeling indicates that the central portion of the tube flow is not in diffusive contact with the tube walls during the reaction time, on the order of 10 s, for a linear flow velocity of about 7 cm/s. Sample times of 100 s yielded maximum daily HO concentrations ranging from about 2×10^6 to 9×10^6 cm^{-3} with standard deviations in the counting statistics around 2×10^5 cm^{-3}.

A third configuration of the characteristic tracer measurement (CTM) technique was developed for aircraft HO measurements during the NASA CITE 1 experiments (83). The high air velocity associated with aircraft motion caused the investigators to attempt a direct HO titration with ^{14}CO rather than try to decelerate the air to velocities used in the photostationary reactor. A 0.2-s reaction time was coupled with higher ^{14}CO concentrations to produce a 5 to 30% conversion of HO to $^{14}CO_2$. However, in the titration mode, very low impurities in the ^{14}CO and very high $^{14}CO_2$ separation ratios are required [2×10^{-9} and 10^9, respectively (83)]. During the CITE 1 experiments the $^{14}CH_4$ contamination problem could not be successfully overcome, and the investigators did not obtain meaningful data.

In spite of prior difficulties, problems associated with ^{14}C impurities have been overcome, and recently reported ambient HO measurements (84) have given HO concentrations with impressively short collection times of 100 s per HO datum. Current implementation of the radiocarbon technique has been described in detail (85), and repetitive measurements at a clean-air site in eastern Washington state have yielded midday HO concentrations of $(5.6 \pm 0.1) \times 10^6$ molecules per cubic centimeter with impressively high precision (86). Instrumental sensitivity fluctuations were estimated to have an upper bound of 16%, and the detection limit was 10^5 cm^{-3} (86).

Ion-Assisted CTM. In contrast to the radiocarbon approach, in which perturbations of ambient HO photochemistry are minimized, an alternative approach is rapid HO titration, exemplified by the ion-assisted mass-spectroscopic measurements of Eisele and Tanner (87). In their method, ambient HO is titrated with added $^{34}O_2$, forming $H_2^{34}SO_4$ (in three rapid steps in the presence of ambient O_2 and H_2O) in a flow tube. In a subsequent stage, NO_3^- ions are added to ionize H_2SO_4 to HSO_4^- for mass-spectrometric analysis. The HO titration is restricted to times shorter than 0.1 s to minimize replenishment of HO by reaction 3, and the ion reaction time is kept similarly short to minimize HSO_4^- production by extraneous paths. Subtraction of the background (arising mainly from ionization of ambient $H_2^{34}SO_4$) is achieved by periodic addition of propane to compete with $^{34}O_2$ for HO removal in the titration stage.

Table III summarizes the performance of LPA, LIDAR, EAPF, FAGE, and CTM as reflected in measurements described in the literature since

Table III. Comparison of Recent HO Measurements

Method	Best Average Time	Random Uncertainty[a]	Altitude[b]	Noise Sources	Calibration	Possible Systematic Error Sources
LPA:						
Jülich[c]	10 min	2	low	turbulence, haze, and unknown absorptions	absolute	underlying absorptions
NOAA[d]	15 min	0.5	2.7			
Frankfurt[e]	1 h	0.3	1.2			
LIDAR[f]	~1 h	3.2	0.6 to 10	solar background, scattering, and air fluorescence	calculated	sloping solar and air fluorescence baseline
EAPF[f]	~1 h	2.2	0.6 to 10	air fluorescence, interference, and scattering	O_3 or H_2O $+ h\nu \rightarrow HO$	$O_3 \rightarrow HO$ rate constants
FAGE2[g]	~1 h	0.4	0	air fluorescence, scattering, and PMT ringing	CT[h]	negative O_3 offset
FAGE3[i]	9 min	1	0	solar background and air fluorescence	CT	varying nozzle transmission
14C[j]	100 s	0.4	low	^{222}Rn contamination and $^{14}CO_2$ counting statistics	absolute	collection and purification
34SO2[d]	300 s	0.1	2.8	ambient $H_2{}^{34}SO_4$	O_3 and/or H_2O $+ h\nu \rightarrow HO$	unrecognized chemical interferences or inefficiencies

NOTE: Sensitivities of individual techniques have improved with time.

[a] The 2σ uncertainty is in units of 10^6 molecules per cubic centimeter ambient HO at the stated averaging time. Uncertainties are stated for the most recent published measurements and were converted from 1σ values as necessary. [b] Altitude in kilometers of recent experiments. [c] From Hofzumahaus et al. (41) for ambient HO data at Jülich on July 14, 1987. [d] From Mount and Eisele (88) ambient data at Fritz Peak, Colorado, August 23 and 24, 1991. [e] From ambient data of Comes et al. (54) at Schauinsland, Germany, August 22, 1991. [f] From spring 1984 daytime measurements reported in Table 5 of Beck et al. (95) for all altitudes. When HO data from below 3 km are excluded, 2σ uncertainties are 1.6×10^6 molecules per cubic centimeter for EAPF and 2.2×10^6 for LIDAR. [g] From ambient data of Hard et al. (80) at a Pacific coastal site, May and August 1987. [h] CT, chemical tracer calibration. [i] From ambient data of Felton et al. (84) at a ground-level rural clean-air site. [j] From calibration results of Chan et al. (81) at a ground-level rural clean-air site.

1987. In the high-altitude techniques, the background from absorbing and fluorescing species is lower and the fluorescence yield of the relatively constant HO concentration (Figure 2A) is higher than in the atmospheric boundary layer. Thus, fluorescence measurements from aircraft should be more sensitive than ground-based measurements.

Fundamental Hurdles in HO Determination

Tropospheric HO determination has involved surmounting formidable barriers in fundamental analytical chemistry—barriers that in retrospect might have been recognized before they were discovered experimentally. Most of the problems of the radiocarbon approach have been associated with achieving high purity of the tracer and the oxidation product. Once these problems are overcome, sensitivities are dependent upon counting statistics of these two species and are therefore related to the counting statistics associated with the spectroscopic measurements. Several considerations in the spectroscopic approach are treated here, including the production of HO by the probe laser and fundamental considerations of signal-to-noise ratio (SNR) and its interrelationship with averaging times.

Reactions 9 and 10 are thought to dominate HO production in the cleanest regions of the troposphere. Thus any approach that utilizes intense laser radiation at wavelengths shorter than about 320 nm should be undertaken with some care. In addition to ozone, H_2O_2 and HONO photolyze in the UV to produce HO and must be considered photolytic precursors to spurious HO. Nevertheless, O_3 has been the only significant known photolytic parent of spurious HO to date. The earliest measurements of tropospheric HO contained significant contributions from spurious HO for which corrections were applied (71).

Subsequent measurements have revisited this problem on a surprising number of occasions. For instance, Davis et al. (96) corrected additional aircraft-borne tropospheric HO measurements for ozone interference with improved data on nascent HO distributions, such as those of Rodgers et al. (113). Furthermore, Hard et al. (79) reported a small negative offset of 2 × 10^5 cm^{-3} in their wintertime measurements but found no evidence for an offset in the summer data. The origin of negative offsets in FAGE was later recognized (107) as ozone photolysis to produce O (^1D), which reacts with the chemical modulation reagent in FAGE, as described later in the chapter.

Two spurious HO cases may be recognized when excitation is with a pulsed laser—both involving ozone photolysis to produce O (^1D)—and its subsequent reaction with ambient water vapor to produce HO. In the first case, this spurious HO is detected by the same laser pulse, whereas in the second case it is detected by a subsequent laser pulse. The latter problem can be more significant, because the spurious HO grows rapidly in time following the initial production of O (^1D). These two types of behavior make laser temporal pulse width, repetition rate, and air velocity important in preventing excessive accumulation of laser-generated HO. The largest potential difficulties exist in the LIDAR and LPA configurations at the earth's surface, where local winds or eddy diffusion is required to remove sampled air in which significant concentrations of spurious HO may accumulate. In aircraft operation, or in sampled airstreams, the sampling velocity makes the two-pulse problem easier to deal with. In the absence of saturation, the production and detection of spurious HO is a phenomenon second-order in the laser flux. Because ambient HO determination is first-order in laser intensity, the relative signal from spurious HO effectively depends upon the first power of the laser flux. Thus, one method of minimizing this problem is laser beam expansion (71), resulting in a more diffuse fluorescence signal that is more difficult to collect geometrically and that is contaminated by higher levels of ambient light. Wang et al. (72) made a preflight measurement of the interference with the beam-expanding telescope and Davis et al. (73) made interference measurements in flight by temporarily removing the telescope to reduce the laser beam diameter 200-fold. From a plot of the interference signal versus the product of $[O_3]$, $[H_2O]$, and laser flux they calculated spurious HO to be less than 1×10^5 cm^{-3} in the expanded beam, except for a value of 4×10^5 for low-altitude runs.

Laser beam expansion was also employed in LPA laser detection of HO (36–39). Hübler et al. (38) calculated an asymptotic laser-generated HO concentration in their quasi-continuous-wave expanded laser beam by assuming a chemical decay lifetime of 1 s for the excess HO. Chemical recycling of this HO was assumed to be slow with respect to the residence time of air in the laser beam.

In the absence of radiative saturation of the HO transition, the processes occurring during excitation—that is, vibrational relaxation (in the $1 \leftarrow 0$ excitation case), rotational relaxation, quenching, and fluorescence—of the ambient HO are straightforward, and the necessary rate parameters have been measured for the evaluation of equations 15 and 16. When O (^1D) is produced by the probe laser, it reacts with water vapor to produce a hot nascent internal energy distribution (113–118). With pulsed laser excitation these nascent distributions relax on time scales that are long with respect to typical laser-pumped dye laser pulse widths but short with respect to the interval between laser pulses. Thus these relaxation processes are inefficient during a single laser pulse, but they can be essentially completed at the arrival of the next laser pulse and thus can lead to significantly larger interferences. Only recently (80, 103) has detailed kinetic modeling of these relaxation processes been used to estimate the magnitude of fluorescence from photolytic HO relative to that from the native radical concentration. In the absence of radiative saturation, this treatment is still relatively straightforward, although experimental uncertainties in the various kinetic parameters used in predicting photolytic HO introduce a fair amount of uncertainty, estimated by Smith and Crosley (103) to be of the same order as the net interference in chemical modulation by isobutane.

If the HO transition is partly saturated, then the detailed kinetic treatment is further complicated, but a detailed solution is still possible. Smith and Crosley (103) used their model to calculate interferences in the HO data of Hard et al. (79). Subsequently, Hard et al. (80) developed a similar model, measured ozone interferences with the instrument used to obtain the ambient HO data in their 1986 paper (79), and applied the model results to both those data and more recent ambient HO data obtained with the same system. In the more recent measurements, ambient ozone data were available, so observed negative nighttime offsets could be compared directly with model calculations.

The two modeling efforts used a similar set of spectroscopic and kinetic processes and associated parameters but employed different excitation conditions. These different conditions were chiefly in laser power, beam cross section, and details of the pulse geometry within the excitation cell. Smith and Crosley (103) used nominal laser energies, whereas Hard et al. (80) specifically measured the laser energy for input to their model. The Hard et al. (79) ambient HO measurements employed multiple-pass laser excitation in a White cell with partial overlap of the reflected beams. As both models have shown, most of the interference arises from the growth of spurious HO between the several passes of each laser pulse in the White cell, and the fraction of this HO that is detected depends upon the relative beam alignment. Hard et al. (80) did interference experiments with the multipass arrangement used in ambient measurements and also measured multipass-to-single-pass ratios of the interferences and the instrument's response to ambient HO. Thus they were able to compare their model-calculated spurious HO production to that observed experimentally when measured concentrations of ozone and water vapor were sampled. From this comparison they used the model to estimate a fractional overlap of successive beams. Smith and Crosley, on the other hand, assumed a given level of beam overlap and did not present results for single-pass excitation under the same conditions.

Because most of the interference arises from beam overlap, the absence of single-pass output from the latter model makes it difficult to directly compare the two models. In spite of these differences between the modeling efforts, the major findings of both models are similar and may be summarized as follows:

1. Partial saturation of the HO absorption and use of excitation line widths wider than the HO absorption line width are in general undesirable, because they may produce spurious HO while exciting HO fluorescence with less than 100% efficiency (103). However, the issue is not as simple as matching the HO absorption profile exactly and avoiding all saturation. For instance, a matched laser excitation line width is more difficult to achieve and to keep tuned to the HO absorption, and the

ambient HO signal-to-noise ratio may continue to improve with increasing laser power even when partial saturation of HO occurs. HO saturation can be reduced by expanding the laser beam; there is a gain in signal until either saturation is eliminated or the emission exceeds the field of view of the detection optics. However, in an enclosed volume, light scattering or fluorescence from windows, walls, masks, and so forth is a major background source that can be harder to control with a larger diameter laser beam. The effects of partial saturation on SNR can be quantitatively evaluated by using the model to derive signal-to-noise ratios that include all background sources (*80*).

2. Chemical modulation in FAGE allows the use of higher laser fluxes than can be used with spectral modulation because it reduces nearly all of the photolytic HO fluorescence to a background subtracted out in the two-channel system (*78–80*). However, the isobutane modulating reagent used by Hard and co-workers was not without its problems, at least for excitation at 282 nm with their Nd:YAG-pumped dye laser. All reported FAGE measurements—except for interference tests—chemically modulate HO by alternating ~460-ppm isobutane between the two air-sampling nozzles. Three sources of interferences have been associated with this modulating reagent:

- The O (^1D) precursor to HO reacts partially with the isobutane modulating reagent; this reaction reduces the production of HO from its reaction with water in the background channel. This contributes a positive offset. Shirinzadeh et al. (*106*) first pointed out this effect but overestimated its magnitude about 1000-fold (*80, 107*) by applying a steady-state model to a transient kinetic effect. Hard et al. (*80*) showed that this effect is negligibly small.

- Substitution of isobutane for a small fraction of the airstream increases the quenching of spurious HO fluorescence in the background channel by about 1%, so exact cancellation is not achieved in subtraction and a false positive signal is introduced (*80, 103, 107*). This situation applies not only to ozone photolysis but also to all other photolytic HO sources.

- As a corollary of the first interference, the reaction of O (^1D) with the isobutane modulating reagent produces additional HO in the background channel and contributes a negative offset (*80, 103, 107*).

These effects are summarized in Table IV. Of the three effects, the third is dominant, and it results in a negative offset proportional to ozone with a magnitude estimated by Smith and Crosley as -8×10^6 (units of ambient equivalent HO per cubic centimeter) for 100-ppbv ozone. Hard et al. (80), using less extreme laser conditions they determined to be representative of their system but larger overlap between adjacent beams in their White cell, calculated this offset as -2×10^6 cm^{-3} for 50-ppbv ozone, which is an upper-limit ozone concentration for the downtown urban and coastal locations of their reported HO measurements (78, 79). In the coastal measurements, nighttime negative offsets were observed that are not inconsistent with the values calculated with their model for similar conditions.

Table IV. Model Estimates of Ozone Interference in FAGE2 with 10-torr (1300-Pa) H_2O and 50-ppbv O_3

Interference Effect	Single Pass, Reference 80	Multipass Reference 80	Multipass Reference 103
O (^1D) competition	+0.0007	+0.005	NA
Quenching of spurious HO*	+0.062	+0.47	+0.4
O (^1D) + isobutane $\rightarrow$ HO	−0.34	−2.60	−4
Net	−0.28	−2.1	−3.6

NOTE: Units are the equivalent signal from 1×10^6 HO molecules per cubic centimeter ambient; the symbol " + " indicates positive interference from spurious HO, and "−" indicates a negative interference. Gross resonant background is the sum of backgrounds with and without isobutane; values for single-pass (80) and multipass (80, 103) measurements were 10, 78, and 63, respectively. NA, not available.

SOURCES: Hard et al. (80) and Smith and Crosley (103). Smith and Crosley performed calculations for 100-ppbv O_3; their results are halved here for comparison. Differing predictions from these two references are due largely to Smith and Crosley's use of 1-mJ nominal laser energy and a 0.4-cm beam diameter compared to Hard et al.'s 0.6-mJ measured energy and measured multipass diameters alternating between 0.6 and 0.4 cm.

Smith and Crosley suggested remedies for these problems. The most obvious is the removal of overlap between successive passes in the White cell, because the time delay between these passes allows the conversion of a large amount of photolytic O (^1D) to spurious HO. If the resultant interference levels are still unacceptable, modification of the chemical modulation system can lower them to acceptable levels. A chemical modulator that does not react with O (^1D) to yield HO (they suggest deuterated isobutane or $CClFCF_2$) will eliminate the third effect, and the quenching may be remedied by balancing the quenching in the background channel by the addition of an appropriate quantity of a non-HO-reactive quencher to the signal channel. Neither remedy will influence the first effect, but its magnitude is negligible. The chemical remedies remove the net ozone interference even with the overlapping White cell beams, and the nonreactive quencher in the HO channel is also effective in eliminating interference from direct production of spurious HO, for example, from H_2O_2 photolysis. However, the remedies do not affect the background from laser-produced HO (reactions 7 and 8 still occur), which degrades the signal-to-noise ratio.

3. Because the photolytic HO production is linear in laser intensity, an increase in the laser repetition rate at constant average power has the same effect as beam expansion in reducing interference. This approach, simulated in the models of Smith and Crosley (*103*) and Hard et al. (*80*), has the advantage of not requiring a greatly enlarged beam cross section. However, the necessary increase in fractional duty cycle increases the importance of the scattered ambient light background and the photomultiplier tube (PMT) dark current.

   A pumping laser satisfying these requirements is the Cu vapor laser, first advocated for the same reasons by J. G. Anderson in about 1980. Thus Stimpfle et al. (*93, 94*) used a Cu vapor laser pumped dye laser that was frequency-doubled to 282 nm to determine stratospheric HO during balloon-borne descent in the stratosphere. At the 17-kHz repetition rate of this laser, multiple-pulse photolytic HO accumulation appears to have been avoided by the use of a fan downstream of the excitation zone to increase the air velocity beyond that provided by balloon descent.

   Chan et al. (*81*) likewise reported the use of a Cu vapor laser system in a third-generation instrument (FAGE3), but they used a 6-kHz repetition rate, pumping the HO ($^2\Sigma$ v' = 0) $\leftarrow$ ($^2\Pi$ v'' = 0) excitation at 308 nm. At this repetition rate (and excitation wavelength; *see* the next section) and at the flow velocity used, back diffusion of reacting O (^1D) and its spurious HO product produce negligible net interference, as shown by both experimental measurement (*81*) at very high ozone concentrations and the model (*80*).

4. Major reductions in photolytic background may be obtained by shifting the excitation wavelength from (1 $\leftarrow$ 0) at 282 nm to (0 $\leftarrow$ 0) at 308 nm. Although Smith and Crosley (*103*) did not consider this excitation approach, Hard et al. (*80*) assessed it with their kinetic model, and Chan et al. (*81*) successfully implemented (0 $\leftarrow$ 0) pumping in a third-generation FAGE instrument. At 308 nm the absorption cross section is 4 times

greater and the fluorescence yield is about 30% larger relative to the values at 282 nm. In contrast, the ozone absorption cross section is 24 times less and the O (^1D) quantum yield is 1.2 times less. Hard et al. attempted (0 ← 0) excitation with their YAG-pumped system (FAGE2) but found this detection method unsuccessful because of the very high directly scattered background and the associated ringing of the photomultiplier that comes with the intense 30-Hz YAG-pumped pulse (81). With 6-kHz Cu vapor laser pumping at 308 nm, the less intense directly scattered light may be discriminated against simply by time-gating of the detection electronics.

5. Other photolytic precursors, such as H_2O_2 and HONO, are photolyzed by the laser and thus may yield significant HO relative to that generated by ozone photolysis. The H_2O_2 interference was addressed by both Smith and Crosley (103) and Hard et al. (80) as a possible error in the ambient HO data of Hard et al. (79). For 1-ppbv H_2O_2 in the FAGE2 system, Hard et al. (80) predicted gross HO production, net interference, and isobutane's fractional modulation of the total quenching as 7×10^6 cm^{-3}, 4×10^4 cm^{-3}, and 1.2%, respectively, compared with Smith and Crosley's results of approximately 1.2 $\times$ 10^7 cm^{-3}, 8×10^4 cm^{-3}, and 1.3%. Smith and Crosley's higher values are due to their assumption of fully thermalized nascent HO. Smith and Crosley examined several other photochemical mechanisms for laser production of HO, including photolysis of HNO_3, HONO, and alkyl hydroperoxides, and found that none of them yield more than 0.5% of the HO that O_3 photolysis does.

Signal-to-Noise Considerations

All HO measurement techniques require signal averaging for varying lengths of time, and the improvement of SNR with averaging time is a major goal. An allied issue is the choice of an optimal modulation frequency for switching between background and signal measurements. In fluorescence measurements this choice involves tuning the laser off the absorption wavelength or switching a chemical modulator between channels; in LPA, the modulation frequency is that with which the spectral region near the HO absorption has been scanned. Hübler et al. (38) increased their spectral scan rate from 400 to 6600 Hz to eliminate laser fluctuation noise and achieved SNR limited by "photon statistics", but they mentioned that photon-limited statistics usually were not achieved for path lengths over 10 km. For Poisson statistics as applied to a single-signal measurement (transmitted laser light in LPA, gross fluorescence signal and background in any of the fluorescence techniques, or radiolytic events in CTM), the standard error in the mean signal is expected to decrease and the SNR to increase as the accumulated number of events to the 1/2 power, or as the square root of the averaging time. This $t^{1/2}$ improvement with averaging time can result more fundamentally from Gaussian statistics (discussed later in the chapter). On the other hand, $t^{1/2}$ dependence is not implicit to fluorescence measurements: t^1 dependence can be achieved with methods that have a sufficiently low background and yield signal-limited detection uncertainties, such as those demonstrated in sequential two-photon excitation of NO fluorescence in field experiments (99) and in HO fluorescence in a laboratory simulation (100).

A summary of the instrumental variables (that may be optimized by the designer or experimenter) and their relations with the data-averaging variables is given in Table V. These relations are based on Poisson statistics.

Table V. Exponent of Dependence of SNR, MDC, and MAT on Three Instrumental Variables for Signal-Limited and Background-Limited Detection Cases in the Poisson Noise Limit

	SNR		MDC		MAT	
Variable	*SL[a]*	*BL[a]*	*SL*	*BL*	*SL*	*BL*
R_S	$1/2$[b]	1	−1	−1	−1	−2
R_B	0	1/2	0	1/2	0	1
$R_S[HO] + 2R_B$	1/2	1/2	NA[c]	NA	−1	−1
SNR	NA	NA	2	1	2	2
MDC	1/2	1	NA	NA	−1	−2
t_{av}	1/2	1/2	−1	−1/2	NA	NA

[a]SL, signal-limited case; BL, background-limited case.
[b]The numerical entry represents the power to which the row variable, varied alone, must be raised to be proportional to the column variable to the first power. For example, the column variable SNR is proportional to the row variable $R_S^{1/2}$ in the signal-limited case and to R_S^1 in the background-limited case.
[c]NA, not applicable.

The quantity t_{av} is taken as the averaging time in any convenient time unit, and photon arrival rates are calculated per that time unit. If (1) R_B is the background photon arrival rate, (2) R_S is the net HO signal photon arrival rate per unit HO concentration, (3) the noise source consists of photon fluctuations, and (4) there is no correlation between corresponding values of the measured quantities $S + B = G$ (averaged gross signal) or B (averaged background signal), then the expression for the signal-to-noise ratio SNR $= S/\sigma_S = S/(\sigma_G^2 + \sigma_B^2)^{1/2}$ is

$$SNR = R_S[HO](t_{av}/(R_S[HO] + 2R_B))^{1/2} \qquad (20)$$

After the resulting quadratic equation is solved for [HO] and a solution of negative sign is rejected, a formula for the minimum detectable HO concentration (MDC) corresponding to specified values of SNR and t_{av} is obtained (MDC is defined by substituting the desired level of accuracy, SNR = 1, SNR = 2, and so forth, as desired):

$$\text{MDC} = \text{SNR}^2(1 + (1 + 8R_B t_{av}/\text{SNR}^2)^{1/2})/2R_S t \qquad (21)$$

Equation 20 can be rearranged to find the minimum averaging time (MAT) required to obtain a desired value of SNR:

$$\text{MAT} = \text{SNR}^2(R_S[\text{HO}] + 2R_B)/(R_S[\text{HO}])^2 \qquad (22)$$

$$= (\text{SNR}_{\text{desired}}/\text{SNR}_{t_{av}=1})^2$$

The instrument variables R_S, R_B, and $R_S + 2R_B$ are used in instrument optimization; for example, an improved matching of the laser bandwidth to the HO absorption could increase R_S, a reduction in illumination of walls near the detection zone by ambient light or scattered or diffracted laser light could decrease R_B, and an increase in photon collection efficiency could increase ($R_S + 2R_B$). The remaining quantities t_{av}, MAT, SNR, and MDC may be traded off during data processing, but the choice of their values is restricted by the instrument variables.

The ideal of signal-limited detection, in which the measurement accuracy is independent of the size and fluctuations of the background, has not yet been achieved in any spectroscopic measurement of ambient HO. However, because the signal-limited condition may be achieved in the future, this case is included in Table V.

In LPA and fluorescence experiments, the third source of interference (see bulleted list) has often been weakened by the presence of noise sources other than photon noise: fluctuations in laser power or in ambient concentrations of substances causing nonresonant absorption and fluorescence, for example. In addition, all measurements are influenced at least implicitly by variability in HO during the averaging time. These additional noise sources mean that $(R_S[\text{HO}] + 2R_B)^{1/2}t_{av}^{1/2}$ may not be a satisfactory approximation of $(\sigma_G^2 + \sigma_B^2)^{1/2}$. However, the standard deviation in the net may be less than that given by uncorrelated Poisson statistics because of a significant degree of correlation between G and B (discussed later in the chapter). In this case, equations 20 through 22 may be regarded as conservative estimates of actual instrument performance. In cases where G and B have uncorrelated sources of variation beyond those of Poisson statistics, however, equations 20 through 22 may overestimate the SNR.

In many cases it may be impossible to separate fluctuations in ambient HO from background fluctuations or other sources of noise. The proper uncertainty in the net HO signal, averaged over any chosen time interval, is the standard error of the mean calculated from standard statistical formulas from the net data only. This calculation is independent of the applicability of Poisson statistics but may also yield improvement in SNR with the square root of the averaging time, because the computed standard deviation σ must be divided by $n^{1/2}$ in computing the standard error in the mean σ_m (n refers to the number of measurement intervals lumped together in averaging). However, the presence of nonphoton noise due to fluctuations in (1) laser pulse energy, (2) spectral overlap of the laser with the HO absorption line, (3) laser alignment, (4) ambient sources of nonresonant fluorescence and scattering, or (5) ambient HO concentration typically leads to $1/f$ noise, which yields an improvement of less than $n^{1/2}$ or $t_{av}^{1/2}$ in SNR upon averaging.

On the other hand, Poisson statistics may provide too conservative an estimate of SNR if significant correlation exists between background and signal-plus-background measurements. Such correlation is to be expected in many circumstances and has led Rodgers et al. (77) to use a two-wavelength scheme to measure these two quantities almost simultaneously—one laser excites the HO transition while the second measures the background signal at a similar wavelength. Although effective, the use of two complete laser systems may not be a practical approach for routine measurements. Wang and co-workers (73) minimized the time between signal and background measurement with rapid sequential mode hopping in the excitation laser to maximize the correlation between signal and background measurements. Armerding et al. (53) determined the sampling-frequency dependence of the baseline standard deviation for their DOAS LPA measurement of HO. This dependence is shown in Figure 8.

The early signal uncertainty estimates of Hard et al. (78, 79) employed the standard error from Poisson statistics $(R_S + 2R_B)^{1/2}$, which gives too conservative an estimate of the error in these fluorescence measurements. Thus, the actual standard error in the mean σ_m just discussed was often $1/2$, or less than the value given by $(R_S[HO] + 2R_B)^{1/2}t_{av}^{1/2}$. Although the covariance is implicitly accounted for in the calculation of σ_m from the net signal data, its role in improving SNR may be seen explicitly by expressing the SNR in terms of the individual variances and covariance in the background B and the gross signal G:

$$\sigma_{net} = [\text{var}(G) + \text{var}(B) - 2\,\text{cov}(G,B)]^{1/2} \tag{23}$$

In reading various HO measurement reports [including Hard et al. (78)], we find too little attention given to clearly expressing how the uncertainty limits are derived. This negligence may be the result of assuming such uncertainties are trivial to calculate. However, future reports should state explicitly whether 1σ, 2σ, 90% confidence limits, and so forth employ Gaussian or Poisson statistics and whether the quoted uncertainties are internal to the ambient HO data or include calibration uncertainties.

This discussion deals with random errors and their propagation in reported HO concentrations. Equal attention should be given, of course, to systematic errors of calibration or instrument drift.

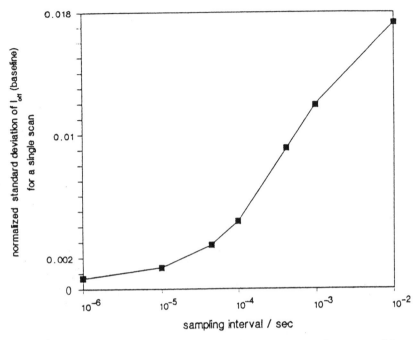

Figure 8. Dependence of baseline standard deviation on sampling interval for LPA–DOAS determination of HO near 308 nm. (Reproduced with permission from reference 51. Copyright 1990.)

Advantages and Limitations of HO Measurement Techniques

LIDAR. *Advantages.* A small air volume is sampled that is removed from local surfaces, especially in aircraft implementation.

Disadvantages. The method uses calculated calibration with associated uncertainties in input parameters—however, this problem is resolvable with the increasingly accurate information on relevant energy-transfer processes. It is difficult, but not impossible, to calibrate against a common HO source of "known" concentration. The method is very sensitive to solar background at the detection wavelength, the fluorescence signal is effectively coincident in time with direct scattering and fluorescence, and LIDAR is subject to accumulation of photolytic HO in stagnant air. It may be amenable to sequential 2-λ excitation, which has apparently not been proposed; calibration in a 2-λ mode might be difficult, however.

EAPF. *Advantages.* A large-diameter sample tube excludes most ambient light and provides minimum perturbation to ambient HO. Past sensitivities have been marginal with single-photon excitation. Sequential

2-λ excitation could greatly reduce photolytic interference, nonresonant fluorescence, and scattering backgrounds. The method can be calibrated by fundamental calculation or by ozone photolysis by a shortwave UV source. The successful 2-λ determination of NO provides confidence in the fundamental principles of this technique and shows its adaptability for nonsimultaneous measurements of NO, NO_2, and HNO_3.

Disadvantages. Scattering and fluorescence are not separated in time if performed at ambient pressure; however, sequential two-step excitation will reduce detected scattering and nonresonant background to negligible levels. Reproducibility of calibrations and field measurements may be compromised by the need to monitor two laser energies and two spectral overlaps between the laser lines and the respective HO lines, although the reproducibility of the calibration is enhanced by a reference HO cell. In ozone (or H_2O_2) photolysis calibration, uncertainties in the absorption cross sections will be carried over into the calibration. With ozone, rate constant uncertainties for the reaction of O (^1D) with water and other atmospheric species also propagate into calibration uncertainties. Uncertainties in concentration measurements for H_2O_2 or O_3–H_2O also propagate into the calibration. However, agreement within 20% has been reported between the two photolytic parent calibrations. (Because both calibrations used the same 266-nm photolysis laser, this agreement is independent of the absolute power of the photolysis source.) Relative vibrational and rotational distributions of photolytic HO will not present a problem as long as sufficient time is allowed for the HO products to thermalize. The large sampling rate would make it difficult for this instrument to be intercompared with instruments drawing samples from a closed volume of known HO concentration. The method requires two complete laser systems.

FAGE. *Advantages.* A small air volume is sampled. The method stretches the HO fluorescence lifetime out for a simple separation from directly scattered laser light (Rayleigh, Raman, Mie, and container wall scattering). FAGE is easily calibrated from known HO sources, could be easily intercompared with most other point measurements, and can be implemented with either chemical or spectral modulation, giving independent measurements of same quantity. In chemical modulation, simultaneous measurement of signal and background in two channels may improve the signal-to-noise ratio because backgrounds in each may be correlated. FAGE can be configured for HO_2 measurement as HO, the calibration factor of which can be directly compared with that of ambient HO. Flow tube kinetics and dynamics have been studied for several decades by chemical kineticists. The system can be configured to simultaneously determine several other chemical parameters with a single pump laser, including NO_2 and NO (*104*).

Disadvantages.   FAGE passes sample through a nozzle and down a flow tube, the HO transmission of which must be incorporated into the overall calibration, and it adds the complexity of a vacuum apparatus to the laser system.

LPA.   *Advantages.*   This is an absolute technique, requiring no "calibration" per se, as long as absorption cross sections are accurately treated. An actual spectrum of several closely spaced lines can be obtained that can unambiguously identify HO as the absorber if there is an adequate signal-to-noise ratio. Nearby absorption bands may allow other species concentrations to be compared with "point" measurements by independent instruments. The sampled area is well removed from surfaces, including the ground. Reported sensitivities that use DOAS have been very good.

Disadvantages.   Unless a folded path is used, concentrations averaged over several kilometers make comparison with other HO measurements or modeled calculations difficult unless these quantities are also determined over the same path as that of HO. Even then, the effects of volume averaging are difficult to determine. The uncertainties associated with volume averaging are shared with "point" HO measurements requiring significant averaging times, for instance, from a moving aircraft. Measurements are only possible in relatively clean air. The method is difficult to implement from aircraft, but it could conceivably sample an atmospheric region subsequently probed by a low-flying aircraft. Absorptions by unidentified species may not always be separable from HO absorption. The maximum possible number of HO transitions should be monitored.

Radiocarbon.   *Advantages.*   This is a high-sensitivity, absolute method that can increase sample precision by longer laboratory counting until counting times become prohibitively long. The method has shown very good reproducibility (precision) in repeated measurements.

Disadvantages.   Real-time data cannot be obtained for use in modifying the experimental protocol during the experiment. Although impressively short sampling times have been reported (100 s), the ability to obtain repetitive samples for a quasi-continuous measurement is not currently possible. Frequent sampling would require a large number of sample canisters, in contrast to continuous measurements that have routinely run 24 h in some cases. Contamination of $^{14}CO_2$ by ^{14}CO would result in spurious HO concentrations when none were expected (such as at night) and would be easily recognized as a problem. Failure to capture all $^{14}CO_2$ or its loss during purification would result in a concentration measurement below the true ambient level and might be much harder to recognize as a problem. Further characterization of this technique—for instance, by comparing signals at varying sampling times for linearity—would be worthwhile.

Titration and Mass Spectrometry.   *Advantages.*   This method allows high-sensitivity, low background, quasi-continuous measurements, is amenable to all present external calibration methods, and has been successfully compared with LPA.

Disadvantages.   Further study may reveal possible positive or negative interferences in the chemical titration and ion generation reactions.

HO Intercomparison

Intercomparison of atmospheric chemistry analytical techniques is a fairly recent phenomenon, but one that has received wide acceptance. These efforts have sometimes been called "shoot-outs"; it has been stated that "most of the intercomparison field campaigns, particularly those done fairly rigorously, have been rather humbling experiences" (119). Two HO intercomparisons have been carried out under the sponsorship of NASA as one element of the CITE 1 campaign (95). One joint study involved ground measurements at Wallops Island in Virginia, and the second was an airborne comparison of five individual flights. Three research groups [LIDAR (74), EAPF (77), and radiocarbon (83)] participated in each phase of this study. As summarized by Beck et al. (95), "the ground results were inconclusive because operational problems with the instrumentation over the 10 days of testing provided no overlapping measurements with a signal-to-noise ratio (SNR) greater than 1."

Although CTM experienced instrumental difficulties during the airborne comparison (e.g., $^{14}CH_4$ contamination of the ^{14}CO), the other two techniques reported complementary measurements with SNR exceeding 1 on six occasions. The reported 1σ uncertainties of the two techniques were comparable, averaging 1.1×10^6 cm^{-3} (EAPF) versus 1.6×10^6 cm^{-3} (LIDAR); the latter value is higher possibly because of the presence of a stronger scattered solar signal in the open atmosphere data. Because no "OH standard" was possible in this study, the success of the intercomparison could only be based upon the level of agreement between the two techniques. Although the 1σ error bars were above zero for the 12 data points considered, higher confidence levels were achieved far less frequently, and agreement among the two sets was poor. The overview (95) concluded as follows:

> The OH intercomparison results were not of sufficient quantity to allow one to conclude that OH measurements in the clean, remote troposphere can be made with sufficient accuracy or reliability. The near-zero nighttime results from the two LIF techniques, however, do indicate that there are no major problems that can be attributed to artifacts or interference effects for clean, remote tropospheric measurements. Finally, it is of some significance that within the accuracy of the ensemble of OH measurements reported during

CITE 1, no disagreements between measurements were observed and current predictions of OH concentration in a "clean" environment do not disagree with the general OH level indicated by the data.

Considerably greater success has recently been achieved in an intercomparison of local and long-path HO measurements in Colorado. In this intercomparison (88) Mount (LPA) and Eisele (CTM titration–mass spectrometry) compared 10-km path-averaged HO concentrations with a local CTM measurement near the retroreflector. Agreement between the two techniques (Figure 9) was in general good, especially because there is reason to believe that actual atmospheric HO concentrations would not necessarily be the same for each technique, given their quite divergent averaging volumes.

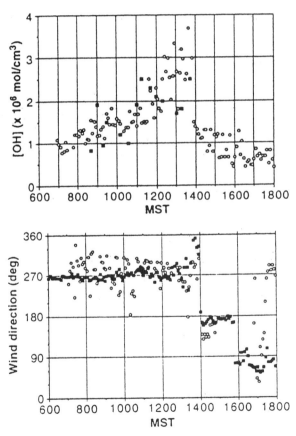

Figure 9. Intercomparison of ○, LPA and ■, CTM titration–mass spectrometric determinations of [HO] at Fritz Peak, Colorado. (Reproduced with permission from reference 88. Copyright 1992 American Association for the Advancement of Science.)

Generation of Reference HO Concentrations for Calibration and Instrument Intercomparison

Generation of reference HO concentrations during the measurement process is an extremely valuable procedure. This approach was developed for in-flight calibration of the EAPF sampling system (77) and employed O_3–H_2O photolysis by a second laser just prior to HO measurement. For laser delay times greater than 10 μs, HO thermalization will be complete. This calibration is subject to the uncertainties in concentration measurements of O_3–H_2O–N_2–O_2 and to uncertainties in the relevant rate constants, although the latter appear as systematic errors that can be corrected at any time provided that more accurate rate constants are known. If HO concentrations much higher than ambient are generated, reference HO measurements will have better precision, but extrapolation of a calibration curve outside its determined range then carries its own uncertainty. Still, the EAPF system's self-calibration possibilities present a major advantage currently available with no other technique.

Future intercomparisons of HO instruments should incorporate measurements of "known" or "standard" HO concentrations (the norm with less reactive analytes) as well as blind comparisons of ambient measurements. The simplest "known" HO source is a large-volume continuously stirred tank reactor (CSTR)—with volume flow sufficient to satisfy instrumental sampling rates—that is illuminated by sunlight. This source is equivalent to the CSTR used to calibrate FAGE and could similarly deliver flow to any CTM experiment.

Calibration of FAGE1 from a static reactor (a Teflon film bag that collapses as sample is withdrawn) has been reported (78). In static decay, HO reacts with a tracer T that has a loss that can be measured by an independent technique; T necessarily has no sinks other than HO reaction (*see* Table I) and no sources within the reactor. From equation 17, the instantaneous HO concentration is calculated from the instantaneous slope of a plot of ln[T] versus time. The presence of other reagents may be necessary to ensure sufficient HO; however, the mechanisms by which HO is generated and lost are of no concern, because the loss of the tracer by reaction with whatever HO is present is what is observed. Turbulent transport must keep the reactor's contents well mixed so that the analytically measured HO concentration is representative of the volume-averaged HO concentration reflected by the tracer consumption. If the HO concentration is constant, the random error in [HO] calculated from the tracer decay slope can be obtained from the slope uncertainty of a least squares fit. Systematic error would arise from uncertainties in the rate constant for the T + HO reaction, but several tracers may be employed concurrently. In general, HO may be nonconstant in the reactor, so its concentration variation must be separated from noise associated with the [T] measurement, which must therefore be determined separately.

Two dynamic alternatives to the static approach have been used in HO calibration and measurement. In the CSTR (continuously stirred tank reactor) approach, air containing the tracer or tracers flows into the reactor to balance the bulk flow out to the HO measuring devices, and the contents are stirred by a fan or other means. The HO chemical tracer is measured in the inlet flow to obtain $[T]_0$ and in the outlet flow to obtain $[T]$. Mass balance requires

$$\frac{d[T]}{dt} = \frac{F}{V}[T]_0 - (k[HO][T] + F/V)[T] \tag{24}$$

where F/V is the ratio of flow rate to volume, which is the inverse of the reactor residence time τ. At steady state, equation 24 reduces to

$$[HO] = (k\tau)^{-1}([T]_0/[T] - 1) \tag{25}$$

Errors in $[HO]$ due to $[T]$ measurement can be estimated by error propagation in equation 25, which yields

$$\sigma_{|HO|} = \frac{1.4([HO] + (k\tau)^{-1})\sigma_{|T|}}{[T]} \tag{26}$$

Here, the uncertainty $\sigma_{|HO|}$ introduced by the relative uncertainty $\sigma_{|T|}/[T]$ applies to a single hydrocarbon measurement, and the presence of this uncertainty in both $[T]$ and $[T]_0$ introduces the factor 1.4. If $[T]_0$, other reagents, or illumination of the reactor are not constant, $[HO](t)$ can be obtained from the time derivative of equation 25:

$$H' = \tau^{-1} - (k[HO] + r^{-1})H \tag{27}$$

in which $H \equiv [T]/[T]_0$ and the prime represents differentiation with respect to time.

Although Hard et al. (78) used the static method to calibrate the FAGE1 instrument's response to HO, they adopted the CSTR method to calibrate FAGE2 (79) and FAGE3 (81). In the static method, the large-volume flow rate into the FAGE sampling nozzles deflated conveniently sized reactors (250-L Teflon bags) rapidly and left inadequate time for signal averaging at very low HO concentrations. In contrast, CSTR allows longer averaging times and is less sensitive to wall sources and sinks of HO and its precursors that would be associated with collapse of the reactor. The CSTR calibrator could also deliver sample for EAPF, perhaps in the sequential 2-λ mode, if the sample flow rate were reduced to that used in the radiocarbon measurements. Before such an intercomparison is considered, radical wall loss and generation should be modeled for proposed geometries to investigate effects that the walls might have on HO concentrations at the sampling ports versus effects in the reactor interior.

The second dynamic approach is the atmospheric pressure flow tube, in which an organized two-dimensional flow field replaces the bulk mixing of the CSTR, as has been used by Davis and co-workers (77) for in-flight calibration of their EAPF system. It might be difficult to adapt this method to generate "known" HO for a instrument intercomparison, however.

An early intercomparison of LPA and CTM measurements of HO within a chemical reactor has been reported (120, 121). The measurements were made by directing the beam transverse to the detection axis within the reactor at atmospheric pressure; N_2 was replaced by He to improve fluorescence efficiency. A folded-path LPA measurement within a calibration chamber would have obvious advantages in an instrumental intercomparison and seems within the realm of possibility.

Summary

Tropospheric HO measurements published to date include chemical tracer techniques and several spectroscopic approaches, including both absorption and fluorescence. Each of these approaches is now capable of providing ambient HO data, but their sampling systems, sensitivities, averaging volumes, and potential systematic errors are widely divergent. Past ambient HO determinations have been the subject of varying controversy, which could be reduced in future measurements if existing methods were intercompared, as has become the norm in other areas of atmospheric analytical chemistry. A chemical tracer approach to generate a "known" HO concentration in such an intercomparison is feasible and should ultimately employ three or more successful methods to allow for potential inaccuracy in one of the methods. Provision of ancillary measurements of chemical and photochemical quantities would be useful so that the HO results and potential inconsistencies between them may be interpreted with the use of tropospheric photochemical models.

Acknowledgments

We gratefully acknowledge support from the National Science Foundation (ATM–8615163), the National Aeronautics and Space Administration (NAG–1–697), and the U.S. Environmental Protection Agency (R81–3012).

References

1. Logan, J.; Prather, M. J.; Wofsy, S. C.; McElroy, M. B. *J. Geophys. Res.* **1981**, *86*, 7210–7254.
2. Khalil, M. A. K.; Rasmussen, R. A. *Atmos. Environ.* **1987**, *21*, 2445–2452.
3. Khalil, M. A. K.; Rasmussen, R. A. *Chemosphere* **1982**, *11*, 877–883.

4. Khalil, M. A. K.; Rasmussen, R. A. *Atmos. Environ.* **1985**, *19*, 397–407.
5. Craig, H.; Chou, C. C. *Geophys. Res. Lett.* **1982**, 9, 477–481.
6. Blake, D. R.; Rowland, F. S. *J. Atmos. Chem.* **1986**, *4*, 43–62.
7. Blake, D. R.; Rowland, F. S. *Science (Washington, D.C.)* **1988**, *239*, 1129–1132.
8. Khalil, M. A. K.; Rasmussen, R. A. *Atmos. Environ.* **1987**, *21*, 2445–2452.
9. Levine, J. S.; Rinsland, C. P.; Tennille, G. M. *Nature (London)* **1985**, *318*, 254–257.
10. Khalil, M. A. K.; Rasmussen, R. A. *Chemosphere* **1990**, *20*, 222–242.
11. Finlayson-Pitts, B. J.; Pitts, J. N., Jr. *Atmospheric Chemistry: Fundamentals and Experimental Techniques*; Wiley–Interscience: New York, 1986.
12. Calvert, J. G.; Stockwell, W. R. *Can. J. Chem.* **1983**, 983–992.
13. Atkinson, R.; Balch, D. L.; Cox, R. A.; Hampson, R. F.; Kerr, J. A.; Troe, J. *J. Phys. Chem. Ref. Data* **1989**, *Monograph 1*, 881–1097.
14. Leighton, P. *Photochemistry of Air Pollution*; Academic Press: New York, 1961.
15. O'Brien, R. J.; George, L. A. In *The Science of Global Change: The Impact of Human Activities on the Environment*; Dunnette, D. A.; O'Brien, R. J., Eds; ACS Symposium Series 483; American Chemical Society: Washington, DC, 1992; pp 64–116.
16. Weinstock, B. *Science (Washington, D.C.)* **1969**, *166*, 224–225.
17. McFarland, M.; Kley, D.; Drummond, J. W.; Schmeltekopf, A. H.; Winkler, R. H. *Geophys. Res. Lett.* **1979**, *6*, 605–608.
18. Liu, S. C.; Kley, D.; McFarland, M.; Mahlman, J. D.; Levy, H. *J. Geophys. Res.* **1980**, *85*, 7546–7552.
19. Crutzen, P. J.; Gidel, L. T. *J. Geophys. Res.* **1983**, *88*, 6641–6661.
20. Levy, H. *Science (Washington, D.C.)* **1971**, *173*, 141–143.
21. Levy, H. *Planet. Space Sci.* **1972**, *20*, 919–935.
22. Levy, H. *Adv. Photochem.* **1974**, *9*, 369–524.
23. DeMore, W. B.; Sander, S. P.; Molina, M. J.; Golden, D. M.; Hampson, R. F.; Kurylo, M. J.; Howard, C. J.; Ravishankara, A. R. *Chemical Kinetics and Photochemical Data for Use in Stratospheric Modeling*; Evaluation No. 9; Jet Propulsion Laboratory: Pasadena CA, 1988; JPL Publication 90–1.
24. Atkinson, R.; Darnall, K. R.; Lloyd, A. C.; Winer, A. M.; Pitts, J. N., Jr. *Adv. Photochem.* **1979**, *11*, 375–488.
25. O'Brien, R. J.; Green, P. J.; Doty, R. A. *J. Phys. Chem.* **1979**, *83*, 3302–3305.
26. Zellner, R.; Exner, M.; Herrmann, H. *J. Atmos. Chem.* **1990**, *10*, 411.
27. Greenberg, J. P.; Zimmerman, P. R.; Chatfield, R. B. *Geophys. Res. Lett.* **1985**, *3*, 113–116.
28. Watanabe, T.; Yoshida, M.; Fujiwara, S.; Abe, K.; Onoe, A.; Hirota, M.; Igarashi, S. *Anal. Chem.* **1982**, *54*, 2470–2474.
29. Dieke, G. H.; Crosswhite, H. M. *J. Quant. Spectrosc. Radiat. Transfer* **1962**, *2*, 97–199.
30. Burnett, C. R. *Geophys. Res. Lett.* **1976**, *3*, 319–322.
31. Burnett, C. R.; Burnett, E. H. *J. Geophys. Res.* **1981**, *86*, 5185–5202.
32. Burnett, E. H.; Burnett, C. R.; Minschwaner, K. R. *Geophys. Res. Lett.* **1989**, *16*, 1285–1288.
33. Kendall, D. J. W.; Clark, T. A. *Nature (London)* **1988**, *283*, 57–58.
34. Carli, B.; Carlotti, M.; Dinelli, B.; Mencaraglia, F.; Park, J. H. *J. Geophys. Res.* **1989**, *94*, 11,049–11,058.
35. Carli, B.; Park, J. H. *J. Geophys. Res.* **1988**, *93*, 3851–3865.
36. Perner, D.; Ehhalt, D. H.; Pätz, H. W.; Platt, U.; Roth, E. P.; Volz, A. *Geophys. Res. Lett.* **1976**, *3*, 466–468.

37. Perner, D.; Platt, U.; Trainer, M.; Hübler, G.; Drummond, J. W.; Ehhalt, D. H.; Helas, G.; Junkermann, W.; Rudolph, R.; Schubert, B.; Rumpel, K. J.; Volz, A. *J. Atmos. Chem.* **1987**, *5*, 185.
38. Hübler, G.; Perner, D.; Platt, U.; Tonnisen, A.; Ehhalt, D. H. *J. Geophys. Res.* **1984**, *89*, 1309–1319.
39. Platt, U.; Rateike, M.; Junkermann, W.; Rudolph, J.; Ehhalt, D. H. *J. Geophys. Res.* **1988**, *93*, 5159–5166.
40. Dorn, H. P.; Callies, J.; Platt, U.; Ehhalt, D. H. *Tellus* **1988**, *40B*, 437–445.
41. Hofzumahaus, A.; Dorn, H. P.; Callies, J.; Platt, U.; Ehhalt, D. H. *Atmos. Environ.* **1991**, *25A*, 2017–2022.
42. Mount, G. H. *J. Geophys. Res.* **1992**, *97*, 2427–2444.
43. Baardsen, E. L.; Terhune, R. W. *Appl. Phys. Lett.* **1972**, *21*, 209–211.
44. Anderson, J. G. *J. Geophys. Res.* **1971**, *76*, 7820–7824.
45. Crutzen, P. J.; Isaksen, I. S. A.; McAfee, J. R. *J. Geophys. Res.* **1978**, *83*, 345–363.
46. Goldman, A.; Gillis, J. R. *J. Quant. Spectrosc. Radiat. Transfer* **1981**, *25*, 111–135.
47. Chidsey, I. L.; Crosley, D. R. *J. Quant. Spectrosc. Radiat. Transfer* **1980**, *23*, 187–199.
48. McGee, T. J.; McIlrath, T. J. *J. Quant. Spectrosc. Radiat. Transfer* **1984**, *32*, 179–184.
49. Killinger, D. K.; Wang, C. C. *Chem. Phys. Lett.* **1977**, *52*, 374–376.
50. Bakalyar, D. M.; James, J. V.; Wang, C. C. *Appl. Opt.* **1982**, *21*, 2901–2905.
51. Armerding, W.; Herbert, A.; Schindler, T.; Spiekermann, M.; Comes, F. J. *Ber. Bunsen-Ges. Phys. Chem.* **1990**, *94*, 776–781.
52. Armerding, W.; Herbert, A.; Spiekermann, M.; Walter, J.; Comes, F. J. *Fresenius' Z. Anal. Chem.* **1991**, *340*, 654–660.
53. Armerding, W.; Spiekermann, M.; Grigonis, R.; Walter, J.; Herbert, A.; Comes, F. J. *Ber. Bunsen-Ges. Phys. Chem.* **1992**, *96*, 314–318.
54. Comes, F. J.; Armerding, W.; Grigonis, R.; Herbert, A.; Spiekermann, M.; Walter, J. *Ber. Bunsen-Ges. Phys. Chem.* **1992**, *96*, 284–286.
55. Doyle, G. J.; Lloyd, A. C.; Darnall, K. R.; Winer, A. M.; Pitts, J. N., Jr. *Environ. Sci. Technol.* **1975**, *9*, 237–241.
56. Wu, C. H.; Japar, S. M.; Niki, H. *J. Environ. Sci. Health, Part A* **1976**, *11*, 191–200.
57. Wu, C. H.; Wang, C. C.; Japar, S. M.; Davis, L. I.; Hanabusa, M.; Killinger, D. K.; Niki, H.; Weinstock, B. *Int. J. Chem. Kinet.* **1976**, *8*, 765–776.
58. Calvert, J. G. *Environ. Sci. Technol.* **1976**, *10*, 256.
59. Chang, T. Y.; Norbeck, J. M.; Weinstock, B. *Environ. Sci. Technol.* **1979**, *13*, 1534–1537.
60. Lovelock, J. E. *Nature (London)* **1977**, *267*, 32.
61. Singh, H. B. *Geophys. Res. Lett.* **1977**, *4*, 453–456.
62. Singh, H. B.; Salas, L. J.; Shigeishi, H.; Scribner, E. *Science (Washington, D.C.)* **1979**, *203*, 899–903.
63. Prinn, R. G.; Cunnold, D.; Rasmussen, R. A.; Simmonds, P.; Alyea, F.; Crawford, A.; Fraser, P.; Rosen, R. *Science (Washington, D.C.)* **1987**, *238*, 945–950.
64. Donahue, N. M.; Prinn, R. G. *J. Geophys. Res.* **1990**, *95*, 18,387–18,411.
65. Volz, A.; Ehhalt, D. H.; Derwent, R. G. *J. Geophys. Res.* **1981**, *86*, 5163–5171.
66. McKeen, S. A.; Trainer, M.; Hsie, E. Y.; Tallamraju, R. K.; Liu, S. C. *J. Geophys. Res.* **1990**, *95*, 7493–7501.

67. Prinn, R. G. *Geophys. Res. Lett.* **1985**, *12*, 597–600.
68. *Assessment of Techniques for Measuring Tropospheric H_xO_y*; Hoell, J. M., Ed.; National Aeronautics and Space Administration: Washington, DC, 1984; NASA Conference Publication 2332.
69. Crosley, D. R.; Hoell, J. M. *Future Directions of H_xO_x Detection*; National Aeronautics and Space Administration: Washington, DC, 1986; NASA Conference Publication 2488.
70. Crosley, D. R., to be published as a NASA Conference Publication; National Aeronautics and Space Administration: Washington, DC, 1986.
71. Wang, C. C.; Davis, L. I.; Wu, C. H.; Japar, S. M.; Niki, H.; Weinstock, B. *Science (Washington, D.C.)* **1975**, *189*, 797–800.
72. Wang, C. C.; Davis, L. I.; Seltzer, P. M.; Munoz, R. *J. Geophys. Res.* **1981**, *86*, 1181–1186.
73. Davis, L. I.; Guo, C.; James, J. V.; Morris, P. T.; Postiff, R.; Wang, C. C. *J. Geophys. Res.* **1985**, *90*, 12,835–12,842.
74. Davis, L. I.; James, J. V.; Wang, C. C.; Guo, C.; Morris, P. T.; Fishman, J. *J. Geophys. Res.* **1987**, *92*, 2020–2024.
75. Shirinzadeh, B.; Wang, C. C.; Deng, D. Q. *Geophys. Res. Lett.* **1987**, *14*, 123–126.
76. Davis, D. D.; Heaps, W. S.; McGee, T. J. *Geophys. Res. Lett.* **1976**, *3*, 331–333.
77. Rodgers, M. O.; Bradshaw, J. D.; Sandholm, S. T.; KeSheng, S.; Davis, D. D. *J. Geophys. Res.* **1985**, *90*, 12,819–12,834.
78. Hard, T. M.; O'Brien, R. J.; Chan, C. Y.; Mehrabzadeh, A. A. *Environ. Sci. Technol.* **1984**, *18*, 768–777.
79. Hard, T. M.; Chan, C. Y.; Mehrabzadeh, A. A.; Pan, W. H.; O'Brien, R. J. *Nature (London)* **1986**, *322*, 617–620.
80. Hard, T. M.; Mehrabzadeh, A. A.; Chan, C. Y.; O'Brien, R. J. *J. Geophys. Res.* **1992**, *97*, 9795–9817.
81. Chan, C. Y.; Hard, T. M.; Mehrabzadeh, A. A.; George, L. A.; O'Brien, R. J. *J. Geophys. Res.* **1990**, *95*, 18,569–18,576.
82. Campbell, M. J.; Sheppard, J. C.; Au, B. F. *Geophys. Res. Lett.* **1979**, *6*, 175–178.
83. Campbell, M. J.; Farmer, J. C.; Fitzner, C. A.; Henry, M. N.; Sheppard, J. C.; Hardy, R. J.; Hopper, J. F.; Muralidhar, V. *J. Atmos. Chem.* **1986**, *4*, 413–427.
84. Felton, C. C.; Sheppard, J. C.; Campbell, M. J. *Nature (London)* **1988**, *335*, 53–55.
85. Felton, C. C.; Sheppard, J. C.; Campbell, M. J. *Environ. Sci. Technol.* **1990**, *24*, 1841–1847.
86. Felton, C. C.; Sheppard, J. C.; Campbell, M. J. *Atmos. Environ.* **1992**, *26A*, 2105–2109.
87. Eisele, F. L.; Tanner, D. J. *J. Geophys. Res.* **1991**, *96*, 9295–9308.
88. Mount, G. H.; Eisele, F. L. *Science (Washington, D.C.)* **1992**, *256*, 1187–1190.
89. Anderson, J. G. *Geophys. Res. Lett.* **1976**, *3*, 165–168.
90. Heaps, W. S.; McGee, T. J.; Hudson, R. D.; Caudill, L. O. *Appl. Opt.* **1982**, *21*, 2265–2274.
91. Heaps, W. S.; McGee, T. J. *J. Geophys. Res.* **1983**, *88*, 5281–5289.
92. Heaps, W. S.; McGee, T. J. *J. Geophys. Res.* **1985**, *90*, 7913–7921.
93. Stimpfle, R. M.; Anderson, J. G. *Geophys. Res. Lett.* **1988**, *15*, 1503–1506.
94. Stimpfle, R. M.; Lapson, L. B.; Wennberg, P. O.; Anderson, J. G. *Geophys. Res. Lett.* **1989**, *16*, 1433–1436.

95. Beck, S. M.; Bendura, R. J.; McDougal, D. S.; Hoell, J. M.; Gregory, J. L.; Curfman, H. J.; Davis, D. D.; Bradshaw, J. D.; Rodgers, M. O.; Wang, C. C.; Davis, L. I.; Campbell, M. J.; Torres, A. L.; Carroll, M. A.; Ridley, B. A.; Sachse, G. W.; Hill, G. F.; Condon, E. P.; Rasmussen, R. A. *J. Geophys. Res.* **1987**, *92*, 1977–1985.

96. Davis, D. D.; Rodgers, M. O.; Fischer, S. D.; Asai, K. *Geophys. Res. Lett.* **1981**, *8*, 69–72.

97. Davis, D. D.; Rodgers, M. O.; Fischer, S. D. *Geophys. Res. Lett.* **1981**, *8*, 73–76.

98. Bradshaw, J. D.; Davis, D. D. *Opt. Lett.* **1982**, *7*, 224–226.

99. Bradshaw, J. D.; Rodgers, M. O.; Sandholm, S. T.; KeSheng, S.; Davis, D. D. *J. Geophys. Res.* **1985**, *90*, 2861–2873.

100. Bradshaw, J. D.; Rodgers, M. O.; Davis, D. D. *Appl. Opt.* **1984**, *23*, 2134–2145.

101. Hard, T. M.; O'Brien, R. J.; Cook, T. B.; Tsongas, G. J. *Appl. Opt.* **1979**, *18*, 3216–3217.

102. Hard, T. M.; O'Brien, R. J.; Cook, T. B. *J. Appl. Phys.* **1980**, *51*, 3459–3464.

103. Smith, G. P.; Crosley, D. R. *J. Geophys. Res.* **1990**, *95*, 16,427–16,442.

104. George, L. A.; O'Brien, R. J. *J. Atmos. Chem.* **1991**, *12*, 195–209.

105. Heaps, W. S. *Appl. Opt.* **1981**, *20*, 583–587.

106. Shirinzadeh, B.; Wang, C. C.; Deng, D. Q. *Appl. Opt.* **1987**, *26*, 2102–2105.

107. Hard, T. M.; Chan, C. Y.; Mehrabzadeh, A. A.; O'Brien, R. J. *Appl. Opt.* **1989**, *28*, 26–27.

108. Lonneman, W. A.; Bufalini, J. J.; Kuntz, R. J.; Meeks, S. A. *Environ. Sci. Technol.* **1981**, *15*, 99–103.

109. Kelly, N. A. *Environ. Sci. Technol.* **1980**, *16*, 763–770.

110. Sakamaki, F.; Hatakeyama, S.; Akimoto, J. *Int. J. Chem. Kinet.* **1983**, *15*, 1013–1029.

111. Pitts, J. N., Jr.; Sanhueza, E.; Atkinson, R.; Carter, W. L. P.; Winer, A. M.; Harris, G. W.; Plum, C. N. *Int. J. Chem. Kinet.* **1984**, *16*, 919–939.

112. Zafonte, L.; Reiger, P. L.; Holmes, J. R. *Environ. Sci. Technol.* **1977**, *11*, 483–487.

113. Rodgers, M. O.; Asai, K.; Davis, D. D. *Chem. Phys. Lett.* **1981**, *78*, 246–252.

114. Butler, J. E.; Talley, L. D.; Smith, G. K.; Lin, M. C. *J. Chem. Phys.* **1981**, *74*, 4501–4508.

115. Gericke, K. H.; Comes, F. J. *Chem. Phys. Lett.* **1980**, *74*, 63–66.

116. Gericke, K. H.; Comes, F. J. *Chem. Phys.* **1982**, *65*, 113–121.

117. Gericke, K. H.; Comes, F. J.; Levine, R. D. *J. Chem. Phys.* **1981**, *74*, 6106–6112.

118. Cleveland, C. B. Ph.D. thesis, Cornell University, Ithaca, NY, 1988.

119. Albritton, D. L.; Fehsenfeld, F. C.; Tuck, A. F. *Science (Washington, D.C.)* **1990**, *250*, 75–81.

120. Wu, C. H.; Japar, S. M.; Niki, H. *J. Environ. Sci. Health, Part A* **1976**, *11*, 191–200.

121. Wu, C. H.; Wang, C. C.; Japar, S. M.; Davis, L. I.; Hanabusa, M.; Killinger, D. K.; Niki, H.; Weinstock, B. *Int. J. Chem. Kinet.* **1976**, *8*, 765–776.

RECEIVED for review March 20, 1991. ACCEPTED revised manuscript September 25, 1992.

13

Measurement of Personal Exposure to Air Pollution: Status and Needs

Paul J. Lioy

Environmental and Occupational Health Sciences Institute, University of Medicine and Dentistry of New Jersey, Robert Wood Johnson Medical School, 681 Frelinghuysen Road, Piscataway, NJ 08854

Research on air pollution monitoring has expanded its scope of inquiry from characterization of the ambient atmosphere and identification of its chemical and aerosol constituents to determination of an individual's total indoor and outdoor air pollution exposure. (Exposure is the integral of a time-varying concentration over a specified interval of contact.) This new emphasis has spurred the development and evaluation of personal air monitors for applications within populations at risk to high exposure and within the general population. The types of pollutants presently requiring or being considered for personal monitoring are discussed. The associated technological issues and problems are described and illustrated by examples. The criteria for a design of a personal monitor are reviewed, as are the scientific approaches currently being used for personal monitoring and plausible approaches for the future. Activity logs, which are needed to ensure proper allocation of the sources of significant exposure, are briefly discussed.

Personal monitoring is a relatively new concept in community air pollution measurement research (*1–3*). This fact is not surprising because most air pollution investigations have been directed toward the characterization of the ambient atmosphere, the observation of pollutant trends, the acquisition of data on chemical kinetic parameters and on the physical properties of aerosols, and the determination of compliance to national and other standards (*4*). Before the late 1970s, research on personal monitors was primarily conducted in industrial settings (*5, 6*) because American Confer-

0065–2393/93/0232–0373$06.00/0
© 1993 American Chemical Society

ence of Governmental Industrial Hygienists (ACGIH) guidelines and Occupational Safety and Health Administration (OSHA) standards for workplace contaminants are an 8-h time-weighted average or the peak concentration of a particular pollutant or mixture of pollutants (7, 8). The concentration of workplace contaminants may be associated with more than one area and may not vary predictably with distance from a source; therefore, area monitors have been known to underestimate exposures in the workplace that can cause known health effects. Consequently, the hygienist uses personal monitors.

Occupational exposures are usually associated with relatively high concentrations. In fact, the concentration of many substances (but not ozone) exceeds ambient air concentrations by 2 to 3 orders of magnitude. A multitude of samplers are used to detect the inorganic or organic compounds encountered in industrial settings, but these samplers collect material primarily to determine an 8-h time-weighted average (5, 6). Even with long sampling times, the total quantities of contaminants collected in nonindustrial microenvironments (e.g., a living room, a park, or a library) are small, so sensitive techniques are needed (1–3, 9, 10).

Frequently, community air environments are complex, and most outdoor exposures do not occur near the actual source of air contaminants, a situation that is the norm for the workplace (4). The concentrations in community air are an average of the emissions dispersed within the atmosphere by a number of sources, by the same source in a number of different locations, or by a single large [>100 lb/year (>45 kg/year)] or small [<5 lb/year (<2.25 kg/year)] source in one location (11). In other situations the ambient concentrations are an average of the secondary products formed in a defined area or large region (12). In addition, indoor air exposures result from outdoor air penetrating indoors and from emissions by indoor sources (e.g., tobacco smoke or solvent evaporation). Indoor emissions have some features similar to occupational settings because the person can be located adjacent to or can pass near the indoor source (1). As a consequence of the needs in air exposure research, improved instrumentation for personal monitoring must be developed; this instrumentation is the focus of this chapter.

Rationale

The personal monitoring of community air pollutants is required for four basic reasons; these reasons are associated with a need for a more accurate description of an individual's contact with a pollutant that can affect health.

1. For specific air pollutants, even some air pollutants in the National Ambient Air Quality Standard (e.g., nitrogen dioxide and particulate matter), the highest exposures may not occur outdoors.

2. The concentration of an air pollutant varies from location to location, so a stationary monitor may not be representative of the major exposures to many pollutants.

3. A person's activities alter the patterns of exposure to contaminants throughout a day.

4. A person's exposure occurs all day long; this situation indicates the need to have the device accompany an individual throughout an entire day or, for some pollutants, during that portion of the day when peak exposures may occur.

The development of personal monitoring techniques and their application to a community setting are essential in exposure studies designed for epidemiology, risk assessment, and clinical intervention (10). Further, there is little information on the formation, transformation, accumulation, and fate of pollutants in locations where the population spends its time (e.g., office buildings or residences). Thus, studies are required that couple measurements of pollutants at levels of environmental concern with the places where people spend time or conduct significant activities. Some traditional and new applications of personal monitoring include the following (3, 10):

1. Outdoor Air Pollution. Applications include monitoring neighborhoods near a small local source; municipal incinerators; photochemical smog episodes and their impact on outdoor athletics and recreation; urban traffic congestion; and dust resuspension from hazardous wastes.

2. Indoor Air Pollution. This category includes monitoring high source emissions or ubiquitous sources; air emissions from contaminated water, which can come from bathroom showers, basement seepage, or pesticide contamination; and tight buildings, which, because of a lack of dispersal, may have high concentrations of many chemicals.

3. Commuter Transit. Applications include monitoring automobile cabin pollution and self-service gasoline refueling.

Each situation requires an evaluation of the hypotheses to be tested before the personal monitor can be designed and the identification of the types and duration of exposure that may occur. Once inhaled, the compound may be rapidly metabolized in the body; short-term measurements would be appropriate for such processes. The compound may instead have a long residence time at a specific site in the lung or may be stored in an organ or tissue. Thus, the inhaled compound or one or more of its adducts or metabolites could eventually deliver a biologically effective dose to a target organ or cell (13). The monitor must be developed with consideration of the nature, concentration, time of contact, and biological effects of the compound.

Criteria and Techniques

Once the inhalation exposure questions have been identified, the specifications for each personal monitor must be determined and the monitor must be validated for the contaminant being measured. Table I, updated from Samet et al. (14), identifies currently available personal monitors, and Table II, taken from an Environmental Protection Agency (EPA) report (15), shows the projected needs in the 1990s. There are a number of opportunities for research on personal monitors; Table II indicates that relatively few commercial units are currently available for either particulate or gas-phase species. For compounds such as polycyclic aromatic hydrocarbons (PAHs), a two-stage sampler is required because some PAHs exist simultaneously in the gaseous and particulate phase (16). Consequently, research must be ranked with respect to the significance of the air pollution problem, and the technological developments required to provide reliable samplers must be defined.

After a personal monitor is developed, the first level of use would be within a target population potentially having high exposures, and the second level would be the introduction of a streamlined monitoring package for applications within larger segments of the general population. New personal monitors must address these six criteria (10):

1. Sensitivity.   The monitor should detect analytes at levels below those causing adverse health effects, be sensitive to changes that are one-tenth of the level of interest, have precision of ±5%, and be easy to calibrate accurately.

2. Selectivity.   The monitor should have no response to other compounds that might also be present.

3. Rapidity.   Sampling and analysis times should be short compared with biological response times, response time in 90% of samples should be less than 30 s, and output should be RS232 or the equivalent.

4. Comprehensiveness.   The monitor should be sensitive to all contaminants that could result in adverse health effects and adaptable to several analytes.

5. Portability.   The sampling and analysis device should be rugged and should not interfere with the normal behavior of the individual. It should have low power consumption, a stabilization time of less than 15 min, a temperature range of 20–

40 °C, and a humidity range of 0–100%, and it should be
battery powered.

6. Cost.   Sampling and analysis should not be prohibitively expensive. The monitor should have few components that are
consumed by analysis and should require little maintenance.

Five of the criteria are normally considered when any air pollution
monitor is designed; however, the fifth criterion—portability—is essential
in a personal monitor. Obviously, this requirement has an impact on the
other criteria because it establishes a challenge for achieving specificity for
a chemical or suite of chemicals, for low detection limits for adequate determination of the concentration and exposure, and for units that are not
cost-prohibitive. Sensitivity of current personal monitors is less than that
for stationary monitoring techniques. Thus, entirely new approaches appear
to be necessary for detection of a contaminant in a personal monitor.

The current methodologies for personal sampling include two major
types (*17, 18*):

1. Passive Sampling.   These techniques provide for the accumulation of a contaminant on a substrate on the basis of the
principles of diffusion, sedimentation, adsorption, or absorption.

2. Active Sampling.   These techniques use the dynamic passage
of the sampled air at a specified rate through a substrate (e.g.,
a filter), an absorbant (e.g., Tenax (diphenylphenylene oxide)
or activated charcoal), or a detector (e.g., a photometer) that
measures a parameter that is proportional to detectable quantities of a contaminant.

Applications of different personal monitors have increased over the past
10 years; passive monitoring techniques are used primarily for long-term
sampling, which provides data to quantify exposures associated with chronic
health effects. One of the most well known passive techniques is the Palmes
tube, Figure 1 (*19*). It has been used to study the magnitude of indoor
nitrogen dioxide exposures for a variety of time periods, ranging from a day
to greater than a week. Other samplers are now available with variations on
this approach, including devices that can be read by an individual who has
a color chart (*20*). Passive monitoring techniques were considered for the
first EPA Total Exposure Assessment Methodology (TEAM) studies for volatile organic species; however, TEAM investigations have used active samplers because the personal measurements were made only for 12-h durations
(*21*).

Table I. Representative Monitoring Equipment for Particulate Matter for Indoor Air Quality Samples

Pollutant Sampler	Manufacturing Company	Sensitivity and Integrating Time	Approximate Cost
Organic vapors	Industrial Scientific Corporation, Oakdale, PA 15071	NA	NA
Organic vapors: hydrocarbon chemical reaction tubes	National Draeger Inc., P.O. Box 120, Pittsburgh, PA 15230	100–3000 ppm for 4–8 h[a]	$3/tube; $900 for pump and accessories
Organic vapors: charcoal badges	3M Corporation, Technical Service Department, 3M Center, St. Paul, MN 55144	depends on vapors and sampling times; minimum level, 10 mg	$10/badge; $50– $300 for analysis by GC or GC/MS
Formaldehyde: Pro-Tek adsorption badge	Du Pont, Applied Technical Division, P.O. Box 110, Kennett Square, PA 19348	1.6–54 ppm/h up to 7 days or 0.2–6.75 ppm/8 h TWA	$20/badge; $25– $80 for analysis
Formaldehyde: diffusion monitor	3M Corporation, Technical Service Department, Building 260-3-2, 3M Center, St. Paul, MN 55144	0.1 ppm for 8 h	$37/monitor and $37/analysis
NO_2: personal and alarm	MDA Scientific, Lincolnshire, IL 60069	2–3 ppm; 1/3 TLV; electrochemical cell based 15 min–8 h TWA	$800/detector; $100/output; $2,075/dosimeter; $1,045/readout unit
NO_2 and SO_2: Diffusion tubes	Environmental Sciences and Physiology, Harvard School of Public Health, Boston, MA 02115	500 ppb/h integrated	$10/tube, research only
Mercury vapor (and others)	SKC, Inc., Eighty Four, PA 15330	0.05 mg/m³ integrated	NA
Ozone	Environmental Sciences and Physiology, Harvard School of Public Health, Boston, MA 02115	>25 ppb at 8 h	NA
NO_2: Diffusion badge	Environmental Sciences and Physiology, Harvard School of Public Health, Boston, MA 02115	50 ppb/h	$15/badge, research only

CO: Passive badge	Lab Safety Supply Company, P.O. Box 1368, Janesville, WI 53547	50 ppm for 8 h produces color change	$3/holder; $12.75 per 10 indicating papers
CO: Detector tube (integrated)	National Draeger Inc., P.O. Box 120, Pittsburgh, PA 15230	2.5 ppm for 8 h	$255 for pump and accessories; $3/tube
Integrated gravimetric; particles <3.5 μm in diameter	Several manufacturers	1.7 L/m	pumps, $200–$700; filters, $2; cyclones, $20–$100
Integrated gravimetric; particles with 10- to 3-μm or with less than 3-μm diameters	National Bureau of Standards under EPA contract, U.S. EPA, Research Triangle Park, NC 27711	6 L/m; separates by using filters in series	NA
Instantaneous (0.2 s); TSP or RSP; 0.1 to 10 μm; forward light-scattering	GCA-Mini-RAM (personal aerosol monitor), GCA Corporation, Bedford, MA 01730	NA	NA
Continuous; RSP submicrometer light-scattering multi-sensor monitor	Handheld Aerosol Monitor (HAM), PPM Inc., Knoxville, TN 37922	>10 μg/m³ mass concentration; 1.5 L/s	$3,000–$10,000
Acid aerosols	Environmental Sciences and Physiology, Harvard School of Public Health, Boston, MA 02115	H^+ and O_4^{2-}; nmol/m³	NA

NOTES: Particles can be measured with a variety of techniques. With cyclone or impactor separators, smaller fractions can be collected on filters. Mass can also be measured by using the optical properties of particles. For the most part, measuring particles requires equipment costing several hundred to a few thousand dollars. Filters must be pre- and post-weighed in a temperature- and humidity-controlled room; TWA, time-weighted average; TLV, threshold limit value; TSP, total suspended particles; and RSP, respirable particles. NA, not available.

[a]Reference 20.

SOURCE: Adapted from reference 14.

Table II. Status of Personal Monitor Development

Pollutants	Monitor Needed D	I	Monitor Under Development D	I	Prototype Under Development D	I	Tested and Evaluated D	I	Used in Pilot Studies D	I	Used in Large Field Studies D	I	Ready for Routine Use D	I
CO	Y	Y	Y	Y	Y	Y	Y	Y	Y	Y	Y	Y	Y	Y
NO$_2$	Y	Y	Y	Y	Y	Y	Y	Y		Y		Y		Y
Breathable particles (>10 µm diameter)	Y	Y	Y	Y	Y									
Formaldehyde	Y	Y		Y	Y	Y		Y		Y		Y		Y
VOCs	Y	Y		Y		Y		Y		Y		Y		Y
Polar VOCs	NA	Y		Y		Y		Y		Y				
Pesticides	NA	Y		Y		Y		Y		Y		Y		
Radon	Y	Y	Y	Y	Y	Y	Y	Y	Y	Y	Y	Y	Y	Y
PAH	NA	Y		Y		Y		Y		Y				
Biological aerosols	NA	Y		Y		Y		Y		Y		Y		Y
House dust	NA	Y		Y		Y		Y		Y		Y		Y
O$_3$	Y	Y		Y		Y		Y		Y				

NOTE: Y denotes yes; D, direct readout; I, integrated collection of samples; NA, not applicable; and VOCs, volatile organic compounds.
SOURCE: Adapted from reference 15.

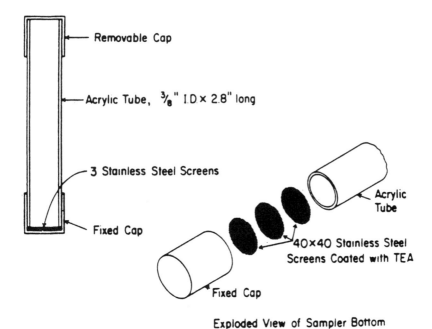

Removable Cap

Acrylic Tube, ⅜" I.D × 2.8" long

3 Stainless Steel Screens

Fixed Cap

Acrylic Tube

40×40 Stainless Steel Screens Coated with TEA

Fixed Cap

Exploded View of Sampler Bottom

Figure 1. Schematic diagram of the Palmes personal NO_2 sampler. (Reproduced with permission from reference 8. Copyright 1989.)

The obvious advantages of a passive sampler, as seen in Figure 1, are its simplicity, total portability, and lack of any ancillary equipment needed to collect the sample. The primary drawback is the necessity to collect an integrated sample over an extended period of time. This drawback may not be important if the response time for a biological effect is not significantly shorter than the minimum sampling duration.

Active sampling has always been preferred in traditional air pollution studies because a substance can be concentrated on a particular substrate and because continuous measurements can be taken. These samplers have been placed at fixed monitoring sites on a roof or in a trailer (4). The use of active sampling, however, has not been without problems. For instance, the use of substrates such as filters and sorbents can affect the measured concentration by artifact formation, breakthrough, and blow-off associated with individual compounds or classes (22).

Personal monitors that use active sampling have technical problems that must be addressed during the design phase because

- there are smaller component parts,
- the size of the pump is limited,
- the volume or surface of the collection medium or detector is reduced, and

- the energy needed to power the device is from a self-contained source.

Each of these considerations must be resolved before a personal monitor can be applied to air pollution research and characterization studies. Some advances have been made for carbon monoxide, volatile organic compounds, acid aerosols, and particulate matter (PM–10 and RSP; these represent the masses of all particles collected in samplers with 50% cut sizes of 10 and 25 μm, respectively) and its components (12, 22–31). Each advance is still undergoing development, and further advances can be anticipated for these as well as for other pollutants. The next generation of monitors will probably include devices for some pollutants that incorporate the use of microsensors. Currently, microsensors are being examined for detection of nitrogen dioxide and ozone, but the range of sensors available suggests that they can be used for a number of compounds (10, 32):

- Biosensors
- Electrochemical sensors
  Potentiometric devices
  Amperometric devices
  Elements sensitive to contact potential
- Thermal sensors
  Thermistor and resistance thermometer elements
  Thermoelectric–bolometric sensors
  Semiconductor-based elements
  Elements sensitive to piezoelectric thermal oscillation
  Pyroelectric sensors
  Black-body radiation sensors
- Stress and pressure sensors
  Photoacoustic elements
  Mass-sensitive elements
    Bulk piezoelectric elements (thickness monitors)
    Surface acoustic-wave (SAW) elements
    Plate-mode oscillators
    Interface impedance elements
  Fiber optic elements sensitive to elastic constants
- Electromagnetic sensors: passive
  Solid-state conductivity (chemiresistance) sensors
  Dielectrometric sensors
  Dielectric sensors
    Absorptivity elements
    Index of refraction elements
    Phase-shift and interface impedance (e.g., ellipsometry) elements
    Spectral "fingerprint" elements
  Surface-enhanced Raman spectrometers

- Electromagnetic sensors: active
  Nonlinear behavior, including frequency doubling, elements
  Fluorescence elements

Each of the passive and active samplers mentioned has a specific sampling rate. However, they do not reflect the respiration rate of an individual. For an estimate of the dose this respiration rate must be estimated from literature values or collection devices.

Example of Personal Monitor Development and Application

The difficulties encountered in attempts to apply personal monitoring can be illustrated by the developments required to collect PM–10 samples during the Total Human Environmental Exposure Study (THEES) (25). The results of the THEES are described in a number of research articles (25, 33–36). The following is a synopsis of the approach used to collect daily samples from individuals for 2 weeks a year over the course of 2 years (a total of 28 days). A major problem with the collection of particulate matter is the size of the pump. The pump needs to provide a sufficient flow rate to ensure that the inlet and collection medium components are operating efficiently (e.g., correct cut size must be provided) and that the system can obtain a large enough sample to achieve the detection limits for the measurement of the compound in question. In THEES the target chemical was benzo[a]pyrene, which had a lower detection limit of 0.1 ng/m^3. After a series of laboratory experiments it was determined that a PM–10 sampler operating for 24 h with a 4 L/min flow rate was required to collect adequate mass (>0.5 ng) for analysis. An impactor with a sharp cut size at 10 μm and a 25-mm filter was developed by V. Marple and used as the sampler (37). It was evaluated in my laboratories for collection efficiency in an intercomparison with a dichotomous sampler and a stationary indoor air sampling impactor (IASI) (38). The results of the study were excellent for all samplers; a slope of 1.0 and 1.08 was found for the regression between mass collected by IASI and by the dichotomous sampler, respectively. The impactor was adequate for our needs; unfortunately, the only pump available that operated at the correct flow rate and had a flow controller was designed for occupational hygiene (28). The pump presented three logistical problems. The first was a bulky design that was unresolvable because of insufficient time to redesign the unit. A sturdy shoulder harness was constructed to facilitate carrying of the monitor by the participants; however, the pump eventually should be separated into a battery and a pump assembly. This setup was tested in the particle TEAM investigations, although the system was still bulky (39).

The second problem was pump noise. A worker may not have the same sensitivities about wearing a pump and collector as would a member of the general public. For the general public, the noise levels of greater than 72 dB generated by a typical pump are intolerable, especially when the pump is worn for 14 days. At such noise levels relaxation or conversation is not possible (40). A sound control package was constructed to minimize the noise level; the pump was housed in a box with attenuation material, and the pump mounts were muffled. The noise was reduced to approximately 50 dB, which was found to be a reasonable value for all participants.

The third concern was that the pump batteries only operated for 12 h. Therefore, it was essential to devise a sampling strategy that allowed for the changing the batteries halfway through a sampling period. This situation led to a 6–8 p.m. start time for the samples and a 6–8 a.m. battery change for each 24-h sample. Obviously, a battery that maintains a charge for greater than 24 h is needed for future exposure studies. The final personal monitoring system used is shown in Figure 2. The approach worked; we successfully collected greater than 95% of the personal samples, and no potential participant refused to wear the sampler.

During the day the participants wore the sampler attached to the lapel of an article of clothing, but at night this was not feasible. Therefore, at night it was placed in a convenient location near the sleeping participant. Participants could not wear the sampler during specific types of exercise, and they were instructed to place it in a convenient location at such times.

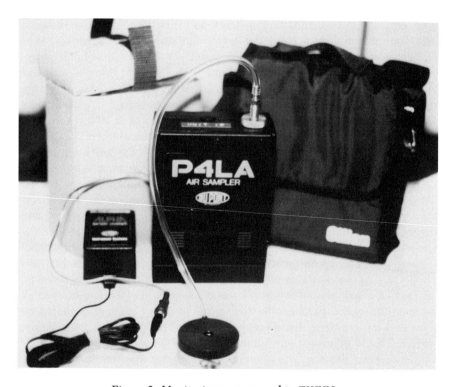

Figure 2. Monitoring system used in THEES.

Obviously the methodology was not optimal, and more research is required on PM–10 and other particle samplers to produce a more compact and efficiently operated system for consistent use in large populations. The technological problems will be the most direct to resolve. However, a long-term concern is whether the participants are altering usual activities because they are wearing a personal monitor. In the THEES this situation did not happen often because the participants were told that in such circumstances the device should be removed and placed in a convenient location. Furthermore, each person was provided with a sports bag in which to place the pump, allowing each person to carry the personal sampler when he or she participated in social functions. In the strict terms of industrial hygiene practice, the approach used did not result in all sampling times being associated with breathing zone air. The logistics were physically impossible and not reasonable because the participants could not wear the samplers to bed. Future attempts at miniaturization or the use of a continuous sensor may solve some of these methodological issues. However, personal monitors will probably continue to sample within the personal zone of an individual but not within the breathing zone.

The THEES conclusions indicated the importance of personal samplers. Results illustrated in Figure 3 show that estimates of the benzo[a]pyrene

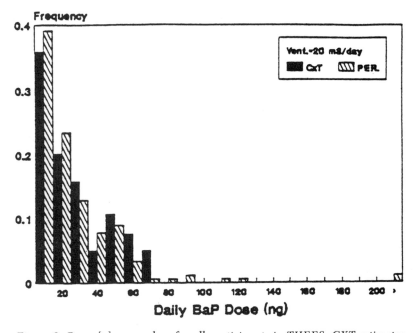

Figure 3. *Benzo[a]pyrene dose for all participants in THEES; CXT estimates from indoor and outdoor measurements (solid bars) and personal air measurements (cross-hatched bars) are shown. (Reproduced with permission from reference 36. Copyright 1991.)*

inhalation dose from time-weighted indoor and outdoor environmental samples was not equivalent to the exposures measured by the personal samples (36). The estimated exposures were similar for the mean values, but the time-weighted approach missed the highest exposures in the distribution. In THEES most participants were nonsmokers, so the indoor levels were primarily caused by the outdoor air and indicated that the personal activities contributed primarily to the extreme values. The importance of personal activities was verified by a daily activity log that was filled out by each participant. The log was designed to ensure that sources of benzo[a]pyrene could be accounted for; specific questions on location and operation of the source and passive contact with sources were used.

For smokers, the personal mean values for particulate matter, in this instance respirable particle concentrations, were driven by the indoor air passive smoke (23). The studies summarized in Table III show similar information for other pollutants, and for many the outdoor levels are much lower than the indoor levels that result from the presence of specific indoor sources or activities (40).

Discussion

The types of samplers required for personal air monitoring in the future will use a mix of passive and active techniques because not all situations lend themselves to the use of both and because the biological effect being studied may require one sampling protocol in favor of another. In the case of ozone, which is one of the major outdoor pollutant problems in the United States (41), exposure is estimated from the results from continuous monitors located at stationary sites away from sources of scavenging compounds like NO_2. These types of sites have provided adequate information for human health effects field studies because the daily peak or other daytime average concentrations in places away from these sources are correlated with decreases in lung function. However, to obtain detailed information on exposure profiles for individuals participating in outdoor activity, an investigator needs a personal device that can record the daily maximum or integrate over a number of hours between 10 a.m. and 9 p.m. Thus, a continuous monitor would be most desirable. A passive monitor to determine integrated exposure might work, but it must be capable of at least providing an 8-h average for the concentration range of 20 to >200 parts per billion (ppb) (42). Concurrently, it may become necessary to monitor personal exposure to other smog constituents such as formaldehyde and fuel constituents such as methanol if the United States promotes the use of alternative fuels to operate a large fraction of motor vehicles in metropolitan areas (43). Because of the lack of information on the chemical constituents of alternative fuels (or their emissions), there could be an opportunity for chemical characterization and exposure research to link and enhance efforts to identify the impact of toxic compounds on segments of the population.

Table III. Summary of Selected Personal Monitoring Studies Carried Out in the United States

Pollutant	Reference	Number of Subjects	Summary of Findings
RSP	44	20	Personal exposure did not correlate with outdoor measurements; in most cases it was substantially higher than outdoor concentrations would predict. Exposure to passive tobacco smoke was a major determinant of RSP exposure.
	23	37	
	45	48	
	46	101	
CO	47	66	Time spent in transit was the primary determinant of personal exposure. Highest CO exposures were due primarily to motor vehicle exhaust.
	48	98	
	49	3	
	22	1083	
NO_2	50	9	Outdoor monitors overestimated exposures for people not exposed to indoor sources but underestimated exposures for people who resided in homes with unvented combustion appliances.
	51	350	
VOCs	21	355	Outdoor measurements did not correlate well with personal exposures. Personal exposure and in-home concentrations tended to be higher than outdoor concentrations for many volatile organic compounds.
	30		
Pb	52	150	Highest Pb exposures were experienced by taxi drivers. All subjects except office workers experienced highest exposures at work.
PM–10	25	10–18	Highest personal exposures were due to individual activities; nonsmokers' exposures correlated to exposure to ambient air.
PM–10	25	9	Highest personal exposures were due to personal activities.

SOURCE: Adapted from reference 1.

This discussion also indicates the need for research on strategies for implementation of personal monitoring. The current monitoring techniques are in a state of evolution, and in some cases the devices are still primitive. To obtain conditions favorable for the development of personal monitors, a major effort must be directed toward defining the strategies that will yield information on targeted populations or populations with high exposures. These definitions can be accomplished through purposeful (focused on a small number of people to detail the feature of air pollution exposure) or statistically representative sampling of segments of the general public. Concurrently, the time–activity logs must be refined for the pollutant and for situations of concern to ensure that the personal samples can be placed into a meaningful perspective. Work is now under way to develop a hand-held microprocessor data logger that can be used by a participant to sequentially record each activity, the time spent in that activity, or the time spent in contact with a source or a type of pollution (e.g., smog).

Summary

The technological advances in the instrumentation required to conduct personal monitoring in the general population are occurring but at a relatively slow rate. The reasons for the delays are not associated with a lack of important questions to be addressed but with the need for a more systematic approach to answering the questions. This approach will provide the required market for manufacturers to invest development costs in an instrument. The net result will be better integration of the criteria for personal monitors with the instrumentation needs of a particular study. In the past equipment not originally suited for personal monitors was modified to address exposure and exposure–health effects problems. Many of the modified units were successfully used; however, for future air pollution exposure assessments, basic research on passive and active monitors for a ranked set of chemicals should be completed before the design of a study. This foundation will lead to the availability of monitors as the research and assessment needs arise. One main requirement associated with future research is that the investigators developing equipment must recognize that the devices are to be worn by people and that the monitor must be tested to ensure that the participants in an exposure study can wear it comfortably. In addition, air pollution exposure research must be integrated with research on the characterization of the atmospheres in various indoor and outdoor environments. This integration will provide a basis for selecting air pollutants or mixtures with biological significance.

Acknowledgment

I thank Malti Patel for work in preparing the final manuscript. This effort was completed as part of the National Institute of Environmental Health Sciences Center of Excellence Award ESOSO22.

References

1. Sexton, K.; Ryan, P. B. *Assessment of Human Exposure to Air Pollution: Methods, Measurements and Models*; Air Pollution, the Automobile and Public Health; Health Effects Institute, National Academy Press: Washington, DC, 1988; pp 207–238.

2. Ott, W. R. *Environ. Int.* **1982**, *7*, 179–196.
3. Ott, W. R. *Environ. Sci. Technol.* **1985**, *19*, 880–885.
4. Lioy, P. J. In *Air Sampling Instruments for Evaluation of Atmospheric Contaminants*, 7th ed.; Hering, S., Ed.; American Conference of Governmental Industrial Hygienists: Cincinnati, 1989; pp 33–50.
5. *Threshold Limit Values and Biological Indices, 1990–1991*; American Conference of Governmental Industrial Hygienists: Cincinnati, 1990.
6. *Code of Federal Regulations*; 29 CFR 19101000, 1989, Occupational Safety and Health Administration: Washington, DC, 1989; OSHA 3112.
7. Rose, V. E.; Perkins, J. L. *Am. Ind. Hyg. Assoc. J.* **1982**, *43*, 605–621.
8. Saltzman, B. E.; Caplan, P. E. In *Air Sampling Instruments for Evaluation of Atmospheric Contaminants*, 7th ed.; Hering, S., Ed.; American Conference of Governmental Industrial Hygienists: Cincinnati, 1989; pp 449–476.
9. National Research Council. Committee on the Epidemiology of Air Pollution. *Epidemiology and Air Pollution*; National Academy Press: Washington, DC, 1985; pp 89–124.
10. National Research Council. *Air Pollution Exposure Assessment: Advances and Opportunities*; National Academy Press: Washington, DC, 1991; pp 1–294.
11. *Compilation of Air Pollution Emission Factors*, 4th ed.; U.S. Environmental Protection Agency: Research Triangle Park, NC, 1985; EPA–AP 42.
12. Wolff, G. T.; Lioy, P. J. *Environ. Sci. Technol.* **1980**, *14*, 1257–1261.
13. Lioy, P. J. *Environ. Sci. Technol.* **1990**, *24*, 938–945.
14. Samet, J.; Marbury, M. C.; Spengler, J. D. *Am. Rev. Respir. Dis.* **1988**, *137*, 221–242.
15. *Research Needs in Human Exposure: A 5-Year Comprehensive Assessment (1990–1994)*; Total Human Exposure Research Council. Office of Research and Development. U.S. Environmental Protection Agency: Washington, DC, 1988.
16. Cautreels, W.; Van Cauwenberghe, K. *Atmos. Environ.* **1978**, *12*, 1134–1141.
17. American Conference of Governmental Industrial Hygienists. *Advances in Air Sampling*; Lewis Publishers: Chelsea, MI, 1988; pp 1–409.
18. Hering, S. V. *Air Sampling Instruments for Evaluation of Atmospheric Contaminants*; American Conference of Governmental Industrial Hygienists: Cincinnati, 1989; pp 1–612.
19. Palmes, E. D.; Gunnison, A. F.; DiMattio, J.; Tomczyk, C. *Am. Ind. Hyg. Assoc. J.* **1976**, *37*, 570–577.
20. Woebkenberg, M. L. *Am. Ind. Hyg. Assoc. J.* **1982**, *43*, 553–561.
21. Wallace, L. A.; Pellizzari, E. D.; Hartwell, T. D.; Sparacino, C. M.; Sheldon, L. S.; Zelon, H. *Atmos. Environ.* **1985**, *19*, 1651–1661.
22. Akland, G. G.; Hartwell, T. D.; Johnson, T. R.; Whitmore, R. W. *Environ. Sci. Technol.* **1985**, *19*, 911–918.
23. Dockery, D. W.; Spengler, J. D. *J. APCA* **1981**, *31*, 153–159.
24. Geisling, K. L.; Tashima, M. K.; Girman, J. R.; Miksch, R. R.; Rappaport, S. M. *Environ. Int.* **1982**, *8*, 153–158.
25. Lioy, P. J.; Waldman, J. M.; Buckley, T.; Butler, J.; Pietarinen, C. *Atmos. Environ.* **1990**, *24B*, 57–66.
26. Vo-Dinh, T. *Environ. Sci. Technol.* **1985**, *19*, 997–1003.
27. Seifert, B.; Abraham, H. J. *Int. J. Environ. Anal. Chem.* **1983**, *13*, 237–253.
28. Tosteson, T.; Spengler, J. D.; Weker, R. *Environ. Int.* **1984**, *8*, 265–268.
29. Koutrakis, P.; Fasano, A. M.; Slater, J. L.; Spengler, J. D.; McCarthy, J. F.; Leaderer, B. P. *Atmos. Environ.* **1989**, *23*, 2767–2774.
30. Wallace, L. A.; Pellizzari, E. D.; Hartwell, T. D.; Whitmore, R.; Zelon, H.; Perritt, R.; Sheldon, L. *Atmos. Environ.* **1988**, *22*, 2141–2164.
31. Hammond, S. K.; Leaderer, B. P. *Environ. Sci. Technol.* **1987**, *21*, 494–497.

32. Zaromb, S.; Stetter, J. R. *Sens. Actuators* **1984**, *6*, 225–243.
33. Lioy, P. J.; Waldman, J.; Greenberg, A.; Harkov, R.; Pietarinen, C. *Arch. Environ. Health* **1988**, *43*, 304–312.
34. Waldman, J. M.; Buckley, T. J.; Greenberg, A.; Butler, J.; Pietarinen, C.; Lioy, P. J. *Polycyclic Aromatic Compounds* **1990**, *1*, 137–149.
35. Lioy, P. J.; Greenberg, A. *J. Toxicol. Ind. Health: Exposure Anal. Sect.* **1990**, *6*, 206–233.
36. Waldman, J.; Lioy, P. J.; Greenberg, A.; Butler, J. *J. Exposure Anal. Environ. Epidemiol.* **1991**, *1*, 197–226.
37. Buckley, T. J.; Waldman, J. M.; Lioy, P. J.; Marple, V. A.; Turner, W. *Aerosol Sci. Technol.* **1991**, *14*, 380–387.
38. Marple, V. A.; Rubow, K. L.; Turner, W.; Spengler, J. D. *J. APCA* **1987**, *37*, 1303–1306.
39. *Particle Total Exposure Assessment Methodology: Pilot Study, Workplan*; Research Triangle Institute: Research Triangle Park, NC, 1990.
40. Glorig, A. In *Patty's Industrial Hygiene and Toxicology*, 2nd ed.; Cralley, L. J., Ed.; Wiley Interscience: New York, 1985; Vol. 31, Chapter 13.
41. *National Air Quality and Emission Trends Report: 1988*; U.S. Environmental Protection Agency. Office of Air Quality Planning and Standards: Washington, DC, 1990; EPA 450/4–90–002.
42. Lioy, P. J.; Dyba, J. *J. Toxicol. Ind. Health* **1989**, *5*, 493–504.
43. *Report Alternate Fuel Research Strategy*; U.S. Environmental Protection Agency. ECAO: Research Triangle Park, NC, 1989.
44. Binder, R. E. *Arch. Environ. Health* **1976**, *36*, 277–279.
45. Sexton, K.; Spengler, J. D.; Trietman, R. D. *Atmos. Environ.* **1984**, *18*, 1371–1383.
46. Spengler, J. D.; Trietman, R. D.; Tosteson, T. D.; Mage, D. T.; Soczek, M. L. *Environ. Sci. Technol.* **1985**, *19*, 700–707.
47. Cortese, A. D.; Spengler, J. D. *J. APCA* **1976**, *26*, 1144–1150.
48. Jabara, J. W.; Keefe, T. J.; Beaulieu, H. J.; Buchan, R. M. *Arch. Environ. Health* **1980**, *35*, 198–204.
49. Ziskind, R. A.; Fite, K.; Mage, D. T. *Environ. Int.* **1982**, *8*, 283–293.
50. Dockery, D. W.; Spengler, J. D.; Reed, M. P.; Ware, J. *Environ. Int.* **1981**, *5*, 101–107.
51. Quackenboss, J. J.; Spengler, J. D.; Kancrek, M. S.; Letz, R.; Duffy, C. P. *Environ. Sci. Technol.* **1986**, *20*, 775–783.
52. Azar, A.; Snee, R. D.; Habini, K. In *Lead*; Griffen, T. F.; Knelson, J. H., Eds.; Academic Press: New York, 1975.

RECEIVED for review March 20, 1991. ACCEPTED revised manuscript August 10, 1992.

Indexes

Author Index

Affiliation Index

Subject Index

393

Copy editing and indexing: Steven Powell
Production: Paula M. Bérard
Acquisition: Cheryl Shanks and Barbara C. Tansill
Cover design: Ronna Hammer

Typeset by Techna Type, Inc., York, PA
Printed and bound by United Book Press, Inc., Baltimore, MD

Bestsellers from ACS Books

The ACS Style Guide: A Manual for Authors and Editors
Edited by Janet S. Dodd
264 pp; clothbound, ISBN 0–8412–0917–0; paperback, ISBN 0–8412–0943–X

Chemical Activities and Chemical Activities: Teacher Edition
By Christie L. Borgford and Lee R. Summerlin
330 pp; spiralbound, ISBN 0–8412–1417–4; teacher ed. ISBN 0–8412–1416–6

Chemical Demonstrations: A Sourcebook for Teachers,
Volumes 1 and 2, Second Edition
Volume 1 by Lee R. Summerlin and James L. Ealy, Jr.;
Vol. 1, 198 pp; spiralbound, ISBN 0–8412–1481–6;
Volume 2 by Lee R. Summerlin, Christie L. Borgford, and Julie B. Ealy
Vol. 2, 234 pp; spiralbound, ISBN 0–8412–1535–9

Writing the Laboratory Notebook
By Howard M. Kanare
145 pp; clothbound, ISBN 0–8412–0906–5; paperback, ISBN 0–8412–0933–2

Developing a Chemical Hygiene Plan
By Jay A. Young, Warren K. Kingsley, and George H. Wahl, Jr.
paperback, ISBN 0–8412–1876–5

Introduction to Microwave Sample Preparation: Theory and Practice
Edited by H. M. Kingston and Lois B. Jassie
263 pp; clothbound, ISBN 0–8412–1450–6

Principles of Environmental Sampling
Edited by Lawrence H. Keith
ACS Professional Reference Book; 458 pp;
clothbound; ISBN 0–8412–1173–6; paperback, ISBN 0–8412–1437–9

Biotechnology and Materials Science: Chemistry for the Future
Edited by Mary L. Good (Jacqueline K. Barton, Associate Editor)
135 pp; clothbound, ISBN 0–8412–1472–7; paperback, ISBN 0–8412–1473–5

Personal Computers for Scientists: A Byte at a Time
By Glenn I. Ouchi
276 pp; clothbound, ISBN 0–8412–1000–4; paperback, ISBN 0–8412–1001–2

Polymers in Aqueous Media: Performance Through Association
Edited by J. Edward Glass
Advances in Chemistry Series 223; 575 pp;
clothbound, ISBN 0–8412–1548–0

For further information and a free catalog of ACS books, contact:
American Chemical Society
Distribution Office, Department 225
1155 16th Street, NW, Washington, DC 20036
Telephone 800–227–5558

Highlights from ACS Books

Good Laboratory Practices: An Agrochemical Perspective
Edited by Willa Y. Garner and Maureen S. Barge
ACS Symposium Series No. 369; 168 pp; clothbound, ISBN 0–8412–1480–8

Silent Spring Revisited
Edited by Gino J. Marco, Robert M. Hollingworth, and William Durham
214 pp; clothbound, ISBN 0–8412–0980–4; paperback, ISBN 0–8412–0981–2

Insecticides of Plant Origin
Edited by J. T. Arnason, B. J. R. Philogène, and Peter Morand
ACS Symposium Series No. 387; 214 pp; clothbound, ISBN 0–8412–1569–3

Chemistry and Crime: From Sherlock Holmes to Today's Courtroom
Edited by Samuel M. Gerber
135 pp; clothbound, ISBN 0–8412–0784–4; paperback, ISBN 0–8412–0785–2

Handbook of Chemical Property Estimation Methods
By Warren J. Lyman, William F. Reehl, and David H. Rosenblatt
960 pp; clothbound, ISBN 0–8412–1761–0

The Beilstein Online Database: Implementation, Content, and Retrieval
Edited by Stephen R. Heller
ACS Symposium Series No. 436; 168 pp; clothbound, ISBN 0–8412–1862–5

Materials for Nonlinear Optics: Chemical Perspectives
Edited by Seth R. Marder, John E. Sohn, and Galen D. Stucky
ACS Symposium Series No. 455; 750 pp; clothbound; ISBN 0–8412–1939–7

Polymer Characterization:
Physical Property, Spectroscopic, and Chromatographic Methods
Edited by Clara D. Craver and Theodore Provder
Advances in Chemistry No. 227; 512 pp; clothbound, ISBN 0–8412–1651–7

From Caveman to Chemist: Circumstances and Achievements
By Hugh W. Salzberg
300 pp; clothbound, ISBN 0–8412–1786–6; paperback, ISBN 0–8412–1787–4

The Green Flame: Surviving Government Secrecy
By Andrew Dequasie
300 pp; clothbound, ISBN 0–8412–1857–9

For further information and a free catalog of ACS books, contact:
American Chemical Society
Distribution Office, Department 225
1155 16th Street, NW, Washington, DC 20036
Telephone 800–227–5558